T0332528

Analytic Pseudo-Differential Operators and their Applications

Mathematics and Its Applications (*Soviet Series*)

Analytic Pseudo-Differential Operators and their Applications

by

Julii A. Dubinskii

Department of Mathematical Modelling,
Moscow Power Engineering Institute,
Moscow, U.S.S.R.

KLUWER ACADEMIC PUBLISHERS
DORDRECHT / BOSTON / LONDON

Library of Congress Cataloging-in-Publication Data

Dubinskii, IU. A.
   Analytic pseudo-differential operators and their applications / by
Julii A. Dubinskii.
      p.   cm. -- (Mathematics and its applicatins. Soviet series ;
   v. 68)
   Translated from Russian.
   Includes bibliographical references and index.
   ISBN 0-7923-1296-1 (HB) printed on acid free paper
   1. Pseudodifferential operators.   I. Title. II. Series:
Mathematics and its applications (Kluwer Academic Publishers).
Soviet series ; 68.
QA329.7.D83   1991
515'.7242--dc20   91-20313

ISBN 0-7923-1296-1

Published by Kluwer Academic Publishers,
P.O. Box 17, 3300 AA Dordrecht, The Netherlands.

Kluwer Academic Publishers incorporates
the publishing programmes of
D. Reidel, Martinus Nijhoff, Dr W. Junk and MTP Press.

Sold and distributed in the U.S.A. and Canada
by Kluwer Academic Publishers,
101 Philip Drive, Norwell, MA 02061, U.S.A.

In all other countries, sold and distributed
by Kluwer Academic Publishers Group,
P.O. Box 322, 3300 AH Dordrecht, The Netherlands.

Printed on acid-free paper

Printed in the Netherlands

SERIES EDITOR'S PREFACE

'Et moi, ..., si j'avait su comment en revenir,
je n'y serais point allé.'

                 Jules Verne

The series is divergent; therefore we may be
able to do something with it.

                 O. Heaviside

One service mathematics has rendered the
human race. It has put common sense back
where it belongs, on the topmost shelf next
to the dusty canister labelled 'discarded non-
sense'.

                 Eric T. Bell

Mathematics is a tool for thought. A highly necessary tool in a world where both feedback and non-linearities abound. Similarly, all kinds of parts of mathematics serve as tools for other parts and for other sciences.

Applying a simple rewriting rule to the quote on the right above one finds such statements as: 'One service topology has rendered mathematical physics ...'; 'One service logic has rendered computer science ...'; 'One service category theory has rendered mathematics ...'. All arguably true. And all statements obtainable this way form part of the raison d'être of this series.

This series, *Mathematics and Its Applications*, started in 1977. Now that over one hundred volumes have appeared it seems opportune to reexamine its scope. At the time I wrote

> "Growing specialization and diversification have brought a host of monographs and textbooks on increasingly specialized topics. However, the 'tree' of knowledge of mathematics and related fields does not grow only by putting forth new branches. It also happens, quite often in fact, that branches which were thought to be completely disparate are suddenly seen to be related. Further, the kind and level of sophistication of mathematics applied in various sciences has changed drastically in recent years: measure theory is used (non-trivially) in regional and theoretical economics; algebraic geometry interacts with physics; the Minkowsky lemma, coding theory and the structure of water meet one another in packing and covering theory; quantum fields, crystal defects and mathematical programming profit from homotopy theory; Lie algebras are relevant to filtering; and prediction and electrical engineering can use Stein spaces. And in addition to this there are such new emerging subdisciplines as 'experimental mathematics', 'CFD', 'completely integrable systems', 'chaos, synergetics and large-scale order', which are almost impossible to fit into the existing classification schemes. They draw upon widely different sections of mathematics."

By and large, all this still applies today. It is still true that at first sight mathematics seems rather fragmented and that to find, see, and exploit the deeper underlying interrelations more effort is needed and so are books that can help mathematicians and scientists do so. Accordingly MIA will continue to try to make such books available.

If anything, the description I gave in 1977 is now an understatement. To the examples of interaction areas one should add string theory where Riemann surfaces, algebraic geometry, modular functions, knots, quantum field theory, Kac-Moody algebras, monstrous moonshine (and more) all come together. And to the examples of things which can be usefully applied let me add the topic 'finite geometry'; a combination of words which sounds like it might not even exist, let alone be applicable. And yet it is being applied: to statistics via designs, to radar/sonar detection arrays (via finite projective planes), and to bus connections of VLSI chips (via difference sets). There seems to be no part of (so-called pure) mathematics that is not in immediate danger of being applied. And, accordingly, the applied mathematician needs to be aware of much more. Besides analysis and numerics, the traditional workhorses, he may need all kinds of combinatorics, algebra, probability, and so on.

In addition, the applied scientist needs to cope increasingly with the nonlinear world and the extra mathematical sophistication that this requires. For that is where the rewards are. Linear models are honest and a bit sad and depressing: proportional efforts and results. It is in the non-linear world that infinitesimal inputs may result in macroscopic outputs (or vice versa). To appreciate what I am hinting at: if electronics were linear we would have no fun with transistors and computers; we would have no TV; in fact you would not be reading these lines.

There is also no safety in ignoring such outlandish things as nonstandard analysis, superspace and anticommuting integration, p-adic and ultrametric space. All three have applications in both electrical engineering and physics. Once, complex numbers were equally outlandish, but they frequently proved the shortest path between 'real' results. Similarly, the first two topics named have already provided a number of 'wormhole' paths. There is no telling where all this is leading - fortunately.

Thus the original scope of the series, which for various (sound) reasons now comprises five subseries: white (Japan), yellow (China), red (USSR), blue (Eastern Europe), and green (everything else), still applies. It has been enlarged a bit to include books treating of the tools from one subdiscipline which are used in others. Thus the series still aims at books dealing with:

- a central concept which plays an important role in several different mathematical and/or scientific specialization areas;
- new applications of the results and ideas from one area of scientific endeavour into another;
- influences which the results, problems and concepts of one field of enquiry have, and have had on the development of another.

Pseudodifferential operators, as the name indicates, generalize differential operators. Their study constitutes a relatively young effort in mathematics; yet they show great promise as a unifying idea in the rather scattered field of differential operators and differential equations, as this volume seeks to make clear.

Partly, no doubt, this is because the class ΨDO is closed under certain natural operations; e.g. the resolvant and complex powers of an elliptic differential operator on a complex manifold are pseudodifferential operators.

But there is no more to it and, again, I can do no better than refer the curious reader to the main text of this very welcome volume.

The shortest path between two truths in the real domain passes through the complex domain.

> J. Hadamard

La physique ne nous donne pas seulement l'occasion de résoudre des problèmes ... elle nous fait pressentir la solution.

> H. Poincaré

Never lend books, for no one ever returns them; the only books I have in my library are books that other folk have lent me.

> Anatole France

The function of an expert is not to be more right than other people, but to be wrong for more sophisticated reasons.

> David Butler

Bussum, May 1991

Michiel Hazewinkel

Contents

Preface

The main purpose of the present book is to give a systematic account of the foundations of the theory of pseudo-differential operators (PD-operators), the symbols of which are arbitrary analytic functions of complex arguments or, as we shall say, the foundations of the theory of analytic PD-operators.

For the author the starting point for constructing this theory was (even from student days) the desire to solve differential equations

$$A(D)u(x) = h(x), \ x \in \mathbf{R}^n,$$

by some natural operator method, that is, in the form

$$u(x) = \frac{I}{A(D)} h(x). \tag{1}$$

Already from this it is clear that in order to give sense to the above formula one needs some algebra of PD-operators at the minimum with meromorphic symbols, because the functions $A(\zeta)$ and $A^{-1}(\zeta)$ are not, as a rule, entire at the same time.

But this idea would most probably have remained merely as a reminiscence of student days if it were not for an important circumstance. Namely, it happens that many classical problems of analysis and differential equations, the investigation of which was considered to be practically complete, in essence cannot be fully understood without using PD-operators with arbitrary analytic symbols. It concerns first of all the theory of the Cauchy problem in a complex domain and the Fourier transformation of analytic functions. In particular, within the bounds of the theory of analytic PD-operators one finds a close connection between the Cauchy-Kovalevskaya theory and the exponential theory of the Cauchy problem which, it would have seemed, have nothing in common.

The reader can find a detailed account of these questions in the first two parts of the present book.

It should be further noted that, despite the fact that the theory of analytic PD-operators is, in essence, part of complex analysis, it allows one to investigate in a new way many real problems of mathematical physics. The point is that many classical problems of mathematical physics are by their nature analytic in the sense that having real analytic data their solutions present the action of some PD-operators with analytic symbols.

Thus it is clear that the question of the existence of a solution for the corresponding problem in the space of functions with finite smoothness is exactly the question of the existence of an extension of analytic PD-operators to operators with a non-analytic domain of definition. The very fact of the existence or non-existence of a "finitely smooth" extension of PD-operators with analytic symbols determines the classical well-posedness or ill-posedness of the initial problem in the Hadamard-Petrovskii sense.

The reader will find an account of these questions in the third part of the book.

These then are the first words we wish to say before embarking on a direct discussion of the material. With this in mind, if that is not sufficient to arouse the reader's interest, we recommend that, after reading this preface, he should turn to the introduction to each part of the book where the contents are described in more detail.

In conclusion we point out that the theory outlined here is very young (not more than 10 years old) and ought to be regarded as an elementary introduction to the theory of analytic PD-operators the fruitful development of which, we hope, still lies ahead.

Moscow, 1990   Julii A. Dubinskii

Part I. PD-Operators with Complex Arguments

Introduction

The first part is devoted to the construction of the algebra of PD-operators with complex arguments and its applications.

First we introduce certain spaces of test functions and distributions. Then we give the main definition of a PD-operator with constant analytic symbol.

We now discuss the ideas of this definition in more detail.

As is well known, the construction of the theory of real PD-operators is based on the theory of the Fourier-Plancherel transformation, which enables one to convert this problem to that of the construction of the algebra of symbols in the space of the dual variables.

In the construction of PD-operators with complex variables one would like to make similar use of the well known Borel transform of exponential functions. This method, however, presents certain difficulties because (by contrast with the real Fourier transformation) the Borel transformation does not define a unique measure in the dual space. This leads to the necessity of constructing PD-operators with complex variables in the original variables $z \in \mathbb{C}^n$ directly. This method is in fact used in the first chapter.

As a result of this approach, the PD-operator $A(D)$ with the symbol $A(\zeta)$ analytic in a domain $\Omega \subset \mathbb{C}^n_\zeta$ acts "locally" as a differential operator of infinite order in direct conformity with Weierstrass's definition of an analytic function, that is, by means of its locally analytic elements.

It is important to note that one uses the space $\mathrm{Exp}_\Omega(\mathbb{C}^n_z)$ as the domain of definition of a PD-operator. This space is associated with the domain of analyticity of the symbol $A(\zeta)$ and contains all entire functions of exponential type that "grow" over the domain Ω. It is this space $\mathrm{Exp}_\Omega(\mathbb{C}^n_z)$ that is invariant under the action of an arbitrary PD-operator whose symbol is analytic in Ω.

1

From this it follows that there is an isomorphism

$$\mathcal{A}(\Omega) \leftrightarrow \mathcal{O}(\Omega),$$

where $\mathcal{A}(\Omega)$ is the algebra of PD-operators with symbols analytic in Ω and $\mathcal{O}(\Omega)$ is the algebra of analytic functions in Ω.

It is necessary to emphasize that the idea of the construction of the space $\mathrm{Exp}_\Omega(\mathbf{C}_z^n)$ associated with Ω is quite closely connected with the classical question about the connection between the growth of the exponential functions and the distribution of the singularities of their Borel transforms. Pólya's well known theorem [1] describes this connection for any convex domain in $\mathbf{C}^1$. In the case of a convex domain in $\mathbf{C}^n$, $n \geq 2$, the functions $u(z) \in \mathrm{Exp}_\Omega(\mathbf{C}_z^n)$ are described in the terms of certain global estimates due to Martineau [1] and Ehrenpreis [1]; moreover the space $\mathrm{Exp}_\Omega(\mathbf{C}_z^n)$ is the Laplace image of the space of all compact measures in Ω. In this latter form, the spaces $\mathrm{Exp}_\Omega(\mathbf{C}_z^n)$ have been used more than once by Hörmander [1], Steinberg [1], Napalkov [1] and others.

We note also that an operator version of the space $\mathrm{Exp}_\Omega(\mathbf{C}_z^n)$ has been given in the recent papers by Radyno [1], [2].

These spaces are used, in particular, in the asymptotic theory in the papers by Kachalov and Lomov [1], [2].

The spectral approach to spaces of exponential type was developed by Umarov [1], [2].

Further it is clear that if

$$A(D) : \mathrm{Exp}_\Omega(\mathbf{C}_z^n) \to \mathrm{Exp}_\Omega(\mathbf{C}_z^n)$$

is a continuous map, then the dual map

$$A(-D) : \mathrm{Exp}'_\Omega(\mathbf{C}_z^n) \to \mathrm{Exp}'_\Omega(\mathbf{C}_z^n)$$

is also continuous. (Here $\mathrm{Exp}'_\Omega(\mathbf{C}_z^n)$ is the space of exponential functionals, that is, the space of all bounded linear functionals on $\mathrm{Exp}_\Omega(\mathbf{C}_z^n)$.) It provides the possibility of defining the Fourier transform of any analytic function $u(z)$ as the exponential functional

$$\tilde{u}(\zeta) = u(-\partial)\delta(\zeta), \ \zeta \in \mathbf{C}^n,$$

$(\partial = (\partial/\partial\zeta_1, \ldots, \partial/\partial\zeta_n))$, where

$$u(-\partial) : \mathrm{Exp}'_G(\mathbf{C}^n_\zeta) \rightarrow \mathrm{Exp}'_G(\mathbf{C}^n_\zeta)$$

is the adjoint of the PD-map

$$u(\partial) : \mathrm{Exp}_G(\mathbf{C}^n_\zeta) \rightarrow \mathrm{Exp}(\mathbf{C}^n_\zeta).$$

The Fourier transformation defined in this way possesses all the natural properties of the "usual" Fourier transformation, moreover, the inversion formula has the standard form

$$u(z) = \langle \tilde{u}(\zeta), \exp z\zeta \rangle, \ z \in G.$$

Finally, there is the diagram

$$\begin{array}{ccc}
\mathcal{O}(G) & \xrightarrow{\ F\ } & \mathrm{Exp}'_G(\mathbf{C}^n_\zeta) \\
\downarrow * & & \uparrow * \\
\mathcal{O}'(G) & \xleftarrow{\ F\ } & \mathrm{Exp}_G(\mathbf{C}^n_\zeta)
\end{array}$$

where * is the operation of forming the adjoint. Moreover, the formula of the complex unitarity

$$\langle \tilde{u}(\zeta), v(\zeta) \rangle = \langle \tilde{v}(z), u(z) \rangle,$$

where $u(z) \in \mathcal{O}(G)$, $v(\zeta) \in \mathrm{Exp}_G(\mathbf{C}^n_\zeta)$ holds.

The "new" property of the Fourier transformation is the formula

$$[A(D)u(z)]\tilde{\ }(\zeta) = A(\zeta)\tilde{u}(\zeta),$$

where $A(\zeta) \in \mathcal{O}(\Omega)$, $u(z) \in \mathrm{Exp}_\Omega(\mathbf{C}^n_z)$ are arbitrary.

The above properties guarantee the well-posedness of the complex Fourier method for the solution of partial differential equations and PD-equations (Chapter 2).

The final Chapter 3 is devoted to PD-operators with more general symbols, namely, symbols that are a formal series

$$A(\zeta) \sim \sum_{|\alpha|=0}^{\infty} a_\alpha \zeta^\alpha.$$

It turns out that in this case the natural domain of definition of the PD-operator with such formal series is the space of entire functions $u(z)$ with the growth of order less than unity. After this we consider various problems for partial differential equations.

In conclusion, I should point out that the results of Chapter 3 were obtained by my post-graduate student O. Odinokov [1], [2] and are published here with his permission.

Chapter 1. PD-Operators with Constant Analytic Symbols

1.1. Spaces of entire functions of exponential type

1. THE SPACE $\mathrm{Exp}_R(\mathbf{C}_z^n)$.

Let $\mathrm{Exp}(\mathbf{C}_z^n)$ be the space of all entire functions $u(z) : \mathbf{C}^n \to \mathbf{C}^1$ of exponential type, that is,

$$\mathrm{Exp}(\mathbf{C}_z^n) = \{u(z) : |u(z)| \leq M \exp(r_1|z_1| + \ldots + r_n|z_n|)\},$$

where $M > 0$ is a constant and $r = (r_1, \ldots, r_n)$, $r_1 \geq 0, \ldots, r_n \geq 0$, is a vector (both depend, in general, on $u(z)$).

For brevity we shall write $r_1|z_1| + \ldots + r_n|z_n| \equiv rz_+$, so that

$$\mathrm{Exp}(\mathbf{C}_z^n) = \{u(z) : |u(z)| \leq M \exp rz_+, \ z \in \mathbf{C}^n\}$$

where the vectors r (vector-types) are arbitrary.

Further, let us fix a vector-type r and set

$$\overline{\mathrm{Exp}}_r(\mathbf{C}_z^n) = \{u(z) \in \mathrm{Exp}(\mathbf{C}_z^n) : |u(z)| \leq M \exp rz_+\}$$

where $M > 0$ is a constant (depending also on $u(z)$).

It is easy to see that $\overline{\mathrm{Exp}}_r(\mathbf{C}_z^n)$ is a Banach space with norm

$$\|u\|_r = \sup_{z \in \mathbf{C}^n} |u(z)| \exp(-rz_+).$$

Proposition 1.1.1. *Let* $r_1 \leq r_2$, *that is,* $r_{1j} \leq r_{2j}$ $(j = 1, \ldots, n)$. *Then*

$$\overline{\mathrm{Exp}}_{r_1}(\mathbf{C}_z^n) \subset \overline{\mathrm{Exp}}_{r_2}(\mathbf{C}_z^n). \tag{1.1.1}$$

Moreover, if $r_1 < r_2$, *then this inclusion is compact.*

5

Proof. The inclusion (1.1.1) is obvious. We prove that for $r_1 < r_2$ this inclusion is compact. Thus, let

$$B_1 = \{u(z) \in \overline{\mathrm{Exp}}_{r_1}(\mathbf{C}_z^n) : \|u\|_{r_1} \leq 1\}$$

be the unit ball in $\overline{\mathrm{Exp}}_{r_1}(\mathbf{C}_z^n)$. Then for every compact $K \subset \mathbf{C}_z^n$ we have the inequality

$$|u(z)| \leq M(K), \ z \in K,$$

where $u(z) \in B_1$ is arbitrary and $M(K) = \max_{z \in K} \exp r z_+$. Then (by Montel's theorem and using a diagonal process) we can find a sequence $u_\nu(z) \in B_1$ ($\nu = 1, 2, \ldots$) and an entire function $u(z)$ such that $u_\nu(z) \to u(z)$ uniformly on any compact set $K \subset \mathbf{C}_z^n$. Further, since $\|u_\nu(z)\|_{r_1} \leq 1$, it follows that for all $z \in \mathbf{C}^n$

$$|u_\nu(z)| \exp(-r_1 z_+) \leq 1,$$

therefore, letting ν tend to $+\infty$, we obtain

$$|u(z) \exp(-r_1 z_+)| \leq 1,$$

that is, $u(z) \in B_1$ and $u(z) \in \overline{\mathrm{Exp}}_{r_2}(\mathbf{C}_z^n)$.

We now show that $u_\nu(z) \to u(z)$ in the norm of the space $\overline{\mathrm{Exp}}_{r_2}(\mathbf{C}_z^n)$. In fact, for every compact $K \subset \mathbf{C}_z^n$

$$\|u - u_\nu\|_{r_2} \equiv \sup_{z \in \mathbf{C}^n} |u(z) - u_\nu(z)| \exp(-r_2 z_+) \leq$$

$$\leq \sup_{z \in K} |u(z) - u_\nu(z)| \exp(-r_2 z_+) + \sup_{z \in \mathbf{C}^n \backslash K} |u(z) - u_\nu(z)| \exp(-r_2 z_+) \leq$$

$$\leq \sup_{z \in K} |u(z) - u_\nu(z)| + \sup_{z \in \mathbf{C}^n \backslash K} |u(z) - u_\nu(z)| \exp(-r_1 z_+) \exp(r_1 - r_2) z_+ \leq$$

$$\leq \sup_{z \in K} |u(z) - u_\nu(z)| + \|u - u_\nu\|_{r_1} \sup_{z \in \mathbf{C}^n \backslash K} \exp(r_1 - r_2) z_+ \leq$$

$$\leq \sup_{z \in K} |u(z) - u_\nu(z)| + 2 \sup_{z \in \mathbf{C}^n \backslash K} \exp(r_1 - r_2) z_+.$$

It is clear that by choosing K and ν sufficiently large, we find that $\|u - u_\nu\|_{r_2} < \epsilon$, where $\epsilon > 0$ is arbitrarily small. Q.E.D.

Now let $R = (R_1, \ldots, R_n)$ be some "positive" vector, that is, $R_1 > 0, \ldots,$ $R_n > 0$.

Definition 1.1.1. We define the complete topological space $\mathrm{Exp}_R(\mathbb{C}_z^n)$ as

$$\mathrm{Exp}_R(\mathbb{C}_z^n) = \varliminf_{r \uparrow R} \overline{\mathrm{Exp}}_r(\mathbb{C}_z^n)$$

(here $r \uparrow R$ means that $r_j \to R_j$ and is monotonically increasing $(j = 1, \ldots, n)$).

In conformity with Proposition 1.1.1 and the standard topological terminology (see, for example, the books of Schaefer [1], A. Robertson and W. Robertson [1]), $\mathrm{Exp}_R(\mathbb{C}_z^n)$ is an LN^*-space. From this it follows, in particular, that $u_\nu(z) \to u(z)$ $(\nu \to \infty)$ if and only if:

1) there exists a vector $r < R$ such that for all $\nu = 1, 2, \ldots$ the functions $u_\nu(z) \in \overline{\mathrm{Exp}}_r(\mathbb{C}_z^n)$ and, in addition, the sequence $\{u_\nu(z)\}$, $\nu = 1, 2, \ldots$, is a bounded set in $\overline{\mathrm{Exp}}_r(\mathbb{C}_z^n)$;

2) $u_\nu(z) \to u(z)$ locally uniformly on $\mathbb{C}_z^n$, that is, for any compact $K \subset \mathbb{C}_z^n$ the convergence $u_\nu(z) \to u(z)$ is uniform.

Remark. One can allow $R = +\infty$, that is, $R_1 = +\infty, \ldots, R_n = +\infty$. In this case,

$$\varliminf_{r \uparrow \infty} \overline{\mathrm{Exp}}_r(\mathbb{C}_z^n) = \mathrm{Exp}(\mathbb{C}_z^n),$$

that is, the space of all functions of exponential type with the $\lim$ ind-topology.

Example. The elementary functions $\exp \lambda z$, $P_N(z) \exp \lambda z$, $\sin \lambda z$, $P_N(z) \sin \lambda z$, $\cosh \lambda z$, $P_N(z) \cosh \lambda z$, etc. ($P_N(z)$ is a polynomial) belong to $\mathrm{Exp}_R(\mathbb{C}_z^n)$ if $\lambda = (\lambda_1, \ldots, \lambda_n)$ is a vector such that $|\lambda_j| < R_j$ $(j = 1, \ldots, n)$.

2. ESTIMATES OF DERIVATIVES

Sharp estimates of the derivatives of $u(z) \in \mathrm{Exp}_R(\mathbb{C}_z^n)$ play an essential role in the theory of differential operators of infinite order and (more generally) in the theory of PD-operators with analytic symbols. In this subsection we describe the space $\mathrm{Exp}_R(\mathbb{C}_z^n)$ in terms of such estimates.

Let $\alpha = (\alpha_1, \ldots, \alpha_n)$ be a multi-index of differentiation and set

$$D^\alpha \overset{\mathrm{def}}{=} \partial^{|\alpha|} / \partial z_1^{\alpha_1} \ldots \partial z_n^{\alpha_n},$$

where $|\alpha| = \alpha_1 + \ldots + \alpha_n$.

Proposition 1.1.2 *An entire function $u(z)$ belongs to $\mathrm{Exp}_R(\mathbb{C}_z^n)$ if and only if there exist a vector $r < R$ and a constant $M > 0$ such that for any α and $z \in \mathbb{C}^n$*

$$|D^\alpha u(z)| \leq M r^\alpha \exp r z_+. \tag{1.1.2}$$

Proof. The sufficiency of inequalities (1.1.2) is clear. Let us prove the necessity of the conditions. In fact, by Cauchy's formula we have

$$u(z) = \frac{1}{(2\pi i)^n} \int_{|\zeta_1 - z_1| = a_1} \cdots \int_{|\zeta_n - z_n| = a_n} \frac{u(\zeta) d\zeta_1 \ldots d\zeta_n}{(\zeta_1 - z_1) \ldots (\zeta_n - z_n)},$$

where $a = (a_1, \ldots, a_n)$ is arbitrary. Hence

$$D^\alpha u(z) = \frac{\alpha!}{(2\pi i)^n} \int_{|\zeta_1 - z_1| = a_1} \cdots \int_{|\zeta_n - z_n| = a_n} \frac{u(\zeta) d\zeta_1 \ldots d\zeta_n}{(\zeta_1 - z_1)^{\alpha_1 + 1} \ldots (\zeta_n - z_n)^{\alpha_n + 1}}.$$

Furthermore, since $u(z) \in \mathrm{Exp}_R(\mathbf{C}_z^n)$, there exist a vector $r < R$ and a constant $M > 0$ such that

$$|u(z)| \leq M \exp r z_+.$$

Taking into account this inequality we obtain from the last formula

$$|D^\alpha u(z)| \leq M \frac{\alpha!}{a^\alpha} \exp ra \cdot \exp r z_+.$$

Minimizing the right hand side with respect to a (the minimum is attained at $a = \alpha/r \equiv (\alpha_1/r_1, \ldots, \alpha_n/r_n)$), we obtain the inequality

$$|D^\alpha u(z)| \leq M \Big(\prod_{1 \leq j \leq n} \alpha_j! \Big(\frac{e}{\alpha_j}\Big)^{\alpha_j} \exp r z_+ \Big) r^\alpha.$$

Since (by Stirling's formula) $m!(e/m)^m \leq e(2\pi m)^{1/2}$, $m = 1, 2, \ldots$, it immediately follows that there exist $M_1 > 0$ and $\bar{r}$ ($r < \bar{r} < R$) such that

$$|D^\alpha u(z)| \leq M_1 \exp \bar{r} z_+, \quad z \in \mathbf{C}^n.$$

Returning to the original notations, we obtain the inequalities (1.1.2). The necessity is proved and thereby our proposition is proved.

Remark.   In what follows it is helpful to note that M_1 and $\bar{r}$ depend on M and r, but not on the specific function $u(z)$.

3. THE TEST-FUNCTION SPACE $\mathrm{Exp}_\Omega(\mathbf{C}_z^n)$. TOPOLOGY AND CONVERGENCE

Let Ω be an arbitrary open subset of the complex space $\mathbf{C}_\zeta^n$ of the variables $\zeta = (\zeta_1, \ldots, \zeta_n)$ (in what follows they will be the dual variables). Further, let $\lambda = (\lambda_1, \ldots, \lambda_n) \in \Omega$ be an arbitrary point. Denote by $R(\lambda)$ the radius of the maximal polycylinder

$$U(\lambda) = \{\zeta : |\zeta_j - \lambda_j| < R(\lambda), \ j = 1, \ldots, n\},$$

lying inside the set Ω.

Definition 1.1.2. We set

$$\text{Exp}_\Omega(\mathbf{C}_z^n) = \{u(z) \in \text{Exp}(\mathbf{C}_z^n) : u(z) = \sum_\lambda e^{\lambda z}\phi_\lambda(z)\},$$

where the summation is carried out over all finite sets of $\lambda \in \Omega$ and $\phi_\lambda(z) \in$
$\text{Exp}_{R(\lambda)}(\mathbf{C}_z^n)$ (here by definition $R(\lambda) \equiv (R(\lambda), \ldots, R(\lambda))$).

Definition 1.1.3. We say that $u_\nu(z) \to u(z)$ $(\nu \to \infty)$ in $\text{Exp}_\Omega(\mathbf{C}_z^n)$ if there
exists a collection of polycylinders $U(\lambda)$ such that

1) $\quad u_\nu(z) = \sum_\lambda e^{\lambda z}\phi_{\nu\lambda}(z), \quad \phi_{\nu\lambda}(z) \in \text{Exp}_{R(\lambda)}(\mathbf{C}_z^n),$

2) $\quad \phi_{\nu\lambda}(z) \to \phi_\lambda(z)$ in $\text{Exp}_{R(\lambda)}(\mathbf{C}_z^n).$

The space $\text{Exp}_\Omega(\mathbf{C}_z^n)$ is the main test-function space for the construction
of the theory of PD-operators with analytic symbols. We shall call this space
the space of entire test-functions associated with the domain Ω.

We now describe this space in standard topological terms.

We set

$$e^{\lambda z}\,\text{Exp}_{R(\lambda)}(\mathbf{C}_z^n) = \{u(z) : u(z)e^{-\lambda z} \in \text{Exp}_{R(\lambda)}(\mathbf{C}_z^n)\}.$$

We then let

$$\bigoplus_{\lambda \in \Omega} e^{\lambda z}\,\text{Exp}_{R(\lambda)}(\mathbf{C}_z^n)$$

denote the (algebraic) direct sum of the spaces $e^{\lambda z}\,\text{Exp}_{R(\lambda)}(\mathbf{C}_z^n)$.

By definition (see, for example, the book of A. Robertson and W. Robertson [1]) the elements of this sum are arbitrary finite formal sums $\bigoplus_{\lambda \in \Omega} u_\lambda(z)$,
where $u_\lambda(z) \in e^{\lambda z}\,\text{Exp}_{R(\lambda)}(\mathbf{C}_z^n)$ or, what is the same, families $\{u_\lambda(z)\}$, $\lambda \in \Omega$,
for which only a finite set of the $u_\lambda(z) \not\equiv 0$. We identify the families $\{u_\lambda(z)\}$
and $\{u_\mu(z)\}$ if

$$\sum_\lambda u_\lambda(z) \equiv \sum_\mu u_\mu(z),$$

where $\sum$ denotes the usual sum of functions.

In other words, we consider the factor-sum

$$\bigoplus_{\lambda \in \Omega} e^{\lambda z}\,\text{Exp}_{R(\lambda)}(\mathbf{C}_z^n)/L,$$

where

$$L = \Big\{ \bigoplus_{\lambda \in \Omega} u_\lambda(z) : \sum_\lambda u_\lambda(z) \equiv 0 \Big\}.$$

Then

$$\mathrm{Exp}_\Omega(\mathbf{C}_z^n) \approx \bigoplus_{\lambda \in \Omega} e^{\lambda z} \, \mathrm{Exp}_{R(\lambda)}(\mathbf{C}_z^n)/L,$$

this algebraic isomorphism being defined by the map

$$u(z) \leftrightarrow \{u_\lambda(z)\},$$

where $u(z) \in \mathrm{Exp}_\Omega(\mathbf{C}_z^n)$ and $\{u_\lambda(z)\}$ is the class of all families $u_\lambda(z)$, $\lambda \in \Omega$, such that

$$u(z) \equiv \sum_\lambda u_\lambda(z).$$

One can proceed differently. Namely, let Γ be an arbitrary collection of points in the domain Ω. Consider the direct sum

$$\bigoplus_{\lambda \in \Gamma} e^{\lambda z} \, \mathrm{Exp}_{R(\lambda)}(\mathbf{C}_z^n)$$

and let us again identify the formal sums $\bigoplus_{\lambda \in \Gamma} u_\lambda(z)$ and $\bigoplus_{\lambda \in \Gamma} v_\lambda(z)$ if and only if

$$\sum_\lambda u_\lambda(z) \equiv \sum_\lambda v_\lambda(z).$$

Then

$$\bigoplus_{\lambda \in \Gamma} e^{\lambda z} \, \mathrm{Exp}_{R(\lambda)}(\mathbf{C}_z^n)/L_\Gamma \approx \mathrm{Exp}(\Gamma; \mathbf{C}_z^n) \qquad (1.1.3)$$

where L_Γ is the subspace of the direct sum $\bigoplus_{\lambda \in \Gamma} e^{\lambda z} \, \mathrm{Exp}_{R(\lambda)}(\mathbf{C}_z^n)$, containing all families $u_\lambda(z)$, $\lambda \in \Gamma$, such that

$$\sum_{\lambda \in \Gamma} u_\lambda(z) \equiv 0$$

and

$$\mathrm{Exp}(\Gamma; \mathbf{C}_z^n) = \Big\{ u(z) : u(z) = \sum_{\lambda \in \Gamma} u_\lambda(z), \ u_\lambda(z) \in e^{\lambda z} \, \mathrm{Exp}_{R(\lambda)}(\mathbf{C}_z^n) \Big\}.$$

It is clear that

$$\mathrm{Exp}_\Omega(\mathbf{C}_z^n) \approx \bigcup_\Gamma \mathrm{Exp}(\Gamma; \mathbf{C}_z^n)$$

(algebraically) and, consequently,

$$\text{Exp}_{\Omega}(\mathbf{C}_z^n) \approx \bigcup_{\Gamma} \left\{ \bigoplus_{\lambda \in \Gamma} e^{\lambda z} \, \text{Exp}_{R(\lambda)}(\mathbf{C}_z^n)/L_{\Gamma} \right\}.$$

The topology in $\text{Exp}_{\Omega}(\mathbf{C}_z^n)$ may be introduced in different ways. For example, one may proceed as follows.

Let $\Lambda = \{\lambda_1, \lambda_2, \ldots\}$ be an everywhere dense sequence of points in Ω (it is clear that it is enough to consider such sets of $\lambda \in \Omega$). Set $\Gamma_N = \{\lambda_1, \ldots, \lambda_N\}$ and consider the sequence of spaces $\text{Exp}(\Gamma_N; \mathbf{C}_z^n)$, constructed before. In accordance with formula (1.1.3), the topology in $\text{Exp}(\Gamma_N; \mathbf{C}_z^n)$ is introduced as the natural quotient topology of the left side.

Further, by adding to the representation

$$u(z) = \sum_{\lambda \in \Gamma_N} u_{\lambda}(z)$$

of the function $u(z) \in \text{Exp}(\Gamma_N; \mathbf{C}_z^n)$ the necessary number of items of type $\exp(\lambda_j z) \cdot 0 \; (j > N)$, we obtain the chain of inclusions

$$\text{Exp}(\Gamma_1; \mathbf{C}_z^n) \subset \text{Exp}(\Gamma_2; \mathbf{C}_z^n) \subset \ldots .$$

Furthermore, it is clear that each space $\text{Exp}(\Gamma_N; \mathbf{C}_z^n)$ is a closed subspace of $\text{Exp}(\Gamma_{N+1}; \mathbf{C}_z^n)$. Hence,

$$\text{Exp}_{\Omega}(\mathbf{C}_z^n) = \lim_{N \to \infty} \text{ind} \, \text{Exp}(\Gamma_N; \mathbf{C}_z^n)$$

is a strict inductive limit. From the known properties of such limits it follows that $u_{\nu}(z) \to 0$ in $\text{Exp}_{\Omega}(\mathbf{C}_z^n)$ if and only if there exists a number N such that $u_{\nu}(z) \in \text{Exp}\{\Gamma_N; \mathbf{C}_z^n\}$ for all $\nu = 1, 2, \ldots$ and $u_{\nu}(z) \to 0$ in $\text{Exp}(\Gamma_N; \mathbf{C}_z^n)$. According to subsection 1, this means that every $u_{\nu}(z)$ has a representation

$$u_{\nu}(z) = e^{\lambda_1 z} \phi_{\nu 1}(z) + \ldots + e^{\lambda_N z} \phi_{\nu N}(z), \qquad (1.1.4)$$

where $\phi_{\nu j}(z) \in \text{Exp}_{r_j}(\mathbf{C}_z^n)$, $r_j < R(\lambda_j)$, $j = 1, \ldots, N$, and, furthermore, $\inf \|\phi_{\nu j}(z)\|_{r_j} \to 0$ (here inf denotes the greatest lower bound of all possible representations (1.1.4)).

As is not difficult to see, the convergence obtained in this way is majorized by the convergence given in Definition 1.1.3. For definiteness we shall use below the convergence of Definition 1.1.3.

4. DENSITY OF LINEAR COMBINATIONS OF $\exp \lambda z$, $\lambda \in \Omega$

In what follows, we shall need the fact that the set of all linear combinations of $\exp \lambda z$, $\lambda \in \Omega$, is dense in the space $\text{Exp}_\Omega(\mathbb{C}_z^n)$.

Lemma 1.1.1. *The set of all linear combinations of* $\exp \lambda z$, $\lambda \in \Omega$, *is dense in* $\text{Exp}_\Omega(\mathbb{C}_z^n)$.

Proof. First we note that every function $u(z) \in \text{Exp}_\Omega(\mathbb{C}_z^n)$ can be approximated by linear combinations of exponentials in the sense of locally uniform convergence, that is, in the sense of uniform convergence on any compact $K \subset \mathbb{C}_z^n$. For if $u(z)$ is an entire function, then its Taylor series converges to $u(z)$ locally uniformly. Further, for any multi-index α we have

$$z^\alpha = \partial_\zeta^\alpha e^{\zeta z}\big|_{\zeta=0},$$

therefore any monomial z^α can be approximated locally uniformly by a corresponding finite-difference quotient. These quotients are clearly certain linear combinations of exponentials $\exp \lambda z$, $\lambda \in \Omega$. From this it follows immediately that the function $u(z)$ itself is approximated by linear combinations of $\exp \lambda z$, $\lambda \in \Omega$.

Let us show that this scheme gives a sequence of linear combinations of exponentials

$$u_N(z) = c_{1N} e^{\zeta_1 N z} + \ldots + c_{NN} e^{\zeta_{NN} z},$$
$$\zeta_{jN} \in \Omega; \quad j = 1, \ldots, N; N = 1, 2, \ldots,$$

converging to $u(z)$ in the sense of the space $\text{Exp}_\Omega(\mathbb{C}_z^n)$.

To prove this we need the two propositions given below.

Proposition 1.1.3. *If* $\phi(z) \in \text{Exp}_R(\mathbb{C}_z^n)$, *then its Taylor series tends to* $\phi(z)$ *in the sense of* $\text{Exp}_R(\mathbb{C}_z^n)$ *(see Definition 1.1.1 and below).*

Proof. In fact, the well known formulae for the hypersurface of conjugate types $r_1, \ldots, r_n$ of exponential function imply that the hypersurface of conjugate types $r_1^*, \ldots, r_n^*$ for the function

$$\phi^*(z) = \sum_{|\alpha|=0}^{\infty} |\phi_\alpha| z^\alpha$$

is the same as for the function

$$\phi(z) = \sum_{|\alpha|=0}^{\infty} \phi_\alpha z^\alpha, \quad \phi_\alpha = D^\alpha \phi(0)/\alpha! \qquad (1.1.5)$$

(see, for example, Ronkin [1]).

Thus there exists a vector-type $r^* < R$ and a constant $M > 0$ such that

$$|\phi^*(z)| \le M \exp r^* z_+, \quad z \in \mathbf{C}^n.$$

From this we see that for any $N = 1, 2, \ldots$ the partial sums $S_N(z)$ of the series (1.1.5) satisfy the estimate

$$|S_N(z)| \le \sum_{|\alpha|=0}^{N} |\phi_\alpha| |z_1|^{\alpha_1} \ldots |z_n|^{\alpha_n} \le M \exp r^* z_+,$$

which implies that $S_N(z) \to \phi(z)$ $(N \to \infty)$ in the space $\mathrm{Exp}_\Omega(\mathbf{C}_z^n)$. Q.E.D.

Proposition 1.1.4. *Let* $\Delta^\alpha f(0, z)/\zeta^\alpha$ $(\zeta \in \mathbf{R}^1)$ *be the finite-difference quotients (with respect to ζ) of order α for the function $f(\zeta, z) \equiv \exp \zeta z$. Then*

$$\Delta^\alpha f(0, z)/\zeta^\alpha \to z^\alpha \ (\zeta \to 0)$$

locally uniformly in $\mathbf{C}_z^n$.

Proof. In fact, according to the well-known results on finite-difference approximations, we have

$$\Delta^\alpha f(0, z)/\zeta^\alpha = \partial_\zeta^\alpha f(\xi_\alpha, z), \qquad (1.1.6)$$

where $\xi_\alpha = (\xi_{\alpha_1}, \ldots, \xi_{\alpha_n})$ are certain intermediate points, that is, $\xi_{\alpha_1} \in (0, \alpha_1 \zeta_1), \ldots, \xi_{\alpha_n} \in (0, \alpha_n \zeta_n)$. Therefore, for $f(\zeta, z) \equiv \exp \zeta z$, we immediately obtain the inequality

$$\left| z^\alpha - \frac{\Delta^\alpha f(0, z)}{\zeta^\alpha} \right| \le \prod_{i=1}^{n} \alpha_i |\zeta_i| |z_i|^{\alpha_i + 1} \exp \alpha_i |\zeta_i z_i|, \qquad (1.1.7)$$

which clearly implies that

$$\Delta^\alpha f(0, z)/\zeta^\alpha \to z^\alpha \ (\text{ as } \zeta \to 0)$$

locally uniformly in $\mathbf{C}_z^n$. The proposition is proved.

We now complete the proof of the lemma. Namely, let $u(z) \in \text{Exp}_\Omega(\mathbf{C}_z^n)$, that is,

$$u(z) = \sum_\lambda e^{\lambda z} \phi_\lambda(z),$$

where the $\lambda \in \Omega$ take a finite set of values, and let $\phi_\lambda(z) \in \text{Exp}_{R(\lambda)}(\mathbf{C}_z^n)$.

It clearly suffices to prove the assertion of the lemma for each term $e^{\lambda z} \phi_\lambda(z)$ separately. For we note that, according to Propositions 1.1.3 and 1.1.4, there exists a sequence $\zeta \equiv \zeta_N \to 0$ $(N \to \infty)$ such that the sequence

$$\Lambda_{N,\lambda}(z) \equiv \sum_{|\alpha| \leq N} \phi_{\alpha\lambda} \Delta^\alpha f(0, z)/\zeta^\alpha$$

(the notation is clear) tends to $\phi_\lambda(z)$ locally uniformly in $\mathbf{C}_z^n$.

Let us now show that there exist $r^*(\lambda) < R(\lambda)$ and a constant $M > 0$ such that

$$|\Lambda_{N,\lambda}(z)| \leq M \exp r^*(\lambda) z_+, \quad z \in \mathbf{C}^n,$$

(clearly, in this case the proof will be finished).

In fact, in view of formula (1.1.6) and inequality (1.1.7) we have

$$|\Lambda_{N,\lambda}(z)| \leq |S_N(z)| + \sum_{|\alpha| \leq N} |\phi_{\alpha\lambda}| \left[\prod_{i=1}^n \alpha_i |\zeta_i| \cdot |z_i|^{\alpha_i+1} \cdot \exp \alpha_i |\zeta_i z_i| \right].$$

From this, again using the identity of the hypersurfaces of conjugate types for the function $\phi_\lambda(z)$ and the function

$$\phi_\lambda^*(z) \equiv \sum_{|\alpha|=0}^\infty |\phi_{\alpha\lambda}| z^\alpha,$$

we find that for any $\epsilon = (\epsilon_1, \ldots, \epsilon_n)$, $\epsilon_j > 0$ $(j = 1, \ldots, n)$ and for any $z \in \mathbf{C}^n$

$$|\Lambda_{N,\lambda}(z)| \leq M \left[\exp r z_+ + \exp(r + \epsilon) z_+ \right],$$

if the values $\zeta \equiv \zeta_N$ are sufficiently small.

Since $\epsilon > 0$ may be arbitrarily small and $r(\lambda) < R(\lambda)$, this means that

$$|\Lambda_{N,\lambda}(z)| \leq M \exp r^* z_+,$$

where $r^*(\lambda) < R(\lambda)$ and $M > 0$ is a constant.

It remains to note that for the function $e^{\lambda z} \Lambda_{N,\lambda}(z)$ all the exponents in the corresponding exponentials have the form $\lambda + \beta \zeta_N$, where $|\beta| \leq N$, and can therefore be chosen so that $\lambda + \beta \zeta_N \in \Omega$. Thus

$$e^{\lambda z} \Lambda_{N,\lambda}(z) \to e^{\lambda z} \phi_\lambda(z) \quad (N \to \infty)$$

in the sense of $\text{Exp}_{R(\lambda)}(\mathbf{C}_z^n)$. The lemma is completely proved.

1.2. PD-operators with analytic symbols

1. LOCAL ALGEBRA OF DIFFERENTIAL OPERATORS OF INFINITE ORDER

Let $A(\zeta)$ be an analytic function in the polycylinder $U_R = \{\zeta \in \mathbf{C}^n : |\zeta_j| < R,$ $j = 1, \ldots, n\}$. As is well known, every such analytic function may be represented in U_R by its Taylor series, that is,

$$A(\zeta) = \sum_{|\alpha|=0}^{\infty} a_\alpha \zeta^\alpha,$$

where $a_\alpha = \partial^\alpha A(0)/\alpha!$.

We associate with the function $A(\zeta)$ the differential operator of infinite order (d.o.i.o.) according to the formula

$$A(D) \equiv \sum_{|\alpha|=0}^{\infty} a_\alpha D^\alpha.$$

The function $A(\zeta)$ is called the symbol of the operator $A(D)$.

The following theorem guarantees that the operator $A(D)$ is well defined on functions $u(z) \in \mathrm{Exp}_R(\mathbf{C}_z^n)$ by the above formula.

Theorem 1.2.1. *If $u(z) \in \mathrm{Exp}_R(\mathbf{C}_z^n)$, then the function $A(D)u(z)$ is well-defined and also belongs to $\mathrm{Exp}_R(\mathbf{C}_z^n)$. Moreover, the map*

$$A(D) : \mathrm{Exp}_R(\mathbf{C}_z^n) \to \mathrm{Exp}_R(\mathbf{C}_z^n) \tag{1.2.1}$$

is continuous.

Proof. In fact, if $u(z) \in \mathrm{Exp}_R(\mathbf{C}_z^n)$, then according to Proposition 1.1.2,

$$|D^\alpha u(z)| \le M_1 r^\alpha \exp r z_+,$$

where $M_1 > 0$, $r < R$. Hence for every compact $K \subset \mathbf{C}_z^n$ we have

$$|D^\alpha u(z)| \le M r^\alpha, \quad z \in K,$$

where $M = M(K) > 0$ is some constant.

Further, since the symbol $A(\zeta)$ is analytic in U_R, the series

$$A(D)u(z) \equiv \sum_{|\alpha|=0}^{\infty} a_\alpha D^\alpha u(z)$$

converges uniformly on K (together with all derivatives). Moreover for any $z \in \mathbf{C}_z^n$ we clearly have

$$|A(D)u(z)| \leq M_2 \exp rz_+,$$

where

$$M_2 \equiv \sum_{|\alpha|=0}^{\infty} a_\alpha r^\alpha < \infty.$$

Thus we have proved that $A(D)u(z) \in \operatorname{Exp}_R(\mathbf{C}_z^n)$.

We now prove that the map (1.2.1) is continuous. Thus, let $u_\nu(z) \to u(z)$ in $\operatorname{Exp}_R(\mathbf{C}_z^n)$ ($\nu \to \infty$). Clearly we have

$$|A(D)u(z) - A(D)u_\nu(z)| \leq \left| \sum_{|\alpha|=0}^{N} a_\alpha D^\alpha \Big(u(z) - u_\nu(z) \Big) \right| +$$

$$+ \left| \sum_{|\alpha|=N+1}^{\infty} a_\alpha D^\alpha u(z) \right| + \left| \sum_{|\alpha|=N+1}^{\infty} a_\alpha D^\alpha u_\nu(z) \right|, \qquad (1.2.2)$$

where N is any integer.

By definition of convergence in the space $\operatorname{Exp}_R(\mathbf{C}_z^n)$, there exist a constant $M > 0$ and a vector-type $r < R$ such that for any $\nu = 1, 2, \ldots$

$$|D^\alpha u_\nu(z)| \leq M r^\alpha \exp rz_+ ,$$

therefore, according to the remark at the end of subsection 2 in §1.1, there exist $M > 0$ and $r < R$ (a different one) such that

1. M and r do not depend on $\nu = 1, 2, \ldots,$

2. $|D^\alpha u_\nu(z)| \leq M r^\alpha \exp rz_+,\ z \in \mathbf{C}^n.$ $\qquad (1.2.3)$

From this it follows that for any compact $K \subset \mathbf{C}_z^n$ the last term in (1.2.2) is arbitrarily small, if $N > 1$ is sufficiently large. The same is also true for the second term in (1.2.2). It remains to note that for any fixed N

$$\sum_{|\alpha|=0}^{N} a_\alpha (D^\alpha u(z) - D^\alpha u_\nu(z)) \to 0 \quad (\nu \to \infty)$$

uniformly with respect to $z \in K$. Thus we have proved that

$$A(D)u_\nu(z) \to A(D)u(z)$$

locally uniformly in $\mathbf{C}_z^n$.

Finally, using inequalities (1.2.3) , we immediately obtain

$$|A(D)u_\nu(z)| \leq M_1 \exp r z_+,$$

where the constant

$$M_1 = \sum_{|\alpha|=0}^{\infty} |a_\alpha| r^\alpha < \infty$$

does not depend on $\nu = 1, 2, \ldots$.

We infer from the above arguments that

$$A(D)u_\nu(z) \to A(D)u(z)$$

in the space $\mathrm{Exp}_R(\mathbf{C}_z^n)$. The theorem is proved.

We now denote by $\mathcal{A}(U_R)$ the set of all d.o.i.o.'s $A(D)$ with analytic symbols and with domain of definition $\mathrm{Exp}_R(\mathbf{C}_z^n)$. Then, as a corollary of the previous results, we have

Theorem 1.2.2. *The set $\mathcal{A}(U_R)$ is an algebra which is isomorphic to the algebra $\mathcal{O}(U_R)$ of all analytic functions in U_R. Moreover*

$$A(D) \leftrightarrow A(\zeta),$$
$$\alpha A(D) + \beta B(D) \leftrightarrow \alpha A(\zeta) + \beta B(\zeta)$$

(here $\alpha \in \mathbf{C}^1$, $\beta \in \mathbf{C}^1$),

$$A(D) \circ B(D) \leftrightarrow A(\zeta)B(\zeta).$$

In particular, if $A(\zeta) \neq 0$ in U_R, then the operator $I/A(D)$ with symbol $1/A(\zeta)$ is the operator inverse of $A(D)$.

Remark. Let $R = \infty$, that is, $U_R = \mathbf{C}_\zeta^n$. Then $\mathrm{Exp}_R(\mathbf{C}_z^n)$ is the space of all exponential functions $\mathrm{Exp}(\mathbf{C}_z^n)$ and $\mathcal{A}(U_R) \equiv \mathcal{A}(\mathbf{C}_\zeta^n)$ is the algebra of d.o.i.o.'s with arbitrary entire symbols and with domain of definition $\mathrm{Exp}(\mathbf{C}_z^n)$. Thus

$$\mathcal{A}(\mathbf{C}_z^n) \leftrightarrow \mathcal{O}(\mathbf{C}_\zeta^n),$$

where $\mathcal{O}(\mathbb{C}^n_\zeta)$ is the space of all entire functions in $\mathbb{C}^n_\zeta$.

We conclude this subsection with some examples.

Example 1. Consider the shift operator

$$A(u) : u(z) \rightarrow u(z + a),$$

where $a \in \mathbb{C}^n$ is the shift vector. Clearly,

$$A(u) = \sum_{|\alpha|=0}^{\infty} \frac{D^\alpha u(z)}{\alpha!} a^\alpha = \sum_{|\alpha|=0}^{\infty} \frac{a^\alpha}{\alpha!} D^\alpha u(z)$$

is a d.o.i.o. with symbol $A(\zeta) = \exp a\zeta$.

Example 2. Consider the differential operator with shifts a_β, $|\beta| = 0, 1, \ldots, m$,

$$Au(z) \equiv \sum_{|\beta|=0}^{m} b_\beta D^\beta u(z + a_\beta).$$

It is clear that this operator is a d.o.i.o. with symbol

$$A(\zeta) = \sum_{|\beta|=0}^{m} b_\beta \zeta^\beta \exp \zeta a_\beta.$$

Example 3. Let $\mu(dw)$ be a compactly supported (or "rapidly decreasing") measure in $\mathbb{C}^n_w$. Consider the convolution operator

$$Au(z) = \int_{\mathbb{C}^n} u(z - w)\mu(dw), \quad u(z) \in \mathrm{Exp}(\mathbb{C}^n_z).$$

Applying Taylor's formula, we obtain

$$Au(z) = \sum_{|\alpha|=0}^{\infty} a_\alpha D^\alpha u(z),$$

where

$$a_\alpha = \frac{1}{\alpha!} \int_{\mathbb{C}^n} (-w)^\alpha \mu(dw).$$

Thus the convolution operator is a d.o.i.o., the symbol of which is

$$A(\zeta) = \sum_{|\alpha|=0}^{\infty} a_\alpha \zeta^\alpha.$$

In all these examples the domain of definition of the d.o.i.o. is the space $\mathrm{Exp}(\mathbf{C}_z^n)$ of all entire functions of exponential type. The next example is of a different character.

Example 4. The symbol $A(\zeta) = (1 + \zeta)^{-1}$, $\zeta \in \mathbf{C}^1$, is clearly an analytic function in the circle $U_1 = \{\zeta : |\zeta| < 1\}$. Therefore, the operator

$$A(D) \equiv \frac{I}{I + D}, \; D = d/dz,$$

acts continuously in the space $\mathrm{Exp}_R(\mathbf{C}_z^n)$, where $R = 1$, that is, in the space of exponential functions the type of which are less than 1. In this space the operators $A(D) = I/(I + D)$ and $B(D) = I + D$ are mutually inverse. (The full domain of definition of the operator $I/(I + D)$ will be defined below in subsection 2, Example 2.)

Remark. We have studied the local algebra of d.o.i.o.'s in a neighbourhood of zero but, clearly, a completely analogous theory can be constructed in a neighbourhood of an arbitrary $\lambda \in \Omega$.

Namely, let $U_R(\lambda) = \{\zeta \in \Omega : |\zeta_j - \lambda_j| < R, \; 1 \leq j \leq n\}$, $R \equiv R(\lambda) = \mathrm{dist}(\lambda, \partial\Omega)$ be a polycylinder in Ω centred at λ. We set

$$\mathrm{Exp}_{U_R(\lambda)}(\mathbf{C}_z^n) \equiv e^{\lambda z} \, \mathrm{Exp}_{R(\lambda)}(\mathbf{C}_z^n),$$

that is,

$$\mathrm{Exp}_{U_R(\lambda)}(\mathbf{C}_z^n) = \{u(z) : u(z) = e^{\lambda z} \phi(z), \; \phi(z) \in \mathrm{Exp}_{R(\lambda)}(\mathbf{C}_z^n)\}.$$

In other words, the space $\mathrm{Exp}_{U_R(\lambda)}(\mathbf{C}_z^n)$ is the "exponential shift" of the space $\mathrm{Exp}_{R(\lambda)}(\mathbf{C}_z^n)$ to the point $\lambda \in \Omega$.

In this case it is natural to set

$$A(D)u(z) \stackrel{\mathrm{def}}{=} \sum_{|\alpha|=0}^{\infty} a_\alpha(\lambda)(D - \lambda I)^\alpha u(z),$$

where the $a_\alpha(\lambda) \equiv \partial^\alpha A(\lambda)/\alpha!$ are the Taylor coefficients at the point $\lambda \in \Omega$. Then, taking into account the fact that

$$(D - \lambda I)^\alpha [e^{\lambda z} \phi(z)] = e^{\lambda z} D^\alpha \phi(z),$$

we obtain, for $u(z) \in \mathrm{Exp}_{U_R(\lambda)}(\mathbf{C}_z^n)$,

$$A(D)u(z) = e^{\lambda z} \sum_{|\alpha|=0}^{\infty} a_\alpha(\lambda) D^\alpha \phi(z) \equiv e^{\lambda z} \psi(z),$$

where $|\psi(z)| \le M \exp r(\lambda) z_+$, $r(\lambda) < R(\lambda)$. This means that $A(D)u(z) \in \mathrm{Exp}_{U_R(\lambda)}(\mathbf{C}_z^n)$ as well. Repeating the above estimates, it is easy to see that the action

$$A(D) : \mathrm{Exp}_{U_R(\lambda)}(\mathbf{C}_z^n) \to \mathrm{Exp}_{U_R(\lambda)}(\mathbf{C}_z^n)$$

is continuous. Thus there exists an isomorphism

$$\mathcal{A}(U_R(\lambda)) \leftrightarrow \mathcal{O}(U_R(\lambda)).$$

2. ALGEBRA OF PD-OPERATORS WITH ARBITRARY ANALYTIC SYMBOLS

Let $A(\zeta) \in \mathcal{O}(\Omega)$, that is, $A(\zeta)$ is an arbitrary analytic function on $\Omega \subset \mathbf{C}_\zeta^n$. Further, let $\mathrm{Exp}_\Omega(\mathbf{C}_z^n)$ be the associated space of exponential functions. We recall (see subsection 3 of §1.1) that

$$\mathrm{Exp}_\Omega(\mathbf{C}_z^n) = \left\{ u(z) : u(z) = \sum_\lambda u_\lambda(z),\ u_\lambda(z) \in \mathrm{Exp}_{U_R(\lambda)}(\mathbf{C}_z^n) \right\},$$

where the summation is carried out over all finite sets of $\lambda \in \Omega$.

We shall associate with the function $A(\zeta) \in \mathcal{O}(\Omega)$ the PD-operator $A(D)$ acting continuously on $\mathrm{Exp}_\Omega(\mathbf{C}_z^n)$ in the following way.

Definition 1.2.1. The action of the operator $A(D)$ on $u(z) \in \mathrm{Exp}_\Omega(\mathbf{C}_z^n)$ is defined by

$$A(D)u(z) = \sum_\lambda A(D)u_\lambda(z),$$

where

$$A(D)u_\lambda(z) = \sum_{|\alpha|=0}^{\infty} a_\alpha(\lambda)(D - \lambda I)^\alpha u_\lambda(z), \quad a_\alpha(\lambda) = \frac{\partial^\alpha A(\lambda)}{\alpha!}.$$

It is clear (see the remark on p. 19) that for any $u(z) \in \mathrm{Exp}_\Omega(\mathbf{C}_z^n)$ its image $A(D)u(z) \in \mathrm{Exp}_\Omega(\mathbf{C}_z^n)$ also. Moreover

$$A(D) : \mathrm{Exp}_\Omega(\mathbf{C}_z^n) \to \mathrm{Exp}_\Omega(\mathbf{C}_z^n)$$

is a continuous map.

☐ Remark. Sometimes it can happen that the function $u(z) \in \text{Exp}_\Omega(\mathbf{C}^n_z)$ may be represented in the form

$$u(z) = \sum_\lambda u_\lambda(z), \ u_\lambda(z) \in \text{Exp}_{U_R(\lambda)}(\mathbf{C}^n_z),$$

in various ways. It turns out that if Ω is a Runge domain, then the value $A(D)u(z)$ does not depend on the representation of $u(z)$ in this form. This means that the PD-operator $A(D)$ (Definition 1.2.1) is well defined. The proof of this fact will be given in subsection 3 below. ☐

Thus, taking into account this remark and the previous results, we may assert the following.

Theorem 1.2.3. *The space* $\text{Exp}_\Omega(\mathbf{C}^n_z)$ *is invariant with respect to any PD-operator* $A(D)$ *the symbol of which* $A(\zeta)$ *is analytic in* Ω; *moreover, the map*

$$A(D) : \text{Exp}_\Omega(\mathbf{C}^n_z) \to \text{Exp}_\Omega(\mathbf{C}^n_z) \tag{1.2.4}$$

is continuous.

Furthermore, the collection of operators (1.2.4) *with symbols analytic in the region* Ω *forms an algebra* $\mathcal{A}(\Omega)$ *isomorphic to the algebra of analytic functions* $\mathcal{O}(\Omega)$, *that is,*

$$\mathcal{A}(\Omega) \leftrightarrow \mathcal{O}(\Omega).$$

Here

$$A(D) \leftrightarrow A(\zeta), \ \alpha A(D) + \beta B(D) \leftrightarrow \alpha A(\zeta) + \beta B(\zeta)$$
$$(\alpha, \beta \in \mathbf{C}^1), \ A(D) \circ B(D) \leftrightarrow A(\zeta) \cdot B(\zeta).$$

In particular, if along with $A(\zeta)$ the function $A^{-1}(\zeta)$ is also analytic in Ω, then $I/A(D)$ is the inverse operator to $A(D)$, that is,

$$A(D) \circ \frac{I}{A(D)} = \frac{I}{A(D)} \circ A(D) = I,$$

where I is the identity operator.

Example 1. Let $A(\zeta)$ be analytic in a neighbourhood of $\lambda \in \Omega$ and $u(z) = P(z)e^{\lambda z}$, where $P(z)$ is polynomial. Then

$$A(D)u(z) \equiv \sum_{|\alpha|=0}^{\infty} a_\alpha(\lambda)(D - \lambda I)^\alpha [e^{\lambda z} P(z)] =$$

$$= e^{\lambda z} \sum_{|\alpha|=0}^{\infty} a_\alpha(\lambda) D^\alpha P(z) \equiv e^{\lambda z} Q(z),$$

where $Q(z)$ is also polynomial. Moreover, $\deg P(z) \geq \deg Q(z)$.

Example 2. Let $A(\zeta) = 1/\zeta$, $\zeta \in \Omega = \mathbf{C}^1 \backslash \mathbf{R}_+^1$, where $\mathbf{R}_+^1 = \{x \in \mathbf{R}^1 : x \geq 0\}$. Evidently, Ω is a Runge domain. Consequently, the space $\mathrm{Exp}_\Omega(\mathbf{C}_z^1)$ contains all functions of the form

$$u(z) = \sum_{\lambda \notin \mathbf{R}_+^1} e^{\lambda z} \phi_\lambda(z).$$

Furthermore, the $\phi_\lambda(z)$ are entire functions satisfying the inequality

$$|\phi_\lambda(z)| \leq M_\lambda \exp r_\lambda z_+,$$

where $M_\lambda > 0$, $r_\lambda < R(\lambda) = \mathrm{dist}(\lambda, \mathbf{R}_+^1)$.

According to Theorem 1.2.3, for any $u(z) \in \mathrm{Exp}_\Omega(\mathbf{C}_z^1)$ there exists one (and only one) function $w(z) \in \mathrm{Exp}_\Omega(\mathbf{C}_z^1)$ such that

$$w(z) = \frac{I}{D}u(z). \tag{1.2.5}$$

It is natural to call the function (1.2.5) the "natural" inverse of the function $u(z) \in \mathrm{Exp}_\Omega(\mathbf{C}_z^1)$ and denote it by

$$w(z) = \mathrm{nat} \int u(z)dz.$$

If, for example, $u(z)$ is a quasipolynomial, then the "natural" inverse $w(z)$ may be obtained by usual integration. Here we must "forget" to write down the additive constants during the process of integration.

Example 3.   Consider the operator $\sqrt{I - \Delta}$, where $\Delta \equiv \partial^2/\partial x_1^2 + \ldots + \partial^2/\partial x_n^2$ is the Laplace operator. The operator $\sqrt{I - \Delta}$ occurs in relativistic quantum mechanics in the Schrödinger equation of a relativistic free particle (see Bjorken and Drell [1]).

We consider the complex generalization of this operator, namely,

$$\sqrt{I - \Delta} \equiv \sqrt{I - \partial^2/\partial z_1^2 - \ldots - \partial^2/\partial z_n^2}.$$

The function $A(\zeta) \equiv \sqrt{1 - s^2}$, $s \in \mathbf{C}^1$, has two branch points $s = \pm 1$. Joining these points by a simple contour Γ and making the cut along Γ we can choose a single-valued branch of this function. The resulting single-valued function $A(\zeta) = \sqrt{1 - \zeta^2}$, $\zeta \in \mathbf{C}^n$, will be analytic in $\mathbf{C}^n$ except at those points $\zeta \in \mathbf{C}^n$ for which $\zeta^2 \in \Gamma$. Consequently, the PD-operator $A(D) = \sqrt{I - \Delta}$ is well-defined in the space $\mathrm{Exp}_\Omega(\mathbf{C}_z^n)$ where $\Omega \subset \mathbf{C}_\zeta^n \backslash \{\zeta : \zeta^2 \in \Gamma\}$ is an arbitrary Runge domain.

3. THE CORRECTNESS OF THE DEFINITION OF A PD-OPERATOR

In accordance with the remark in the preceding subsection, we now prove that Definition 1.2.1 is correct. It is useful (from the methodological point of view) to consider the cases $n = 1$ and $n > 1$ separately.

1. *Case $n = 1$.* Let $\Omega \subset \mathbf{C}^1$ be a Runge domain (recall that for $n = 1$ a Runge domain is just a simply connected region). Further, suppose that some entire function $u(z)$ can be represented on the one hand as

$$u(z) = \sum_\lambda e^{\lambda z} \phi_\lambda(z) \tag{1.2.6}$$

and, on the other hand, as

$$u(z) = \sum_\mu e^{\mu z} \phi_\mu(z), \tag{1.2.7}$$

where $\lambda \in \Omega$ and $\mu \in \Omega$ belong to certain (in general, different) finite sets of points in Ω. Here, the functions $\phi_\lambda(z)$ and $\phi_\mu(z)$ are entire functions such that

$$|\phi_\lambda(z)| \leq M_\lambda \exp r_\lambda z_+,$$

$$|\phi_\mu(z)| \leq M_\mu \exp r_\mu z_+,$$

where $M_\lambda > 0$, $M_\mu > 0$, $r_\lambda < R(\lambda) = \mathrm{dist}(\lambda, \partial\Omega)$, $r_\mu < R(\mu) = \mathrm{dist}(\mu, \partial\Omega)$ are constants (see §1, Definition 1.1.2). Then in accordance with Definition 1.2.1 we have, on the one hand,

$$A(D)u(z) \stackrel{\mathrm{def}}{=} \sum_\lambda \left(\sum_{|\alpha|=0}^\infty a_\alpha(\lambda)(D - \lambda I)^\alpha [e^{\lambda z} \phi_\lambda(z)] \right) \tag{1.2.8}$$

and, on the other hand,

$$A(D)u(z) \stackrel{\text{def}}{=} \sum_{\mu} \Big(\sum_{|\alpha|=0}^{\infty} a_\alpha(\mu)(D - \mu I)^\alpha [e^{\mu z} \phi_\mu(z)], \tag{1.2.9}$$

where $a_\alpha(\lambda) = \partial^\alpha A(\lambda)/\alpha!$, $a_\mu(\mu) = \partial^\alpha A(\mu)/\alpha!$.

Our aim is to prove that both values $A(D)u(z)$ are equal, that is, the action of the PD-operator $A(D)$ so defined does not depend on the representation of $u(z)$ in the form (1.2.8) or (1.2.9).

To this end we denote by $Bu_\lambda(\zeta)$ the function associated with $u_\lambda(z)$ in the sense of Borel, or what is the same, the Borel transform of $u_\lambda(z)$. As is well known (see, for example, Leont'ev [1], Ronkin [1]), the function $Bu_\lambda(\zeta)$ is analytic outside the disc $U(\lambda) = \{\zeta : |\zeta - \lambda| < r_\lambda\}$. Moreover, by the inversion formula,

$$u_\lambda(z) = \frac{1}{2\pi i} \int_{\Gamma_\lambda} Bu_\lambda(\zeta) e^{z\zeta} d\zeta,$$

where Γ_λ is some contour enclosing $U(\lambda)$.

Clearly, one can say the same about the function $u_\mu(z)$. Therefore, in accordance with (1.2.6) and (1.2.7), our function $u(z)$ may be represented either by the formula

$$u(z) = \frac{1}{2\pi i} \sum_\lambda \int_{\Gamma_\lambda} Bu_\lambda(\zeta) e^{z\zeta} d\zeta$$

or by the formula

$$u(z) = \frac{1}{2\pi i} \sum_\mu \int_{\Gamma_\mu} Bu_\mu(\zeta) e^{z\zeta} d\zeta.$$

Hence, in accordance with (1.2.8) and (1.2.9) we have, in the first case,

$$A(D)u(z) = \frac{1}{2\pi i} \sum_\lambda \int_{\Gamma_\lambda} A(\zeta) Bu_\lambda(\zeta) e^{z\zeta} d\zeta \tag{1.2.10}$$

and, in the second case,

$$A(D)u(z) = \frac{1}{2\pi i} \sum_\mu \int_{\Gamma_\mu} A(\zeta) Bu_\mu(\zeta) e^{z\zeta} d\zeta \tag{1.2.11}$$

We now note that the union of circles $U(\lambda)$ and the union of circles $U(\mu)$ are relatively compact in Ω, so that there exists a contour $\Gamma \subset \Omega$ contained in these unions. Consequently (and in view of the Runge property), formulae (1.2.10) and (1.2.11) may be written as

$$A(D)u(z) = \frac{1}{2\pi i} \int_\Gamma A(\zeta) \sum_\lambda Bu_\lambda(\zeta) e^{z\zeta} d\zeta$$

and

$$A(D)u(z) = \frac{1}{2\pi i} \int_\Gamma A(\zeta) \sum_\mu Bu_\mu(\zeta) e^{z\zeta} d\zeta$$

(we recall that $A(\zeta)$ is analytic in Ω).

Now in view of the linearity of the Borel transformation, (1.2.6) and (1.2.7) yield

$$\sum_\lambda Bu_\lambda(\zeta) = \sum_\mu Bu_\mu(\zeta) = Bu(\zeta),$$

where $Bu(\zeta)$ is the Borel transform of $u(z)$.

Thus, in both cases,

$$A(D)u(z) = \frac{1}{2\pi i} \int_\Gamma Bu(\zeta) e^{z\zeta} d\zeta$$

which clearly implies the desired well-definedness.

2. *Case $n > 1$.* Let us suppose that the region $\Omega \subset \mathbf{C}^n$, $n > 1$, is a Runge domain. By definition, this means that every function $u(z) \in \mathcal{O}(\Omega)$ may be represented as the limit of some sequence of linear combinations of exponentials in the sense of locally uniform convergence.

Again suppose that the function $u(z) \in \mathrm{Exp}_\Omega(\mathbf{C}_z^n)$ can be written in two forms:

$$u(z) = \sum_\lambda e^{\lambda z} \phi_\lambda(z) \qquad (1.2.12)$$

and

$$u(z) = \sum_\mu e^{\mu z} \phi_\mu(z), \qquad (1.2.13)$$

where λ and μ belong to some finite set of points in Ω; the functions $\phi_\lambda(z)$ and $\phi_\mu(z)$ belong to $\mathrm{Exp}_{R(\lambda)}(\mathbf{C}_z^n)$ and $\mathrm{Exp}_{R(\mu)}(\mathbf{C}_z^n)$ respectively.

For the proof that the values $A(D)u(z)$ are well-defined, we use the Borel inversion formula. Namely, if $u(z)$ is an entire function and

$$u(z) = \sum_{|\alpha|=0}^\infty u_\alpha z^\alpha, \quad z \in \mathbf{C}^n,$$

is its Taylor expansion, then the associated Borel function $Bu(\zeta)$ is defined as

$$Bu(\zeta) = \sum_{|\alpha|=0}^{\infty} \frac{\alpha! u_\alpha}{\zeta^{\alpha+1}},$$

where $\alpha + 1 = (\alpha_1 + 1, \ldots, \alpha_n + 1)$. If $r_1 > 0, \ldots, r_n > 0$ are the types of exponential growth of $u(z)$, then the Borel function $Bu(\zeta)$ is analytic for $|\zeta_j| > r_j$, $j = 1, \ldots, n$, and the inversion formula is given by

$$u(z) = \frac{1}{2\pi i} \int_\Gamma Bu(\zeta) e^{z\zeta} d\zeta, \ z \in \mathbb{C}^n,$$

where Γ is the skeleton of the polycylinder

$$U = \{\zeta : |\zeta_j| < r_j + \epsilon, \ \epsilon > 0, \ j = 1, \ldots, n\}, \ d\zeta = d\zeta_1 \ldots d\zeta_n.$$

From this and (1.2.12) we obtain

$$u(z) = \frac{1}{(2\pi i)^n} \sum_\lambda \int_{\Gamma_\lambda} Bu_\lambda(\zeta) e^{z\zeta} d\zeta, \ z \in \mathbb{C}^n, \tag{1.2.14}$$

where $Bu_\lambda(\zeta)$ is the Borel function associated with $u_\lambda(z) = \exp \lambda z \cdot \phi_\lambda(z)$ and Γ_λ is the skeleton of the polycylinder

$$U_\lambda = \{\zeta : |\zeta_j - \lambda_j| < r_j + \epsilon, \ \epsilon > 0, \ j = 1, \ldots, n\}.$$

It is clear that $\epsilon > 0$ may be chosen so small that all the polycylinders U_λ will lie strictly inside Ω.

Consequently, by Definition 1.2.1,

$$A(D)u(z) = \frac{1}{(2\pi i)^n} \sum_\lambda \left(\sum_{|\alpha|=0}^{\infty} a_\alpha(\lambda)(D - \lambda I)^\alpha \int_{\Gamma_\lambda} Bu_\lambda(\zeta) e^{z\zeta} d\zeta \right) =$$

$$= \frac{1}{(2\pi i)^n} \sum_\lambda \int_{\Gamma_\lambda} \left(\sum_{|\alpha|=0}^{\infty} a_\alpha(\lambda)(\zeta - \lambda)^\alpha Bu_\lambda(\zeta) e^{z\zeta} \right) d\zeta =$$

$$= \frac{1}{(2\pi i)^n} \sum_\lambda \int_{\Gamma_\lambda} Bu_\lambda(\zeta) A(\zeta) e^{z\zeta} d\zeta. \tag{1.2.15}$$

It is clear that, starting from (1.2.13) and repeating the above calculations, we obtain the analogous formulae

$$u(z) = \frac{1}{(2\pi i)^n} \sum_\mu \int_{\Gamma_\mu} Bu_\mu(\zeta) e^{z\zeta} d\zeta \tag{1.2.16}$$

and

$$A(D)u(z) = \frac{1}{(2\pi i)^n} \sum_\mu \int_{\Gamma_\mu} Bu_\mu(\zeta) A(\zeta) e^{z\zeta} d\zeta. \qquad (1.2.17)$$

Finally, it is not difficult to see that the values (1.2.16) and (1.2.17) are the same. In fact, consider the analytic functionals

$$L(v) \overset{\text{def}}{=} \frac{1}{(2\pi i)^n} \sum_\lambda \int_{\Gamma_\lambda} Bu_\lambda(\zeta) v(\zeta) d\zeta$$

and

$$M(v) \overset{\text{def}}{=} \frac{1}{(2\pi i)^n} \sum_\mu \int_{\Gamma_\mu} Bu_\mu(\zeta) v(\zeta) d\zeta,$$

where $v(\zeta) \in \mathcal{O}(\Omega)$ is arbitrary.

Formulae (1.2.14) and (1.2.16) imply that

$$L(e^{z\zeta}) = M(e^{z\zeta})$$

for any $z \in \mathbf{C}^n$, which clearly yields the equality

$$L(v) = M(v)$$

for every linear combination of exponentials. Finally, in view of the Runge condition, the last equality is true for any analytic function $v(\zeta)$ in Ω, that is, $L(v)$ and $M(v)$ are identical analytic functionals on Ω. In particular, setting $v(\zeta) = A(\zeta)e^{z\zeta}$, we obtain the equality

$$L(A(\zeta)e^{z\zeta}) = M(A(\zeta)e^{z\zeta}),$$

which, in accordance with (1.2.15) and (1.2.17), means that the value $A(D)u(z)$ does not depend on the form of the representation of $u(z)$, as required. Q.E.D.

In conclusion, we show that Runge condition is essential.

Counterexample. Let $\Omega = \mathbf{C}^1 \backslash \{0\}$ and $A(D) = I/D$, where $D \equiv d/dz$. Clearly, the symbol $A(\zeta) = 1/\zeta$ is analytic in Ω. Let us show, however, that $A(D)$ is not single-valued. In fact, for any $\lambda \neq 0$ we have, by Cauchy's formula,

$$e^{\lambda z} = \frac{1}{2\pi i} \int_{|\eta|=R} \frac{e^{\eta z}}{\eta - \lambda} d\eta, \quad R > |\lambda|,$$

therefore, representing the contour Γ as a union $\Gamma = \bigcup \Gamma_j$ $(1 \leq j \leq N)$, where the Γ_j are sufficiently small, we obtain

$$e^{\lambda z} = \frac{1}{2\pi i} \sum_{j=1}^{N} \int_{\Gamma_j} \frac{e^{\eta z}}{\eta - \lambda} d\eta = \sum_{j=1}^{N} e^{\lambda_j z} \phi_j(z), \quad \lambda_j \in \Gamma_j, \qquad (1.2.18)$$

where

$$\phi_j(z) = \frac{1}{2\pi i} \int_{\Gamma_j} \frac{e^{(\eta - \lambda_j)z}}{\eta - \lambda} d\eta.$$

It is clear that Γ_j may be taken so small that

$$|\phi_j(z)| \leq M_j \exp r|z|,$$

where $r < |\lambda|$ and $M_j > 0$ are constants. Thus, formula (1.2.18) gives a representation of $u(z) \equiv e^{\lambda z}$ in the form (1.2.6).

In accordance with Definition 1.2.1 we have, on the one hand,

$$A(D)e^{\lambda z} \equiv \frac{I}{D} e^{\lambda z} = \frac{1}{\lambda} e^{\lambda z},$$

while on the other, taking into account (1.2.18),

$$A(D)e^{\lambda z} = \frac{1}{2\pi i} \int_{|\eta|=R} \frac{e^{\eta z}}{\eta(\eta - \lambda)} d\eta = \frac{1}{\lambda}(e^{\lambda z} - 1).$$

Thus the operator $A(D)$ is at least two-valued. It follows that $A(D)$ may, in general, be multivalued if Ω is not a Runge domain.

1.3. The operator method

1. PD-EQUATIONS IN THE WHOLE SPACE $\mathbf{C}^n$

We consider the equation

$$A(D)u(z) = h(z), \quad z \in \mathbf{C}^n, \qquad (1.3.1)$$

where $A(D)$ is a PD-operator with analytic symbol.

Theorem 1.3.1. *Let $A(\zeta)$ be an analytic function in some Runge domain $\Omega \subset \mathbf{C}^n_\zeta$, where $A(\zeta) \neq 0$, $\zeta \in \Omega$. Then for any $h(z) \in \mathrm{Exp}_\Omega(\mathbf{C}^n_z)$ equation (1.3.1) has one and only one solution $u(z) \in \mathrm{Exp}_\Omega(\mathbf{C}^n_z)$. Moreover,*

$$u(z) = \frac{I}{A(D)} h(z).$$

Proof. In fact, we have already noted that if $A(\zeta) \neq 0$ in Ω, then the operator $B(D) \equiv I/A(D)$ is the inverse of the operator $A(D)$. The theorem is proved.

Example 1 (A method of selecting quasipolynomial solutions). Consider equation (1.3.1) with quasipolynomial right hand side $h(z)$, that is,

$$A(D)u(z) = \exp \lambda z \cdot P(z),$$

where λ is a complex number and $P(z)$ is a polynomial. Clearly $e^{\lambda z} P(z) \in \text{Exp}_\Omega(\mathbf{C}_z^n)$, where Ω can be chosen to be any domain containing λ. Therefore, if $A(\lambda) \neq 0$, then the solution

$$u(z) = \frac{I}{A(D)}[e^{\lambda z}P(z)] = \sum_{|\alpha|=0}^{\infty} b_\alpha(\lambda)(D-\lambda I)^\alpha[e^{\lambda z}P(z)] =$$

$$= e^{\lambda z}\sum_{|\alpha|=0}^{\infty} b_\alpha(\lambda)D^\alpha P(z) = e^{\lambda z}Q(z),$$

where $Q(z)$ is a polynomial of degree not greater than that of $P(z)$.

Example 2. Consider the complex Helmholz equation

$$\Delta u + \omega^2 u = h(z), \quad z \in \mathbf{C}^n, \tag{1.3.2}$$

where

$$\Delta \equiv \frac{\partial^2}{\partial z_1^2} + \ldots + \frac{\partial^2}{\partial z_n^2}$$

is the complex Laplace operator and $\omega \in \mathbf{C}^1$ is a parameter.

The symbols $A(\zeta) = \zeta^2 + \omega^2$ and $A^{-1}(\zeta) = 1/(\zeta^2 + \omega^2)$ are clearly analytic in the domain $\Omega_0 = \mathbf{C}_\zeta^n \backslash \{\zeta \in \mathbf{C}^n : \zeta^2 + \omega^2 = 0\}$. Consequently, for any function $h(z) \in \text{Exp}_\Omega(\mathbf{C}_z^n)$, where $\Omega \subset \Omega_0$ is an arbitrary Runge domain, there exists a unique solution of (1.3.2)

$$u(z) = \frac{I}{\Delta + \omega^2 I}h(z).$$

Example 3. Consider the equation

$$u(z + a) + u(z - a) = h(z), \quad z \in \mathbf{C}^1, \tag{1.3.3}$$

where $a \in \mathbf{C}^1$ is a complex shift.

Equation (1.3.3) can be written in the form

$$\left[\exp\left(a\frac{d}{dz}\right) + \exp\left(-a\frac{d}{dz}\right)\right]u(z) = h(z)$$

or, what is the same,

$$2\cosh\left(a\frac{d}{dz}\right)u(z) = h(z).$$

The symbol of the operator $2\cosh\left(a\frac{d}{dz}\right)$ is the entire function $\cosh a\zeta$, which is equal to zero at the points $ia\zeta = \pi/2 + k\pi$, $k = 0, \pm 1, \ldots$. Therefore, for any Runge domain

$$\Omega \subset \mathbf{C}^1 \backslash \{\zeta \in \mathbf{C}^1 : \zeta = \pi/2ia + k\pi/ia, \ k = 0, \pm 1, \ldots\}$$

and any $h(z) \in \mathrm{Exp}_\Omega(\mathbf{C}_z^1)$ equation (1.3.3) has the unique solution

$$u(z) = \tfrac{1}{2}\mathrm{sech}\left(a\frac{d}{dz}\right)h(z).$$

We note, in particular, that if $h(z)$ has growth type $r < \pi/2$, then in accordance with the well known expansion of $\mathrm{sech}\,\zeta$ in a neighbourhood of zero,

$$u(z) = \tfrac{1}{2}\sum_{n=0}^{\infty}\frac{(-1)^n a_n E_{2n}}{(2n)!}\frac{d^{2n}}{dz^{2n}}h(z),$$

where the E_{2n} are the Euler numbers ($E_0 = 1$, $E_2 = -1$, $E_4 = 5\ldots$, see, for example, the handbook by Abramowitz and Stegun [1], p. 810). In particular, for any polynomial $h(z) \equiv P(z)$ there exists only one polynomial solution

$$u(z) = \tfrac{1}{2}\mathrm{sech}\left(a\frac{d}{dz}\right)P(z)$$

whose degree is equal to that of $P(z)$.

2. THE CAUCHY PROBLEM IN THE SPACE OF EXPONENTIAL FUNCTIONS

In the space $\mathbf{C}^{n+1}$ of independent variables $t \in \mathbf{C}^1$ and $z \in \mathbf{C}^n$ we consider the following Cauchy problem of order $m \geq 1$ on t:

$$\frac{\partial^m u}{\partial t^m} + \sum_{k=0}^{m-1} A_k(t, D)\frac{\partial^k u}{\partial t^\lambda} = h(t, z), \tag{1.3.4}$$

$$u(0, z) = \phi_0(z), \ldots, \ \frac{\partial^{m-1} u}{\partial t^{m-1}}(0, z) = \phi_{m-1}(z), \tag{1.3.5}$$

where $A_k(t, D)$ are PD-operators (with respect to z), the symbols $A_k(t, \zeta)$ of which are analytic in $t \in \mathbf{C}^1$ and $\zeta \in \Omega$, where Ω is a Runge domain in $\mathbf{C}_\zeta^n$.

The main result of this subsection is the following.

Theorem 1.3.2. *Let $h(t, z)$ be an analytic function in $t \in \mathbf{C}^1$ and $z \in \mathbf{C}^n$ such that for any fixed t, $h(t, \cdot) \in \mathrm{Exp}_\Omega(\mathbf{C}_z^n)$. Furthermore, let $\phi_j(z) \in \mathrm{Exp}_\Omega(\mathbf{C}_z^n)$, $j = 0, 1, \ldots, n-1$. Then there exists a unique solution $u(t, z)$ of problem (1.3.4), (1.3.5) which is analytic in $t \in \mathbf{C}^1$ and is such that $u(t, \cdot) \in \mathrm{Exp}_\Omega(\mathbf{C}_z^n)$.*

Proof. First we note that Duhamel's well known principle is true in the complex case as well. In fact, let $U(t, s, z)$ be the solution of the Cauchy problem

$$L\left(\frac{\partial}{\partial t}, D\right)U \equiv U^{(m)} + \sum_{k=0}^{m-1} A_k(t, D)U^{(k)} = 0,$$

$$U^{(k)}(0, s, z) = 0 \quad (0 \leq k \leq m - 2), \quad U^{(m-1)}(0, s, z) = h(s, z),$$

where $s \in \mathbf{C}^1$ is arbitrary. Then a simple calculation shows that the solution $u(t, z)$ of the Cauchy problem

$$L\left(\frac{\partial}{\partial t}, D\right)u = h(t, z), \quad u^{(k)}(0, z) = 0 \quad (0 \leq k \leq m - 1)$$

is given by the formula

$$u(t, z) = \int_0^t U(t - s, s, z)\,ds.$$

Thus we can set $h(t, z) \equiv 0$, that is, we can solve the problem

$$L\left(\frac{\partial}{\partial t}, D\right)u(t, z) = 0, \tag{1.3.6}$$

$$u^{(k)}(0, z) = \phi_k(z), \quad 0 \leq k \leq m - 1, \tag{1.3.7}$$

where $\phi_k(z) \in \mathrm{Exp}_\Omega(\mathbf{C}_z^n)$.

For this we formally set $D \leftrightarrow \zeta$ and solve the following set of ordinary Cauchy problems with parameter $\zeta \in \mathbf{C}^n$:

$$L\left(\frac{d}{dt}, \zeta\right)u_j(t, \zeta) \equiv u_j^{(m)} + \sum_{k=0}^{m-1} A_k(t, \zeta)u_j^{(k)} = 0,$$

$$u_j^{(k)}(0, \zeta) = \delta_{jk} \quad (0 \leq k, j \leq m - 1),$$

where δ_{jk} is the Kronecker symbol.

Since the coefficients $A_k(t,\zeta)$ are analytic in $\zeta \in \Omega$, it follows (in accordance with the classical result on the analyticity with respect to the parameter of the solution of the ordinary Cauchy problem) that the solutions $u_j(t,\zeta)$ are analytic in $\zeta \in \Omega$. Thus every $u_j(t,\zeta)$ defines a PD-operator $u_j(t,D)$ acting continuously in $\mathrm{Exp}_\Omega(\mathbf{C}_z^n)$. It is clear (since the PD-operators with analytic symbols form an algebra, see §1.2) that the formula

$$u(t,z) = \sum_{j=0}^{m-1} u_j(t,D)\phi_j(z) \tag{1.3.8}$$

gives the desired solution of the problem (1.3.6), (1.3.7). The uniqueness is obvious, since all the derivatives $u^{(k)}(0,z)$, $k = 0,1,\ldots$ are uniquely defined by the initial conditions (1.3.5) and equation (1.3.4). This completes the proof of the theorem.

Remark. It is clear that the results obtained above are also valid for arbitrary systems of type (1.3.4). The arguments for the proof are the same.

Example 1. Let $a \in \mathbf{C}^1$ and $|a| = 1$. Consider the Cauchy problem

$$u' - a\frac{\partial^2 u}{\partial z^2} = 0 \quad (t \in \mathbf{C}^1,\ z \in \mathbf{C}^1)$$

$$u(0,z) = \phi(z).$$

Clearly, the solution of this problem can be written in the form

$$u(t,z) = \exp\left(at\frac{d^2}{dz^2}\right)\phi(z),$$

where $\phi(z) \in \mathrm{Exp}(\mathbf{C}_z^1)$, that is, is a function of arbitrary exponential type (for this example $\Omega = \mathbf{C}_\zeta^1$).

We show that

$$\exp\left(at\frac{d^2}{dz^2}\right)\phi(z) = \frac{1}{2\sqrt{\pi a t}} \int_{-\sqrt{at_0}\infty}^{\sqrt{at_0}\infty} \exp\left(-\frac{s^2}{4at}\right)\phi(z-s)ds, \tag{1.3.9}$$

where $t_0 = t/|t|$ and the integration is performed over the straight line from $-\sqrt{at_0}$ to $\sqrt{at_0}$.

In fact, using the Taylor expansion of $\phi(z)$ we have

$$\int_{-\sqrt{at_0}\infty}^{\sqrt{at_0}\infty} \exp\left(-\frac{s^2}{4at}\right)\phi(z-s)ds = \int_{-\sqrt{at_0}\infty}^{\sqrt{at_0}\infty} \exp\left(-\frac{s^2}{4at}\right)\sum_{n=0}^{\infty}\frac{\phi^{(n)}(z)}{n!}(-s)^n ds =$$

$$= \sum_{n=0}^{\infty}\frac{\phi^{(n)}(z)}{n!}\int_{-\sqrt{at_0}\infty}^{\sqrt{at_0}\infty} \exp\left(-\frac{s^2}{4at}\right)(-s)^n ds, \qquad (1.3.10)$$

where $\phi^{(n)}(z) \equiv d^n\phi(z)/dz^n$. Next, by substituting $s = 2\sqrt{at}\eta$ we obtain

$$\frac{1}{2\sqrt{\pi at}}\int_{-\sqrt{at_0}\infty}^{\sqrt{at_0}\infty} \exp\left(-\frac{s^2}{4at}\right)(-s)^n ds = \frac{1}{\sqrt{\pi}}(-2\sqrt{at})^n \int_{-\infty}^{\infty} e^{-\eta^2}\eta^n d\eta =$$

$$= \begin{cases} 0, & n = 2m+1; \\ \frac{1}{\sqrt{\pi}}(4at)^m\Gamma(m+\frac{1}{2}), & n = 2m \ (m=0,1,\ldots), \end{cases}$$

where $\Gamma(\cdot)$ is the Euler gamma-function.

Bearing in mind that

$$\Gamma\left(m+\frac{1}{2}\right) = \left(m-\frac{1}{2}\right)\left(m-\frac{3}{2}\right)\cdots\frac{3}{2}\cdot\frac{1}{2} = \frac{(2m-1)!\sqrt{\pi}}{2^{2m-1}(m-1)!},$$

we immediately deduce from the last formulae and (1.3.10) that the equality (1.3.9) holds.

Thus the solution of the Cauchy problem for the "complex heat equation" is given by the complex generalization of the classical Poisson integral

$$u(t,z) = \frac{1}{2\sqrt{\pi at}}\int_{-\sqrt{at_0}\infty}^{\sqrt{at_0}\infty} \exp\left(-\frac{s^2}{4at}\right)\phi(z-s)ds.$$

Example 2. Consider the Cauchy problem

$$\frac{\partial^2 u}{\partial t^2} - a^2\frac{\partial^2 u}{\partial z^2} = 0 \quad (t\in \mathbf{C}^1, z\in \mathbf{C}^1)$$

$$u(0,z) = \phi(z), \quad \frac{\partial u}{\partial t}(0,z) = \psi(z),$$

where $\phi(z)$ and $\psi(z)$ are functions in $\mathrm{Exp}(\mathbf{C}_z^1)$. According to the procedure described above, the solution of this Cauchy problem is

$$u(t,z) = \frac{e^{atD} + e^{-atD}}{2}\phi(z) + \frac{e^{atD} - e^{-atD}}{2aD}\psi(z)$$

$(D \equiv d/dz)$, from which it immediately follows that

$$u(t, z) = \frac{\phi(z + at) + \phi(z - at)}{2} + \frac{1}{2a} \int_{z-at}^{z+at} \psi(\tau)d\tau,$$

which is the complex d'Alembert formula.

Example 3. Consider the Cauchy problem for the equation with shift

$$\frac{\partial u}{\partial t}(t, z) + u(t, z + a) = 0$$

$$u(0, z) = \phi(z), \quad z \in \mathbf{C}^1,$$

or, what is the same,

$$\frac{\partial u}{\partial t} + \exp(aD)u = 0, \quad u(0, z) = \phi(z).$$

Clearly,

$$u(t, z) = \exp\{-t\exp(aD)\}\phi(z),$$

where $\phi(z) \in \mathrm{Exp}(\mathbf{C}_z^1)$. Hence

$$u(t, z) = \sum_{n=0}^{\infty} \frac{(-t)^n}{n!} \exp(naD)\phi(z) = \sum_{n=0}^{\infty} \frac{(-t)^n}{n!} \phi(z + an).$$

Example 4. Here we consider the complex analogue of the Schrödinger equation of the relativistic free particle (see Björken and Drell [1]). Namely, we wish to find the solution of the problem

$$i\frac{\partial u}{\partial t} = \sqrt{I - \Delta}u, \quad t \in \mathbf{C}^1, z \in \mathbf{C}^n, \qquad (1.3.11)$$

$$u(0, z) = \phi(z), \qquad (1.3.12)$$

where $\Delta \equiv \partial^2/\partial z_1^2 + \ldots + \partial^2/\partial z_n^2$.

The PD-operator $\sqrt{I - \Delta}$ was described in §2 (p. 22, Example 3). Therefore, for any $\phi(z) \in \mathrm{Exp}_\Omega(\mathbf{C}_z^n)$ the problem (1.3.11), (1.3.12) has the unique solution

$$u(t, z) = \exp(t\sqrt{I - \Delta})\phi(z),$$

which also belongs to $\mathrm{Exp}_\Omega(\mathbf{C}_z^n)$. Here $\Omega \subset \mathbf{C}_\zeta^n \setminus \{\zeta : \zeta^2 \in \Gamma\}$ is a Runge domain and Γ is a simple contour joining the points $+1$ and -1.

3. Cauchy-Kovalevskaya Theorem (special case)

In this subsection we consider a differential equation of Kovalevskaya type with constant coefficients and show that our previous technique may be applied to the proof of the local solvability of the Cauchy problem in the class of all analytic functions. Namely, we consider the Cauchy problem

$$L\left(\frac{\partial}{\partial t}, D\right)u \equiv \frac{\partial^m u}{\partial t^m} + \sum_{k=0}^{m-1} A_k(D)\frac{\partial^k u}{\partial t^k} = 0 \qquad (1.3.13)$$

$$\frac{\partial^k u}{\partial t^k}(0, z) = \phi_k(z), \quad k = 0, 1, \ldots, m-1, \qquad (1.3.14)$$

where the $A_k(D)$ are differential operators with constant coefficients.

First we recall that the polynomial

$$L(\lambda, \zeta) \equiv \lambda^m + \sum_{k=0}^{m-1} A_k(\zeta)\lambda^k \qquad (1.3.15)$$

is called a Kovalevskaya polynomial (with respect to λ) if

$$k + m_k \leq m, \quad k = 0, 1, \ldots, m-1,$$

where $m_k = \deg A_k(\zeta)$ (the degree of $A_k(\zeta)$). In this case we say that the corresponding equation (1.3.13) (or operator $L(\frac{\partial}{\partial t}, D)$) is a Kovalevskaya equation (or operator).

Let $G \subset \mathbb{C}^n_z$ be some domain and $\phi_k(z) \in \mathcal{O}(G)$, $k = 0, 1, \ldots, m-1$, be analytic functions in G.

Definition 1.3.1. We say that the Cauchy problem (1.3.13), (1.3.14) is locally well-posed in the class of analytic functions if for any initial functions $\phi_k(z) \in \mathcal{O}(G)$ and for any compact subdomain $G_\delta \subset G$ there exists a number $\rho > 0$ (depending, in general, on $\phi_k(z)$) such that in the "polycylinder" $W^\rho_\delta = \{t : |t| < \rho\} \times G_\delta$ there exists a unique analytic solution $u(t, z)$ of this problem. Moreover, the solution $u(t, z)$ depends continuously on the initial functions $\phi_k(z)$, that is, if $\phi_{k\nu}(z) \to \phi_k(z)$ ($\nu \to \infty$) locally uniformly in G, then the corresponding solutions $u_\nu(t, z)$ also tend to $u(t, z)$ locally uniformly in some W^ρ_δ.

Our aim is to prove that the Cauchy problem (1.3.13), (1.3.14) is locally well-posed in the sense of Definition 1.3.1 if and only if equation (1.3.13) is of Kovalevskaya type.

First we need a characterization of the behaviour of the roots of a Kovalevskaya polynomial $L(\lambda, \zeta)$ as $|\zeta| \to \infty$.

Lemma 1.3.1. *Let $\lambda_s = \lambda_s(\zeta)$, $s = 0, 1, \ldots, m-1$, be the roots of the polynomial $L(\lambda, \zeta)$. The formula*

$$\lambda_s(\zeta) = O(|\zeta|) \quad (|\zeta| \to \infty). \tag{1.3.16}$$

is true if and only if polynomial $L(\lambda, \zeta)$ is a Kovalevskaya polynomial.

Proof. The necessity is obvious since the coefficients of the polynomials $A_k(\lambda, \zeta)$ are symmetric functions of order $m-k$ of the roots $\lambda_0(\zeta), \ldots, \lambda_{m-1}(\zeta)$, therefore

$$A_k(\zeta) = O(|\zeta|^{m-k}). \tag{1.3.17}$$

This means that $L(\lambda, \zeta)$ is a Kovalevskaya polynomial.

Sufficiency. Suppose, on the contrary, that (1.3.17) holds but there exists at least one root $\lambda_s(\zeta)$ of $L(\lambda, \zeta)$ not satisfying condition (1.3.16). In other words for any constant $\mu > 0$ there exist some points $\zeta \in \mathbb{C}^n$ ($|\zeta| \to \infty$ as $\mu \to \infty$) such that $|\lambda_s(\zeta)| > \mu|\zeta|$. Then for such ζ

$$|L(\lambda, \zeta)| = |\lambda_s^m(\zeta) \left(1 + \sum_{k=0}^{m-1} \frac{A_k(\zeta)}{\lambda_s^{m-k}(\zeta)}\right)| \geq$$

$$\geq |\lambda_s^m(\zeta)| \left(1 - \sum_{k=0}^{m-1} \frac{O(|\zeta|^{m-k})}{\mu^{m-k}|\zeta|^{m-k}}\right) > 0,$$

if $\mu > 0$ is large enough. This contradiction proves the sufficiency of conditions (1.3.17). The lemma is proved.

We now turn to the question of the local solvability of the Cauchy problem itself.

Theorem 1.3.3. *The Cauchy problem (1.3.3), (1.3.4) is locally well-posed in the class of analytic functions (see Definition 1.3.1) if and only if $L(\lambda, \zeta)$ is a Kovalevskaya polynomial.*

Proof. Sufficiency. Let $L(\lambda, \zeta)$ be a Kovalevskaya polynomial. We prove in this case that formula

$$u(t, z) = \sum_{j=0}^{m-1} u_j(t, D)\phi_j(z) \tag{1.3.18}$$

(see subsection 2) gives the desired solution. In fact, let $u_j(t,\zeta)$, $j = 0,1,\ldots$
$\ldots, m-1$, be the fundamental system of solutions for the ordinary differential
equation

$$L\left(\frac{d}{dt},\zeta\right)u(t,\zeta) = 0,$$

that is,

$$L\left(\frac{d}{dt},\zeta\right)u_j(t,\zeta) = 0$$

$$u_j^{(k)}(0,\zeta) = \delta_{jk} \quad (0 \le k \le m-1)$$

($\zeta \in \mathbf{C}^n$ is the parameter). Then, taking into account Lemma 1.3.1 and the
explicit formula for $u_j(t,\zeta)$ as a function of the roots $\lambda_0(\zeta),\ldots,\lambda_{m-1}(\zeta)$, we
find that the $u_j(t,\zeta)$ are entire functions of exponential type, that is,

$$|u_j(t,\zeta)| \le M_j \exp(r_{1j}|\zeta_1| + \ldots + r_{nj}|\zeta_n|)|t|,$$

where $M_j > 0$ and $r_{1j} > 0,\ldots, r_{nj} > 0$ ($j = 0,1,\ldots,m-1$) are certain
constants. Consequently, the Taylor coefficients $u_{j\alpha}$ of the functions $u_j(t,\zeta)$ in
the expansion

$$u_j(t,\zeta) = \sum_{|\alpha|=0}^{\infty} u_{j\alpha}(t)\zeta^\alpha$$

satisfy the inequalities (see Proposition 1.1.2 in §1.1)

$$|u_{j\alpha}(t)| \le \widetilde{M}_j(\alpha!)^{-1}|t|^{|\alpha|}\tilde{r}_j^\alpha,$$

where $\widetilde{M}_j > 0$ is a constant, $\tilde{r}_j = (\tilde{r}_{1j},\ldots,\tilde{r}_{nj})$, $\tilde{r}_{1j} > r_{1j},\ldots \tilde{r}_{nj} > r_{nj}$.

On the other hand, since $\phi_k(z) \in \mathcal{O}(G)$, for any compact subset $G_\delta \subset G$
there exist $M_\delta > 0$ and $R_\delta = (R_{1\delta},\ldots,R_{n\delta})$ such that

$$|D^\alpha \phi_j(z)| \le M_\delta R_\delta^\alpha \alpha!, \quad z \in G_\delta.$$

Hence for any $z \in G_\delta$

$$|u_j(t,D)\phi_j(z)| \equiv \left| \sum_{|\alpha|=0}^{\infty} u_{j\alpha}(t)D^\alpha \phi_j(z) \right| \le$$

$$\le \sum_{|\alpha|=0}^{\infty} |u_{j\alpha}(t)| \cdot |D^\alpha \phi_j(z)| \le M_\delta \widetilde{M}_j \sum_{|\alpha|=0}^{\infty} |t|^{|\alpha|}\tilde{r}_j^\alpha R_\delta^\alpha < \infty,$$

if $|t| < \rho$, where $\rho = \rho(\delta) > 0$ is sufficiently small. Thus the series $u_j(t, D)\phi_j(z)$ $(0 \leq j \leq m-1)$ converge uniformly in the cylinder $W_\delta^\rho = \{|t| < \rho\} \times G_\delta$ and formula (1.3.18) therefore gives the solution of the original problem (1.3.13), (1.3.14).

Similar calculations show that this solution depends on the initial data $\phi_k(z)$, $k = 0, 1, \ldots, m-1$, in the sense of Definition 1.3.1. This proves the sufficiency.

Necessity. Suppose that the Cauchy problem (1.3.13), (1.3.14) is locally solvable in the class of all analytic functions in the sense of Definition 1.3.1. We claim that in this case, the polynomial $L(\lambda, \zeta)$ is of Kovalevskaya type.

In fact, let $u_0(t, \zeta)$ be one of the "basic" symbols, that is,

$$u_0^{(m)}(t, \zeta) + \sum_{k=0}^{m-1} A_k(\zeta) u_0^{(k)}(t, \zeta) = 0,$$

$$u_0(0, \zeta) = 1, \ u_0'(0, \zeta) = 0, \ldots, \ u^{(m-1)}(0, \zeta) = 0.$$

We prove that $u_0(t, \zeta)$ is a function of exponential type provided that $|t| < \rho$, where $\rho > 0$ is some number. Suppose the contrary. This means that for any small $\rho > 0$ there exists $t \in \mathbf{C}^1$, $|t| < \rho$, such that for any sequences $M_\nu \to +\infty$ and $R_{j\nu} \to +\infty$ $(j = 1, \ldots, n)$ there exists a sequence of values $\zeta = \zeta_\nu \in \mathbf{C}^n$ $(\nu = 1, 2, \ldots)$ for which

$$|u_0(t, \zeta)| > M_\nu \exp(R_{1\nu}|z_1| + \ldots + R_{n\nu}|z_n|). \tag{1.3.19}$$

We set $u_{0\nu}(t, \zeta) \equiv u_0(t, \zeta_\nu)\phi_{0\nu}(z)$, where

$$\phi_{0\nu}(z) = \frac{\exp z\zeta_\nu}{M_\nu \exp(R_{1\nu}|\zeta_{1\nu}| + \ldots + R_{n\nu}|\zeta_{n\nu}|)}.$$

Clearly, the functions $u_{0\nu}(t, \zeta)$ are the solutions of the original problem (1.3.13), (1.3.14) under the initial conditions

$$u_{0\nu}(0, z) = \phi_{0\nu}(z), \quad u_{0\nu}^{(k)}(0, z) = 0, \quad k = 1, \ldots, m-1.$$

Furthermore, for any compact $K \subset \mathbf{C}_z^n$ $\sup_{z \in K} |\phi_{0\nu}(z)| \to 0$ as $\nu \to \infty$, so that, in accordance with Definition 1.3.1 the sequence of solutions $u_{0\nu}(t, z)$ must also tend to zero in some cylinder W_δ^ρ. But this is impossible, since in view of (1.3.19),

$$|u_{0\nu}(t,0)| = |u_0(t,\zeta_\nu)\phi_{0\nu}(0)| \geq 1.$$

This contradiction shows that $u_0(t,\zeta)$ is an exponential function in ζ for all t, $|t| < \rho$, where $\rho > 0$ is some number.

Similarly it can be proved that all the other "basic" functions $u_1(t,\zeta), \ldots$ $\ldots, u_{m-1}(t,\zeta)$ are functions of exponential type in ζ for all t, $|t| < \rho$.

To complete the proof, we note that (after performing the appropriate calculations)

$$u_j(t,\zeta) = \sum_{k=0}^{m-1} c_{jk} t^{r_k} e^{t\lambda_k(\zeta)} \quad (0 \leq r_k = r_k(\zeta) \leq m-1),$$

where the $c_{jk} = c_{jk}(\lambda_0, \ldots, \lambda_{m-1})$ are certain rational functions of the roots $\lambda_0(\zeta), \ldots, \lambda_{m-1}(\zeta)$ of the characteristic polynomial $L(\lambda, \zeta)$. From this we obtain

$$t^{r_k} e^{t\lambda_k(\zeta)} = \sum_{j=0}^{m-1} B_{kj} u_j(t,\zeta),$$

where $B_{kj} = B_{kj}(\lambda_0, \ldots, \lambda_{m-1})$ are also rational functions of $\lambda_0(\zeta), \ldots$ $\ldots, \lambda_{m-1}(\zeta)$. Therefore, if $|\zeta|$ is sufficiently large, then $|\lambda_k(\zeta)| \leq M|\zeta|$, where $M > 0$ is a constant.

Thus, in accordance with Lemma 1.3.1 , the polynomial $L(\lambda, \zeta)$ is a Kovalevskaya polynomial. This proves the necessity. Thus the theorem is completely proved.

Proposition 1.3.1. *If the initial functions $\phi_j(z) \in \mathcal{O}(\mathbf{C}_z^n)$ then the Cauchy problem (1.3.13), (1.3.14) has a unique global solution in $\mathbf{C}_t^1 \times \mathbf{C}_z^n$, that is, $u(t,z) \in \mathcal{O}(\mathbf{C}_{t,z}^{n+1})$.*

For the proof it is enough to note that since $\phi_j(z)$ are entire, it follows that for any compact subset $K \subset \mathbf{C}_z^n$

$$|D^\alpha \phi_j(z)| \leq M\alpha! R^{-|\alpha|}, \quad z \in K,$$

where $R > 0$ is an arbitrary number and $M = M(R, K)$ is a constant. Consequently, the series

$$u(t,z) = \sum_{j=0}^{m-1} u_j(t,D)\phi_j(z)$$

converges uniformly on an arbitrary compact subset of the space $\mathbf{C}^{n+1}$ of the variables $t \in \mathbf{C}^1$ and $z \in \mathbf{C}^n$. This means precisely that the Cauchy problem is globally solvable.

4. A TWO-POINT BOUNDARY VALUE PROBLEM

In this subsection we consider the complex problem of boundary-value type with two "boundary" conditions. Similarly, one can consider the many-point complex boundary-value problem.

We consider in the space $\mathbf{C}^{n+1}$, $n \geq 1$, of the variables $t \in \mathbf{C}^1$ and $z \in \mathbf{C}^n$ the problem

$$\frac{\partial^2 u}{\partial t^2} - \Delta u = 0 \tag{1.3.20}$$

$$u(\tfrac{1}{2}, z) = \phi_+(z), \quad \frac{\partial u}{\partial t}(-\tfrac{1}{2}, z) = \phi_-(z). \tag{1.3.21}$$

Using the operator method, we set $\Delta \leftrightarrow \zeta^2$ and consider the following two-point problems for ordinary differential equations with complex parameter $\zeta \in \mathbf{C}^n$:

$$u_+''(t, \zeta) - \zeta^2 u_+(t, \zeta) = 0$$

with conditions

$$u_+(\tfrac{1}{2}, \zeta) = 1, \quad u_+'(-\tfrac{1}{2}, \zeta) = 0,$$

and

$$u_-''(t, \zeta) - \zeta^2 u_-(t, \zeta) = 0$$

with conditions

$$u_-(\tfrac{1}{2}, \zeta) = 0, \quad u_-'(-\tfrac{1}{2}, \zeta) = 1.$$

It is easy to see that

$$u_+(t, \zeta) = \frac{\cosh(t + \tfrac{1}{2})\zeta}{\cosh \zeta}, \quad u_-(t, \zeta) = \frac{\sinh(t - \tfrac{1}{2})\zeta}{\zeta \cosh \zeta}.$$

Consequently, the solution of the original problem (1.3.20), (1.3.21) may be written in the form

$$u(t, z) = \frac{\cosh(t + \tfrac{1}{2})\sqrt{\Delta}}{\cosh \sqrt{\Delta}} \phi_+(z) + \frac{\sinh(t + \tfrac{1}{2})\sqrt{\Delta}}{\sqrt{\Delta} \cosh \sqrt{\Delta}} \phi_-(z)$$

or, what is the same, in the form

$$u(t,z) = \cosh(t + \tfrac{1}{2})\sqrt{\Delta}u_+(z) + \frac{\sinh(t - \tfrac{1}{2})\sqrt{\Delta}}{\sqrt{\Delta}}u_-(z),$$

where the functions $u_+(z)$ and $u_-(z)$ are the solutions of the infinite order differential equation

$$[\cosh\sqrt{\Delta}]u_\pm(z) \equiv \sum_{n=0}^{\infty} \frac{\Delta^n u_\pm(z)}{(2n)!} = \phi_\pm(z), \quad z \in \mathbf{C}^n.$$

The operator $\cosh\sqrt{\Delta}$ is the PD-operator with symbol $\cosh\sqrt{\zeta^2} \equiv \cosh(\zeta_1^2 + \ldots + \zeta_n^2)^{1/2}$ which equals zero at those points $\zeta \in \mathbf{C}^n$ for which $\zeta^2 = -(\pi/2 + m\pi)^2$, $m = 0, \pm 1, \ldots$. Thus, if we choose the Runge domain

$$\Omega \subset \mathbf{C}_\zeta^n \backslash \{\zeta : \zeta^2 = -\left(\frac{\pi}{2} + m\pi\right)^2, \ m = 0, \pm 1, \ldots\},$$

then for any boundary functions $\phi_\pm(z) \in \mathrm{Exp}_\Omega(\mathbf{C}_z^n)$ we obtain a unique solution $u(t,z)$ which is entire on t and $u(t,\cdot) \in \mathrm{Exp}_\Omega(\mathbf{C}_z^n)$ for all $t \in \mathbf{C}^1$.

1.4. The dual theory

In this section we study the space of the exponential functionals, that is, the space of bounded linear maps of the test-function space $\mathrm{Exp}_\Omega(\mathbf{C}_z^n)$ into $\mathbf{C}^1$. We study also PD-equations and the Cauchy problem in the space of the exponential functionals.

1. EXPONENTIAL FUNCTIONALS. EXAMPLES

Let $\mathrm{Exp}_\Omega(\mathbf{C}_z^n)$ be the test-function space of exponential functions associated with the domain Ω.

Definition 1.4.1. We call any bounded linear map

$$u : \mathrm{Exp}_\Omega(\mathbf{C}_z^n) \to \mathbf{C}^1$$

an exponential functional.

The space of all exponential functionals will be denoted by $\mathrm{Exp}_\Omega'(\mathbf{C}_z^n)$. Here, the value of $u \equiv u(z) \in \mathrm{Exp}_\Omega'(\mathbf{C}_z^n)$ on the test function $\phi(z) \in \mathrm{Exp}_\Omega(\mathbf{C}_z^n)$ is $\langle u(z), \phi(z)\rangle$.

The space $\text{Exp}'_\Omega(\mathbf{C}^n_z)$ has all the standard properties as the dual of a complete topological space. Here we merely point out that it is invariant with respect to the action of arbitrary PD-operators with analytic symbols.

Namely, let $\Omega^- = \{\zeta \in \mathbf{C}^n : -\zeta \in \Omega\}$ and $A(\zeta) \in \mathcal{O}(\Omega^-)$ an arbitrary analytic function. Then, for any $u(z) \in \text{Exp}'_\Omega(\mathbf{C}^n_z)$ we also have $A(D)u(z) \in \text{Exp}'_\Omega(\mathbf{C}^n_z)$. Furthermore,

$$\langle A(D)u(z), \phi(z)\rangle \overset{\text{def}}{=} \langle u(z), A(-D)\phi(z)\rangle,$$

where $\phi(z) \in \text{Exp}_\Omega(\mathbf{C}^n_z)$, $-D = (-\partial/\partial z_1, \ldots, -\partial/\partial z_n)$.

This is an immediate corollary of the same property of the test-function space $\text{Exp}_\Omega(\mathbf{C}^n_z)$.

Example 1. The delta-function $\delta(z)$, acting according to the formula

$$\langle \delta(z), \phi(z)\rangle = \phi(0), \ \ \phi(z) \in \text{Exp}_\Omega(\mathbf{C}^n_z),$$

is clearly an element of $\text{Exp}'_\Omega(\mathbf{C}^n_z)$ for any region Ω.

Example 2. Let $A(\zeta)$ be an analytic function. Then

$$u(z) = A(-D)\delta(z)$$

is the exponential functional; here,

$$\langle u(z), \phi(z)\rangle \equiv \langle A(-D)\delta(z), \phi(z)\rangle \overset{\text{def}}{=} \langle \delta(z), A(D)\phi(z)\rangle = A(D)\phi(0).$$

Example 3. Let $u(z) : \mathbf{C}^n \to \mathbf{C}^1$ be an entire function that is rapidly decreasing as $|x| \to \infty$. More precisely, suppose that for any $y \in \mathbf{R}^n$ and $\delta > 0$

$$u(x + iy) \exp \delta|x| \to 0 \quad (|x| \to \infty).$$

Then such a function defines a regular exponential functional acting according to the formula

$$\langle u(z), \phi(z)\rangle \overset{\text{def}}{=} \int_{\mathbf{R}^n} u(x + iy)\phi(x + iy)dx. \tag{1.4.1}$$

Since $u(x + iy)$ is rapidly decreasing along the real axes, this formula is well-defined. It is clear that, in view of the Cauchy formula, the values (1.4.1) do not depend on $y \in \mathbf{R}^n$, so that this regular functional may be identified with its real "trace" $u(x)$. The last functional is a bounded linear functional on the corresponding space $\mathrm{Exp}_\Omega(\mathbf{R}^n)$, whose elements are entire functions of real variables x having an exponential continuation onto the whole complex space $\mathbf{C}_z^n$.

Thus, simultaneously with (1.4.1) we can set

$$\langle u(z), \phi(z)\rangle \overset{\text{def}}{=} \int_{\mathbf{R}^n} u(x)\phi(x)dx.$$

By analogy one can define regular functionals with respect to an arbitrary "contour" in $\mathbf{C}^n$, in particular, with respect to the purely imaginary space $i\mathbf{R}^n$.

Example 4. Let $u_0(z)$ be the regular exponential functional defined in Example 3 and let $A(\zeta) \in \mathcal{O}(\Omega)$. Then $A(-D)u_0(z)$ is the exponential functional, acting according to the formula

$$\langle A(-D)u_0(z)\phi(z)\rangle \overset{\text{def}}{=} \langle u_0(x), A(D)\phi(x)\rangle.$$

Remark. In the next subsection we prove that every exponential functional may be represented in the form given in Example 2, so that the entire space $\mathrm{Exp}_\Omega'(\mathbf{C}_z^n)$ is determined by functionals of type $A(-D)\delta(z)$, where $A(\zeta) \in \mathcal{O}(\Omega)$.

2. THE GENERAL FORM OF EXPONENTIAL FUNCTIONALS

We prove that every exponential functional may be represented as the action of some PD-operator on the Dirac delta-function.

Theorem 1.4.1. *Let $u(z) \in \mathrm{Exp}_\Omega'(\mathbf{C}_z^n)$. Then there exists a function $A(\zeta) \in \mathcal{O}(\Omega)$ such that*

$$u(z) = A(-D)\delta(z),$$

where $A(-D)$ is the PD-operator with the symbol $A(-\zeta)$.

Proof. Consider the function

$$A(\zeta) \overset{\text{def}}{=} \langle u(z), \exp z\zeta\rangle, \quad \zeta \in \Omega.$$

Clearly, $A(\zeta)$ is analytic in the region Ω, therefore the PD-operator

$$A(-D) : \mathrm{Exp}_\Omega'(\mathbf{C}_z^n) \to \mathrm{Exp}_\Omega'(\mathbf{C}_z^n)$$

with symbol $A(-\zeta)$ exists.

Since for any $\zeta \in \Omega$

$$(-1)^{|\alpha|}\langle(D + \lambda I)^\alpha \delta(z), \exp z\zeta\rangle = (\zeta - \lambda)^\alpha, \quad |\alpha| = 0, 1, \ldots,$$

it follows from the main Definition 1.2.1 of a PD-operator that

$$\langle A(-D)\delta(z), \exp z\zeta\rangle = A(\zeta), \quad \zeta \in \Omega.$$

This means that

$$\langle u(z), \exp z\zeta\rangle = \langle A(-D)\delta(z), \exp z\zeta\rangle, \quad \zeta \in \Omega,$$

that is, $u(z) = A(-D)\delta(z)$ on the set of all exponentials of type $\exp z\zeta$, $\zeta \in \Omega$. However, the set of linear combinations of such exponents is dense in the space $\mathrm{Exp}_\Omega(\mathbf{C}_z^n)$ (see Lemma 1.1.1). Hence

$$u(z) = A(-D)\delta(z) \tag{1.4.2}$$

as elements of $\mathrm{Exp}'_\Omega(\mathbf{C}_z^n)$.

We now prove that the representation (1.4.2) is unique. In fact, if there is an alternative representation

$$u(z) = B(-D)\delta(z),$$

then for any $\zeta \in \Omega$ we have

$$\langle u(z), \exp z\zeta\rangle = \langle A(-D)\delta(z), \exp z\zeta\rangle = A(\zeta)$$

and

$$\langle u(z), \exp z\zeta\rangle = \langle B(-D)\delta(z), \exp z\zeta\rangle = B(\zeta).$$

This means that $A(\zeta) \equiv B(\zeta)$ in Ω, therefore $A(D) = B(D)$ as PD-operators on $\mathrm{Exp}_\Omega(\mathbf{C}_z^n)$. Thus the theorem is proved.

Let us now turn to the representation of an exponential functional by means of PD-operators on some regular functional. Such a representation is a simple corollary of Theorem 1.4.1 which we have just proved, and the formula

$$\delta(z) = (2\sqrt{\pi})^{-n} \exp(-D^2) \circ \exp(-z^2/4),$$

where $\exp(-z^2/4)$ is the regular functional defined by this function and the system of real contours $(-\infty, \infty)$ in each variable. (We remark that the above representation of $\delta(z)$ is the solution of the Cauchy problem for the inverse heat equation for $t = 1$ and initial function $(2\pi)^{-n} \exp(-z^2/4)$.)

Using this formula and Theorem 1.4.1, we obtain $u(z) = B(-D)u_0(z)$, where $B(-D) = A(-D)\exp(-D^2)$ and $u_0(z) = (2\sqrt{\pi})^{-n}\exp(-z^2/4)$ is a regular functional.

Thus we proved the following

Theorem 1.4.2. *Let $u(z) \in \text{Exp}'_\Omega(\mathbf{C}^n_z)$ be an exponential functional. Then there exist an analytic function $B(\zeta) \in \mathcal{O}(\Omega)$ and a regular functional $u_0(z)$ such that*

$$u(z) = B(-D)u_0(z).$$

3. THE ALGEBRA OF PD-OPERATORS IN THE SPACE OF EXPONENTIAL
   FUNCTIONALS

As we showed in §1.2, the algebra $\mathcal{A}(\Omega)$ of PD-operators with analytic symbols in Ω and with domain of definition $\text{Exp}_\Omega(\mathbf{C}^n_z)$ is isomorphic to the algebra of all analytic functions in Ω, that is,

$$\mathcal{A}(\Omega) \leftrightarrow \mathcal{O}(\Omega). \tag{1.4.3}$$

There is also an analogous assertion with respect to the algebras of PD-operators acting in the dual spaces. More precisely, let us denote by $\mathcal{A}'(\Omega^-)$ the space of all PD-operators whose symbols are analytic in Ω^- and whose domain of definition is the dual space $\text{Exp}'_\Omega(\mathbf{C}^n_z)$.

As corollary of (1.4.3) we have the following.

Theorem 1.4.3. *There is an isomorphism*

$$\mathcal{A}'(\Omega^-) \leftrightarrow \mathcal{O}(\Omega^-).$$

Example 1. As has been mentioned (§1.2), the PD-operator I/D in the space $\text{Exp}_\Omega(\mathbf{C}^n_z)$ is the "natural" integration. It is natural, therefore, to call the functional $[I/D]\delta(z)$ the "complex Heaviside θ-function". This function plays a useful role in the formalism of the Fourier transformation and we use the following special notation for it:

$$n(z) \equiv \mathrm{nat} \int \delta(z)dz \stackrel{\mathrm{def}}{=} \frac{I}{D}\delta(z).$$

The functional $n(z)$ is clearly a bounded linear functional on the space $\mathrm{Exp}_\Omega(\mathbf{C}_z^1)$, where $\Omega = \mathbf{C}^1 \backslash \mathbf{R}_+^1$.

The following example will be needed during the study of the dual Cauchy problem.

Example 2. We consider the functional

$$E(t,z) = \frac{\exp\left(at\frac{d}{dz}\right) - \exp\left(-at\frac{d}{dz}\right)}{d/dz}\delta(z), \quad z \in \mathbf{C}^1,$$

where $a \in \mathbf{C}^1$, $|a| = 1$, is a parameter and $t \in \mathbf{C}^1$ is "complex time".

Since the function $\sinh \zeta/\zeta$ is entire, it follows that $E(t,z)$ is a functional on the space $\mathrm{Exp}(\mathbf{C}_z^n)$ of all entire functions of exponential type. In accordance with Example 1, $E(t,z)$, as a functional on $\mathrm{Exp}_\Omega(\mathbf{C}_z^n)$, where $\Omega = \mathbf{C}^1 \backslash \mathbf{R}_+^1$, may be written in the form

$$E(t,z) = n(z+at) - n(z-at).$$

Example 3. We consider the functional

$$E(t,z) = \exp\left(at\frac{d^2}{dz^2}\right)\delta(z),$$

which acts on any test function $\phi(z) \in \mathrm{Exp}(\mathbf{C}_z^1)$ according to the formula

$$\langle E(t,z), \phi(t,z) \rangle = \langle \delta(z), \exp\left(at\frac{d^2}{dz^2}\right)\phi(z) \rangle \qquad (1.4.4)$$

As we know (see Example 1 in §1.3)

$$\exp\left(at\frac{d^2}{dz^2}\right)\phi(z) = \frac{1}{2\sqrt{\pi at}}\int_{-\sqrt{at_0}\infty}^{\sqrt{at_0}\infty} \exp\left(-\frac{s^2}{4at}\right)\phi(z-s)ds,$$

where $t_0 = t/|t|$ is the unit vector along the t-direction. Therefore,

$$\langle E(t,z), \phi(z) \rangle = \langle \delta(z), \frac{1}{2\sqrt{\pi at}}\int_{-\sqrt{at_0}\infty}^{\sqrt{at_0}\infty} \exp\left(-\frac{s^2}{4at}\right)\phi(z-s)ds \rangle =$$

$$= \frac{1}{2\sqrt{\pi at}}\int_{-\sqrt{at_0}\infty}^{\sqrt{at_0}\infty} \exp\left(-\frac{s^2}{4at}\right)\phi(-s)ds. \qquad (1.4.5)$$

Thus, $E(t, z)$ is a regular functional, defined by the function $(2\sqrt{\pi a t})^{-1} \times \exp(-s^2/4at)$ and the contour $(-\sqrt{a t_0}\infty, \sqrt{a t_0}\infty)$. In particular (as we have already mentioned), for $t = 1$ we have from (1.4.5) the formula

$$\delta(z) = \frac{1}{2\sqrt{\pi a}} \exp\left(-a\frac{d^2}{dz^2}\right) \exp\left(-\frac{z^2}{4a}\right),$$

that is, one of the possible representations of $\delta(z)$ as the action of a differential operator of infinite order on a regular functional.

In the same way one can obtain the multidimensional formula

$$\delta(z) = (2\sqrt{\pi})^{-n}(a_1 \ldots a_n)^{-1/2} \exp\left(-a_1\frac{\partial^2}{\partial z_1^2} - \ldots - a_n\frac{\partial^2}{\partial z_n^2}\right) \circ$$

$$\circ \exp\left(-\frac{z_1^2}{4a_1} - \ldots - \frac{z_n^2}{4a_n}\right).$$

4. CAUCHY PROBLEM IN EXPONENTIAL FUNCTIONALS

We consider the Cauchy problem

$$L\left(\frac{\partial}{\partial t}, D\right)u \equiv \frac{\partial^m u}{\partial t^m} + \sum_{k=0}^{m-1} A_k(t, D)\frac{\partial^k u}{\partial t^k} = h(t, z), \qquad (1.4.6)$$

$$\frac{\partial^k u}{\partial t^k}(0, z) = \phi_k(z), \quad k = 0, 1, \ldots, m - 1. \qquad (1.4.7)$$

Here $A_k(t, D)$ are PD-operators whose symbols are analytic in $z \in \Omega$ and $|t| < \delta \le +\infty$, where $\Omega \subset \mathbb{C}^n$ is some region.

We find a solution $u(t, z)$ which is analytic on t, $|t| < \delta$, as an exponential functional in z.

Let us denote by $\mathcal{O}(\delta; \mathrm{Exp}'_{\Omega^-}(\mathbb{C}^n_z))$, where (we recall) $\Omega^- = \{\zeta : -\zeta \in \Omega\}$, the space of elements $u(t, z)$ such that

$$u(t, \cdot) \in \mathrm{Exp}'_{\Omega^-}(\mathbb{C}^n_z);$$

moreover, $u(t, \cdot)$ is analytic in t, $|t| < \delta$.

Theorem 1.4.4. *Let* $h(t, z) \in \mathcal{O}(\delta; \mathrm{Exp}'_{\Omega^-}(\mathbb{C}^n_z))$ *and* $\phi_k(z) \in \mathrm{Exp}'_{\Omega^-}(\mathbb{C}^n_z)$, $k = 0, 1, \ldots, m - 1$. *Then the problem* (1.4.6), (1.4.7) *has a unique solution* $u(t, z) \in \mathcal{O}(\delta; \mathrm{Exp}'_{\Omega^-}(\mathbb{C}^n_z))$.

Proof. We show that the solution $u(t,z)$ of the Cauchy problem (1.4.6), (1.4.7) in the space of exponential functionals is given by the same formula as in the case of exponential functions. Namely, let $u_j(t,\zeta)$ be the fundamental system of the Cauchy problem for the ordinary differential equation:

$$L\left(\frac{d}{dt},\zeta\right)u_j(t,\zeta) = 0,$$

$$u_j^{(k)}(0,\zeta) = \delta_{jk},$$

where δ_{jk} $(0 \le j,k \le m-1)$ is the Kronecker symbol. Then the $u_j(t,\zeta)$ $(j = 0,1,\dots,m-1)$ are analytic in ζ in the region Ω and therefore, these functions define PD-operators $u_j(t,D)$. We set

$$u(t,z) = \sum_{j=0}^{m-1} u_j(t,D)\phi_j(z) \tag{1.4.8}$$

and prove that $u(t,z)$ is the desired solution. (As before, it is sufficient to consider $h(t,z) \equiv 0$.)

Indeed, denoting the conjugate of $U(t,D)$ by $U^*(t,D)$, that is, $U^*(t,D) = U(t,-D)$, it is easy to see that if for any $\phi(z) \in \mathrm{Exp}_\Omega(\mathbf{C}_z^n)$ a function $u(t,z) \equiv U(t,D)\phi(z)$ is a solution of the equation

$$\frac{\partial^m u}{\partial t^m} + \sum_{k=0}^{m-1} A_k(t,D)\frac{\partial^k u}{\partial t^k} = 0,$$

then for any function $v(z) \in \mathrm{Exp}_{\Omega^-}(\mathbf{C}_z^n)$ the function $u^*(t,z) \equiv U^*(t,D)v(z)$ is a solution of the equation

$$\frac{\partial^m u^*}{\partial t^m} + \sum_{k=0}^{m-1} A_k^*(t,D)\frac{\partial^k u^*}{\partial t^k} = 0.$$

Consequently, if the initial data $\phi_j(z) \in \mathrm{Exp}_{\Omega^-}'(\mathbf{C}_z^n)$, then for any $v(z) \in \mathrm{Exp}_{\Omega^-}(\mathbf{C}_z^n)$

$$\left\langle L\left(\frac{\partial}{\partial t},D\right)u(t,z),v(z)\right\rangle \equiv \sum_{j=0}^{m-1}\left[\left\langle\frac{\partial^m u_j}{\partial t^m}(t,D)\phi_j(z),v(z)\right\rangle + \right.$$

$$\left. + \sum_{k=0}^{m-1}\left\langle A_k(t,D)\frac{\partial^k u_j}{\partial t^k}(t,D)\phi_j(z),v(z)\right\rangle\right] =$$

$$= \sum_{j=0}^{m-1}\left[\left\langle\phi_j(z),\frac{\partial^m u_j^*}{\partial t^m}(t,D)v(z)\right\rangle + \sum_{k=0}^{m-1} A_k^*(t,D)\frac{\partial^k u_j^*}{\partial t^k}(t,D)v(z)\right\rangle\right] = 0.$$

This means that $u(t, z)$, defined by (1.4.8), is a solution of the equation (1.4.6).

It only remains to show that $u(t, z)$ satisfies the initial conditions (1.4.7). Indeed, from the construction of the operators $u_j(t, D)$ we have

$$\frac{d^k}{dt^k} u_j(0, \zeta) = \delta_{jk},$$

therefore

$$\frac{d^k}{dt^k} u_j^*(0, D) = \delta_{jk} I,$$

where I is the identity operator in the space $\mathrm{Exp}'_{\Omega-}(\mathbf{C}_z^n)$. From this we obtain

$$\left\langle \frac{d^k}{dt^k} u_j(t, D)\phi_j(z), v(z) \right\rangle \Big|_{t=0} =$$

$$= \left\langle \phi_j(z), \frac{d^k}{dt^k} u_j^*(t, D)v(z) \right\rangle \Big|_{t=0} = \langle \phi_j(z), \delta_{kj} v(z) \rangle,$$

where $v(z) \in \mathrm{Exp}_{\Omega-}(\mathbf{C}_z^n)$ is an arbitrary function. This immediately gives the equality $u^{(k)}(0, z) = \phi_k(z)$ in the space $\mathrm{Exp}'_{\Omega-}(\mathbf{C}_z^n)$.

It remains only to prove that the solution $u(t, z)$ is unique. For this we note that, in accordance with the results of subsection 2 (Theorem 1.4.1), every functional $u(z) \in \mathrm{Exp}'_{\Omega-}(\mathbf{C}_z^n)$ may be represented in the form

$$u(z) = U(D)\delta(z),$$

where $U(\zeta) = \langle u(z), \exp \zeta z \rangle$ is analytic in Ω. Hence the solution $u(t, z)$ of the Cauchy problem (1.4.6), (1.4.7) may be written as

$$u(t, z) = U(t, D)\delta(z),$$

where the function $U(t, \zeta)$, $|t| < \delta$, $\zeta \in \Omega$, satisfies the ordinary differential equation

$$U^{(m)}(t, \zeta) + \sum_{k=0}^{m-1} A_k(t, \zeta)U^{(k)}(t, \zeta) = H(t, \zeta) \tag{1.4.9}$$

$$U^{(k)}(0, \zeta) = \Phi_k(\zeta), \quad 0 \leq k \leq m - 1, \tag{1.4.10}$$

where $H(t, \zeta)$ and $\Phi_k(\zeta)$ are the symbols of the exponential functionals $h(t, z)$ and $\phi_k(z)$. Consequently, the function $U(t, \zeta)$ is uniquely defined as the solution of the problem (1.4.9), (1.4.10). From this it follows that the solution of the original Cauchy problem (1.4.6), (1.4.7) is unique also. The theorem is completely proved.

Remark. The above theorem clearly holds also for systems of PD-equations. The proof is the same.

Example 1. (Fundamental solution of the Cauchy problem). By definition, the solution $E(t, z)$ of the problem

$$L\left(\frac{\partial}{\partial t}, D\right) E(t, z) = 0,$$

$$E(0, z) = 0, \ldots, \quad E^{(m-2)}(0, z) = 0, \quad E^{(m-1)}(0, z) = \delta(z),$$

is fundamental.

Since for any $\Omega \neq \varnothing$ the functional $\delta(z) \in \mathrm{Exp}'_{\Omega^-}(\mathbf{C}^n_z)$, it follows from Theorem 1.4.4 that for any PD-operator $L\left(\frac{\partial}{\partial t}, D\right)$ there exists a unique fundamental solution $E(t, z) \in \mathcal{O}(\delta; \mathrm{Exp}'_{\Omega^-}(\mathbf{C}^n_z))$.

Example 2. Let $a \in \mathbf{C}^1$, $|a| = 1$. We consider the Cauchy problem

$$\frac{\partial u}{\partial t} - a\frac{\partial^2 u}{\partial z^2} = 0, \quad u(0, z) = \phi(z) \quad (t \in \mathbf{C}^1, z \in \mathbf{C}^1).$$

For any $\phi(z) \in \mathrm{Exp}_{\Omega^-}(\mathbf{C}^n_z)$, where $\Omega = \mathbf{C}^1$, the solution is (see Example 1 in §1.3)

$$u(t, z) = \frac{1}{2\sqrt{\pi a t}} \int_{-\sqrt{at_0}\infty}^{\sqrt{at_0}\infty} \exp\left(-\frac{s^2}{4at}\right)\phi(z - s)ds.$$

The fundamental solution of the Cauchy problem for the "complex heat equation" is defined by the function $(2\sqrt{\pi a t})^{-1}\exp(-z^2/4at)$ and the contour $(-\sqrt{at_0}\infty, \sqrt{at_0}\infty$ (see Example 3 in subsection 3).

Example 3. We consider the Cauchy problem

$$\frac{\partial^2 u}{\partial t^2} - a^2\frac{\partial^2 u}{\partial z^2} = 0 \quad (t \in \mathbf{C}^1, z \in \mathbf{C}^1)$$

$$u(0, z) = \phi(z), \quad \frac{\partial u}{\partial t}(0, z) = \psi(z),$$

where $\phi(z) \in \mathrm{Exp}'(\mathbf{C}^1_z)$, $\psi(z) \in \mathrm{Exp}'(\mathbf{C}^n_z)$. Simple calculations give the formula

$$u(t, z) = \frac{e^{atD} + e^{-atD}}{2}\phi(z) + \frac{e^{atD} - e^{-atD}}{2aD}\psi(z)$$

$(D \equiv d/dz)$ from which we immediately obtain the complex d'Alembert formula

$$u(t, z) = \frac{\phi(z + at) + \phi(z - at)}{2} + \frac{1}{2a} \int_{z-at}^{z+at} \psi(s)ds.$$

The fundamental solution for the Cauchy problem for "the complex wave equation" is

$$E(t, z) = \frac{1}{2a}[n(z + at) - n(z - at)],$$

where

$$n(z) = \text{nat} \int \delta(z)dz \overset{\text{def}}{=} [I/D]\delta(z)$$

is the "complex Heaviside θ-function".

Example 4. Let us consider the Cauchy problem for the equation with a shift

$$\frac{\partial u}{\partial t}(t, z) + u(t, z + a) = 0 \quad (t \in \mathbf{C}^1, z \in \mathbf{C}^1)$$

$$u(0, z) = \phi(z), \quad \phi(z) \in \text{Exp}'(\mathbf{C}_z^n),$$

or, what is the same,

$$\frac{\partial u}{\partial t} + \exp(aD)u = 0,$$

$$u(0, z) = \phi(z).$$

Clearly,

$$u(t, z) = \exp(-t \exp(aD))\phi(z),$$

that is,

$$u(t, z) = \sum_{n=0}^{\infty} \frac{(-t)^n}{n!} \exp(naD)\phi(z) = \sum_{n=0}^{\infty} \frac{(-t)^n}{n!} \phi(z + na).$$

In particular, the fundamental solution is

$$E(t, z) = \sum_{n=0}^{\infty} \frac{(-t)^n}{n!} \delta(z + na).$$

Chapter 2. Fourier Transformation of Arbitrary Analytic Functions. Complex Fourier Method

2.1. Fourier transformation

In this section we give the definition of the Fourier transformation of an arbitrary analytic function which is equivalent to the well known Borel transformation for the case of exponential functions. We then construct the complex Fourier method equivalent to the operator method.

1. MAIN DEFINITION. THE INVERSION FORMULA

Let $G \subset \mathbf{C}_z^n$ be a Runge domain, and $u(z)$ an arbitrary analytic function in G. As before, let $\zeta = (\zeta_1, \dots, \zeta_n)$ be the dual variables and $\partial = (\partial/\partial\zeta_1, \dots, \partial/\partial\zeta_n)$ the differentiation symbol.

In accordance with the definition of a PD-operator, an analytic function $u(z) \in \mathcal{O}(G)$ is the symbol of the PD-operator

$$u(\partial) : \operatorname{Exp}_G(\mathbf{C}_\zeta^n) \to \operatorname{Exp}_G(\mathbf{C}_\zeta^n),$$

where $\operatorname{Exp}_G(\mathbf{C}_\zeta^n)$ is the test space of exponential functions $\phi(\zeta)$ associated with the domain G (clearly, the variables z and ζ exchange roles).

It is also clear that

$$u(-\partial) : \operatorname{Exp}'_G(\mathbf{C}_\zeta^n) \to \operatorname{Exp}'_G(\mathbf{C}_\zeta^n)$$

is a PD-operator in the space of exponential functionals $\operatorname{Exp}'_G(\mathbf{C}_\zeta^n)$.

Definition 2.1.1. The Fourier transform $\tilde{u}(\zeta)$ of a function $u(z) \in \mathcal{O}(G)$, $G \subset \mathbf{C}_z^n$, is defined to be the functional

$$\tilde{u}(\zeta) \stackrel{\text{def}}{=} u(-\partial)\delta(\zeta),$$

52

where $\delta(\zeta)$ is the Dirac delta-function.

Thus, the Fourier transform $\tilde{u}(\zeta)$ of a function $u(z) \in \mathcal{O}(G)$ is a linear functional on the space $\mathrm{Exp}_G(\mathbf{C}_\zeta^n)$, acting according to the formula

$$\langle \tilde{u}(\zeta), v(\zeta) \rangle = \langle \delta(\zeta), u(\partial)v(\zeta) \rangle = u(\partial)v(0),$$

where $v(\zeta) \in \mathrm{Exp}_G(\mathbf{C}_\zeta^n)$. (The Fourier transform of $u(z)$ will also be denoted by $[Fu](\zeta)$.)

From Definition 2.1.1 and the properties of PD-operators we have the following.

Theorem 2.1.1. *The map*

$$\mathcal{F} : \mathcal{O}(G) \to \mathrm{Exp}_G'(\mathbf{C}_\zeta^n) \tag{2.1.1}$$

is continuous.

Proof. First we note that it suffices to prove the continuity of (2.1.1) for the case of polycylinder $S_R = \{z : |z_j| < R_j \le \infty, \ j = 1, \ldots, n\}$, since in the case of a general domain G the value $\langle \tilde{u}(\zeta), v(\zeta) \rangle$ may be represented as a finite sum of such terms.

Namely, let $G = S_R$ and $u_\nu(z) \to u(z)$ $(\nu \to \infty)$ locally uniformly in S_R. For any test function $v(\zeta) \in \mathrm{Exp}_R(\mathbf{C}_\zeta^n)$

$$\langle \tilde{u}(\zeta) - \tilde{u}_\nu(\zeta), v(\zeta) \rangle = [u(-\partial) - u_\nu(-\partial)]v(0) = \sum_{|\alpha|=0}^{\infty} (u_\alpha - u_{\alpha\nu})\partial^\alpha v(0),$$

where $u_\alpha = D^\alpha u(0)/\alpha!$ and $u_{\alpha\nu} = D^\alpha u_\nu(0)/\alpha!$ are the Taylor coefficients of the functions $u(z)$ and $u_\nu(z)$.

Since $v(\zeta) \in \mathrm{Exp}_R(\mathbf{C}_\zeta^n)$, it follows from Proposition 1.2.1 (§1.1) that there exist a constant $M > 0$ and vector $r = (r_1, \ldots, r_n)$, $r_j < R_j$ $(j = 1, \ldots, n)$ such that for any α

$$|\partial^\alpha v(\zeta)| \le M \exp(r_1|\zeta_1| + \ldots + r_n|\zeta_n|) \cdot r^\alpha$$

and, in particular, $|\partial^\alpha v(0)| \le Mr^\alpha$.

Further, since $u_\nu(z) \to u(z)$ uniformly on arbitrary compact $K \subset S_R$, there exist a constant $M_1 > 0$ and vector r_1 $(r_j < r_{1j} < R_j)$ such that

$$|D^\alpha(u(z) - u_\nu(z))|_{z=0}| \le \frac{\alpha! M_1}{r_1^\alpha} \max_{z \in \Gamma_1} |u(z) - u_\nu(z)|,$$

where Γ_1 is the skeleton of the polycylinder S_{r_1}.

Taking into account these inequalities we obtain

$$|\langle \tilde{u}(\zeta) - \tilde{u}_\nu(\zeta), v(\zeta)\rangle| \le M_2 \max_{z \in \Gamma_1} |u(z) - u_\nu(z)|,$$

where

$$M_2 = M M_1 \sum_{|\alpha|=0}^{\infty} \frac{r^\alpha}{r_1^\alpha} < \infty$$

is a constant.

The last inequality shows that $\tilde{u}(\zeta) - \tilde{u}_\nu(\zeta) \to 0$ in $\mathrm{Exp}'_R(\mathbb{C}^n_\zeta)$ if $u_\nu(z) \to u(z)$ locally uniformly in S_R. Our theorem is proved.

Theorem 2.1.2. *The mapping* (2.1.1) *is one-to-one, the inverse map being defined by the formula*

$$u(z) = F^{-1}\tilde{u}(\zeta) = \langle \tilde{u}(\zeta), \exp z\zeta \rangle, \quad z \in G. \tag{2.1.2}$$

Proof. Indeed, if $\tilde{u}(\zeta) = u(-\partial)\delta(\zeta)$ is the Fourier transformation of $u(z) \in \mathcal{O}(G)$, then for any function $\exp z\zeta$, $z \in G$, we have

$$\langle \tilde{u}(\zeta), \exp z\zeta \rangle = \langle \delta(\zeta), u(\partial) \exp z\zeta \rangle = u(z)$$

and the inversion formula (2.1.2) is obtained. It remains to show that every exponential functional is represented as the Fourier transform of one and only one analytic function in the domain G. But this is a corollary of Theorem 1.4.1 (§1.4) on the generalized form of an arbitrary exponential functional. The theorem is proved.

2. The Fourier Image of Exponential Functions. The Borel Kernel

In this subsection we shall prove that the Fourier transform of the exponential function is not only an exponential functional but also an analytic functional. This fact is important for constructing the complex Fourier method.

Let

$$u(z) = \sum_{|\alpha|=0}^{\infty} u_\alpha z^\alpha, \ u_\alpha = D^\alpha u(0)/\alpha!,$$

be an entire function of exponential type, that is,

$$|u(z)| \le M \exp(r_1|z_1| + \ldots + r_n|z_n|),$$

where $M > 0$ and $r_1 > 0, \ldots, r_n > 0$ are constants.

Further, let $Bu(\zeta)$ be the Borel function associated with the original function $u(z)$, that is,

$$Bu(\zeta) = \sum_{|\alpha|=0}^{\infty} \frac{\alpha! u_\alpha}{\zeta^{\alpha+1}}, \tag{2.1.3}$$

where $\alpha + 1 = (\alpha_1 + 1, \ldots, \alpha_n + 1)$. (The Borel function is also called the Borel transform.) Clearly, $Bu(\zeta)$ is analytic if $|\zeta_j| > r_j$ $(j = 1, \ldots, n)$, moreover, the series (2.1.3) converges in this domain locally uniformly.

According to Definition 2.1.1, any exponential function $u(z)$ as an entire function has the Fourier transform

$$\tilde{u}(\zeta) = \sum_{|\alpha|=0}^{\infty} u_\alpha (-\partial)^\alpha \delta(\zeta)$$

as a functional on $\mathrm{Exp}(\mathbf{C}^n_\zeta)$. However, in this case the functional $\tilde{u}(\zeta) \in \mathrm{Exp}'(\mathbf{C}^n_\zeta)$ can be defined not only for entire functions of exponential type, but also for any functions analytic in $\mathbf{C}^n_\zeta$. Let us show that this extension, that is, the analytic functional $\tilde{u}(\zeta) \in \mathcal{O}'(\mathbf{C}^n_\zeta)$, is a regular analytic functional with kernel $Bu(\zeta)$.

Indeed, let $v(\zeta)$ be analytic inside the polycylinder U_R, where $R_j > r_j$ $(j = 1, \ldots, n)$. Then by the Cauchy formula

$$\partial^\alpha v(0) = \frac{\alpha!}{(2\pi i)^n} \int_\Gamma \frac{v(\zeta) d\zeta}{\zeta^{\alpha+1}},$$

where Γ is the skeleton of any polycylinder $U_{\bar{r}}$; here $\bar{r}$ satisfies the inequalities $r_j < \bar{r}_j < R_j$ $(j = 1, \ldots, n)$. From this we obtain

$$\langle \tilde{u}(\zeta), v(\zeta) \rangle = \sum_{|\alpha|=0}^{\infty} u_\alpha \langle (-\partial)^\alpha \delta(\zeta), v(\zeta) \rangle = \sum_{|\alpha|=0}^{\infty} u_\alpha \partial^\alpha v(0) =$$

$$= \frac{1}{(2\pi i)^n} \int_\Gamma \sum_{|\alpha|=0}^{\infty} \frac{\alpha! u_\alpha}{\zeta^{\alpha+1}} v(\zeta) d\zeta = \frac{1}{(2\pi i)^n} \int_\Gamma Bu(\zeta) v(\zeta) d\zeta. \tag{2.1.4}$$

This last formula shows that $\tilde{u}(\zeta)$ is a regular analytic functional on the space $\mathcal{O}(U_R)$, moreover, (2.1.4) means that the kernel of this functional is the Borel transform $Bu(\zeta)$ of the original function $u(z)$. It is clear that $\tilde{u}(\zeta)$ is a bounded linear functional on the space of all analytic functions $\mathcal{O}(\mathbf{C}^n_\zeta)$ with locally uniform convergence. Conversely, let $h(\zeta) \in \mathcal{O}'(\mathbf{C}^n_\zeta)$ be an analytic functional. Then (see, for example, Hörmander [1], Napalkov [1], etc.) there exists a compact set $K \subset \mathbf{C}^n_\zeta$ such that

$$|\langle h(\zeta), v(\zeta) \rangle| \le M \max_{\zeta \in K} |v(\zeta)|,$$

where $M > 0$ is a constant. From this it follows immediately that the function

$$u(z) = \langle h(\zeta), \exp z\zeta \rangle, \quad z \in \mathbf{C}^n,$$

is an entire function of exponential type. Since the set of linear combinations of exponents is dense in $\mathcal{O}(\mathbf{C}^n_\zeta)$ ($\mathbf{C}^n_\zeta$ is a Runge domain), there is one-to-one correspondence between $h(\zeta)$ and $u(z)$; moreover, $h(\zeta) = \tilde{u}(\zeta)$.

Finally, we have proved the following theorem (which is well known, see Ehrenpreis [1]).

Theorem 2.1.3. *There is an isomorphism*

$$F : \mathrm{Exp}(\mathbf{C}^n_z) \leftrightarrow \mathcal{O}'(\mathbf{C}^n_\zeta).$$

The inverse map is given by the formula

$$u(z) = \langle \tilde{u}(\zeta), \exp z\zeta \rangle, \quad z \in \mathbf{C}^n.$$

We now turn to the general case of a region $\Omega \subset \mathbf{C}^n_\zeta$. We recall that in this case

$$\mathrm{Exp}_\Omega(\mathbf{C}^n_z) = \Big\{ u(z) : u(z) = \sum_{\lambda \in \Omega} u_\lambda(z) \Big\},$$

where $u_\lambda(z) = \exp \lambda z \cdot \phi_\lambda(z)$, $\phi_\lambda(z) \in \mathrm{Exp}_{R(\lambda)}(\mathbf{C}^n_z)$, $R(\lambda) = \mathrm{dist}(\lambda, \partial\Omega)$. Here the summation is carried out over all finite sets of points $\lambda \in \Omega$.

Theorem 2.1.4. *If Ω is a Runge domain, then the map*

$$F : \mathrm{Exp}_\Omega(\mathbf{C}^n_z) \to \mathcal{O}'(\Omega)$$

is an isomorphism, F^{-1} being defined by the inversion formula.

Proof. Indeed, if $u(z) \in \mathrm{Exp}_\Omega(\mathbf{C}^n_z)$, then

$$\tilde{u}(\zeta) = \sum_{\lambda \in \Omega} \tilde{u}_\lambda(\zeta),$$

where, in accordance with Theorem 2.1.3, the functionals $\tilde{u}_\lambda(\zeta)$ are regular analytic functionals in the polycylinders

$$U_{R(\lambda)} = \{\zeta : |\zeta_j - \lambda_j| < R_j(\lambda), \ j = 1, \ldots, n\}$$

and, moreover, in the whole domain Ω. Hence the Fourier transform $\tilde{u}(\zeta)$ is a finite sum of regular analytic functionals in Ω, that is, $\tilde{u}(\zeta) \in \mathcal{O}'(\Omega)$. It is defined by the formula

$$\langle \tilde{u}(\zeta), v(\zeta) \rangle = \frac{1}{(2\pi i)^n} \sum_{\lambda \in \Omega} \int_{\Gamma_\lambda} Bu_\lambda(\zeta)v(\zeta)d\zeta,$$

where the $Bu_\lambda(\zeta)$ are Borel functions; the Γ_λ are the skeletons of the polycylinders $U_{R(\lambda)}$; $v(\zeta) \in \mathcal{O}(\Omega)$ is arbitrary.

Thus

$$F : \operatorname{Exp}_\Omega(\mathbf{C}_z^n) \to \mathcal{O}'(\Omega).$$

Conversely, let $\Omega \subset \mathbf{C}_\zeta^n$ be a Runge domain and $h(\zeta) \in \mathcal{O}'(\Omega)$. Then, as has already been mentioned, there exists a compact set $K \subset \Omega$ such that

$$|\langle h(\zeta), v(\zeta) \rangle| \le M \max_{\zeta \in K} |v(\zeta)|,$$

where $M > 0$ is a constant, $v(\zeta) \in \mathcal{O}(\Omega)$. This inequality shows that by the Hahn-Banach theorem, the functional $h(\zeta)$ may be extended to a bounded linear functional on the space of all continuous functions $C(K)$. Consequently, by Riesz's theorem there exists a countable additive measure $\mu(d\zeta)$ defined on K such that

$$\langle h(\zeta), v(\zeta) \rangle = \int_K v(\zeta)\mu(d\zeta), \quad v(\zeta) \in C(K). \tag{2.1.5}$$

Since K is compact, it is clear that there exists a finite collection of Borel sets K_i ($i = 1, \ldots, N$) such that:

1) $K_i \bigcap K_j = \varnothing, \quad i \ne j;$

2) $\bigcup_{1 \le j \le N} K_j = K;$

3) each K_j is contained in at least one "polycylinder of analyticity" $U_{R(\lambda_j)}$ (here, different j may give rise to the same polycylinder $U_{R(\lambda_j)}$).

We set $\mu_j(d\zeta) = \mu(d\zeta)\big|_{K_j}$. Then in view of the properties 1) and 2)

$$\mu(d\zeta) = \sum_{1 \leq j \leq N} \mu_j(d\zeta),$$

therefore from (2.1.5), in particular, we obtain

$$u(z) = \langle h(\zeta), \exp z\zeta \rangle = \sum_{1 \leq j \leq N} \int_{K_j} e^{z\zeta} \mu_j(d\zeta) \equiv \sum_{1 \leq j \leq N} u_j(z).$$

Further, the property 3) implies that every $u_j(z) \in \mathrm{Exp}_{R(\lambda_j)}(\mathbf{C}_z^n)$ and therefore, $u(z) \in \mathrm{Exp}_\Omega(\mathbf{C}_z^n)$.

In conclusion we note that under a Runge condition the functional $h(\zeta)$ is defined (one-to-one) by the function $u(z)$, moreover, $h(\zeta) = \tilde{u}(\zeta)$. The theorem is proved.

Remark. It is natural to call the above theorem the complex version of the classical Paley-Wiener theorem, since every analytic functional is defined by a compact measure. Thus, Theorem 2.1.3 may be stated as follows:

The Fourier transformation of an entire function $u(z)$ is defined by a compact measure in a Runge domain Ω if and only if $u(z) \in \mathrm{Exp}_\Omega(\mathbf{C}_z^n)$.

3. COMPLEX UNITARITY

Let $u(z) \in \mathcal{O}(G)$, where $G \subset \mathbf{C}_z^n$, be a domain with a Runge condition. Further, let $v(\zeta) \in \mathrm{Exp}_G(\mathbf{C}_\zeta^n)$. Then, according to the previous results,

$$\tilde{u}(\zeta) \in \mathrm{Exp}_G'(\mathbf{C}_\zeta^n) \text{ and } \tilde{v}(\zeta) \in \mathcal{O}'(G).$$

The complex unitarity property is expressed in the following assertion.

Proposition 2.1.1. *For any $u(z) \in \mathcal{O}(G)$ and $v(\zeta) \in \mathrm{Exp}_G(\mathbf{C}_\zeta^n)$ we have*

$$\langle \tilde{u}(\zeta), v(\zeta) \rangle = \langle \tilde{v}(z), u(z) \rangle. \tag{2.1.6}$$

Proof. In fact, since $v(\zeta) \in \mathrm{Exp}_G(\mathbf{C}_\zeta^n)$,

$$v(\zeta) = \sum_{\lambda \in G} e^{\lambda\zeta} \phi_\lambda(\zeta),$$

where $\phi_\lambda(\zeta) \in \mathrm{Exp}_{R(\lambda)}(\mathbf{C}^n_\zeta)$. Hence

$$\langle \tilde{u}(\zeta), v(\zeta)\rangle \overset{\mathrm{def}}{=} \langle u(-\partial)\delta(\zeta), v(\zeta)\rangle =$$

$$= \sum_{\lambda \in G} \langle \delta(\zeta), u(\partial)e^{\lambda\zeta}\phi_\lambda(\zeta)\rangle =$$

$$= \sum_{\lambda \in G} \left\langle \delta(\zeta), \sum_{|\alpha|=0}^\infty \frac{D^\alpha u(\lambda)}{\alpha!}(\partial - \lambda I)^\alpha[e^{\lambda\zeta}\phi_\lambda(\zeta)] \right\rangle =$$

$$= \sum_{\lambda \in G} \sum_{|\alpha|=0}^\infty \frac{1}{\alpha!}D^\alpha u(\lambda)\partial^\alpha\phi_\lambda(0).$$

On the other hand,

$$\langle \tilde{v}(z), u(z)\rangle = \sum_{\lambda \in G} \langle e^{-\lambda D}\phi_\lambda(-D)\delta(z), u(z)\rangle =$$

$$= \sum_{\lambda \in G} \langle \delta(z), \phi_\lambda(D)e^{\lambda D}u(z)\rangle = \sum_{\lambda \in G} \langle \delta(z), \phi_\lambda(D)u(\lambda + z)\rangle =$$

$$= \sum_{\lambda \in G} \sum_{|\alpha|=0}^\infty \frac{\partial^\alpha\phi_\lambda(0)}{\alpha!}D^\alpha u(\lambda).$$

Comparing these formulae, we immediately obtain (2.1.6). The assertion is proved.

Remark. For the sake of symmetry one can write this formula in the form

$$\langle \tilde{u}(\zeta), v(\zeta)\rangle = \langle u(z), \tilde{v}(z)\rangle.$$

In this form it is a variant of the L_2-unitarity of the usual Fourier L_2-transformation, that is, a variant of the classical Parseval equality.

2.2. Complex Fourier Method

In this section we give some applications of the Fourier transformation to the solution of PD-equations in $\mathbf{C}^n$ and to the solution of the Cauchy problem for PD-equations.

1. TABLE OF DUALITY. EXAMPLES

The Fourier transformation considered in §2.1 has all the standard (or almost standard) properties. We collect some of these properties in the following table.

Table of duality

$u(z)$	$\tilde{u}(\zeta)$
1. Linearity of F	
$au(z) + bv(z);\ a, b \in \mathbb{C}^1$	$a\tilde{u}(\zeta) + b\tilde{v}(\zeta);$
$u(z), v(z) \in \mathcal{O}(G)$	$\tilde{u}(\zeta), \tilde{v}(\zeta) \in \mathrm{Exp}'_G(\mathbb{C}^n_\zeta)$
2. Similarity theorem	
$u(az),\ a \in \mathbb{C}^n$	$a_1^{-n} \ldots a_n^{-n} \tilde{u}(a^{-1}\zeta) \in \mathrm{Exp}'_{a^{-1}G}(\mathbb{C}^n_\zeta)$
$(az \equiv (a_1 z_1, \ldots, a_n z_n))$	$(a^{-1}z = (a_1^{-1}z_1, \ldots, a_n^{-1}z_n),\ a^{-1}G = \{z : az \in G\})$
3. Shift theorem	
$u(z - a),\ a \in \mathbb{C}^n$	$e^{-a\zeta}\tilde{u}(\zeta) \in \mathrm{Exp}'_{G+a}(\mathbb{C}^n_\zeta)$
	$(G + a = \{z + a,\ z \in G\})$
4. Retardation theorem	
$e^{\tau z}u(z),\ \tau \in \mathbb{C}^n$	$\tilde{u}(\zeta - \tau) \in \mathrm{Exp}'_G(\mathbb{C}^n_\zeta)$
5. Multiplication by z	
$z^\alpha u(z)$	$(-\partial)^\alpha \tilde{u}(\zeta) \in \mathrm{Exp}'_G(\mathbb{C}^n_\zeta)$
6. Differentiation theorem	
$D^\alpha u(z)$	$\zeta^\alpha \tilde{u}(\zeta) \in \mathrm{Exp}'_G(\mathbb{C}^n_\zeta)$
7. PD-operator theorem	
$A(D)u(z)$	$A(\zeta)\tilde{u}(\zeta)$

The duality of item 7 needs some explanation. Namely, let $A(\zeta)$ be analytic in Ω. Then for any $h(\zeta) \in \mathcal{O}'(\Omega)$ the "product" $A(\zeta)h(\zeta)$ is clearly well defined as a functional on $\mathcal{O}(\Omega)$. Thus, if $u(z) \in \mathrm{Exp}_\Omega(\mathbb{C}^n_z)$ and $A(D)$ is a PD-operator with symbol $A(\zeta) \in \mathcal{O}(\Omega)$, then $(A(D)u(z))^\sim = A(\zeta)\tilde{u}(\zeta)$, that is,

$$A(D)u(z) \overset{\mathcal{F}}{\to} A(\zeta)\tilde{u}(\zeta).$$

Conversely, if $h(\zeta) \in \mathcal{O}'(\Omega)$ has the form $h(\zeta) = A(\zeta)\tilde{u}(\zeta)$, where

$A(\zeta) \in \mathcal{O}(\Omega)$ and $\tilde{u}(\zeta)$ is the Fourier image of $u(z) \in \mathrm{Exp}_\Omega(\mathbf{C}_z^n)$, then $\mathcal{F}^{-1}h(\zeta) = A(D)u(z)$.

Example 1. Let us calculate the Fourier transform of the function $u(z) = \exp az^2$, where $az^2 = a_1 z_1^2 + \ldots + a_n z_n^2$. By definition, $\tilde{u}(\zeta) = \exp a\partial^2 \delta(\zeta)$. Consequently, using the formula

$$\delta(\zeta) = (2\sqrt{\pi})^{-n}(a_1 \ldots a_n)^{1/2} \exp(-a\partial^2) \exp(-\zeta^2/4a)$$

(see §1.4), we find that

$$[\mathcal{F} \exp az^2](\zeta) = (2\sqrt{\pi})^{-n}(a_1 \ldots a_n)^{-1/2} \exp(-\zeta^2/4a).$$

Thus, the Fourier transformation of the entire function $\exp az^2$ is a regular functional which is defined by the kernel $(2\sqrt{\pi})^{-n}(a_1 \ldots a_n)^{-1/2} \exp(-\zeta^2/4a)$ and the system of contours $(-\sqrt{a_j}\infty, \sqrt{a_j}\infty)$, $1 \leq j \leq n$.

Example. 2. Let us find the Fourier image of the function $u(z) = z^{-m}$, where $z \in \mathbf{C}^1$, m is a positive integer. Clearly,

$$\frac{1}{z^m} = \frac{(-1)^{m-1}}{(m-1)!} D^{m-1}\frac{1}{z}.$$

Hence (see Table of duality),

$$F[z^{-m}](\zeta) = \frac{(-1)^{m-1}}{(m-1)!}\zeta^{m-1}F[z^{-1}](\zeta) = \frac{(-1)^m}{(m-1)!}\zeta^{m-1}n(\zeta), \qquad (2.2.1)$$

where

$$n(\zeta) = \mathrm{nat} \int \delta(\zeta)d\zeta$$

is the natural inverse of the delta function (see §1.4).

Example 3. Let

$$u(z) = \exp\left(-\frac{1}{z-a}\right),$$

where $a \in \mathbf{C}^1$ is a fixed number. In view of the shift theorem (see Table of duality),

$$\left[F \exp\left(-\frac{1}{z-a}\right)\right](\zeta) = e^{-a\zeta} F\left[\exp\left(-\frac{1}{z}\right)\right](\zeta).$$

Further, since

$$\exp\left(-\frac{1}{z}\right) = \sum_{m=0}^{\infty} \frac{(-1)^m}{m!z^m} = 1 + \sum_{m=1}^{\infty} \frac{(-1)^m}{m!z^m},$$

in accordance with formula (2.2.1) we have

$$\left[F \exp\left(-\frac{1}{z}\right)\right](\zeta) = \delta(\zeta) + \sum_{m=1}^{\infty} \frac{\zeta^{m-1}}{m!(m-1)!} n(\zeta).$$

Hence,

$$\mathcal{F}\left[\exp\left(-\frac{1}{z-a}\right)\right](\zeta) = e^{-a\zeta}\left[\delta(\zeta) + \sum_{m=1}^{\infty} \frac{\zeta^{m-1}}{m!(m-1)!} n(\zeta)\right].$$

2. FOURIER METHOD FOR PD-EQUATIONS

We consider the PD-equation

$$A(D)u(z) = h(z), \quad z \in \mathbf{C}^n. \tag{2.2.2}$$

Here $A(D)$ is a PD-operator with symbol $A(\zeta) \in \mathcal{O}(\Omega)$, where $\Omega \subset \mathbf{C}_\zeta^n$ is a Runge domain.

Theorem 2.2.1. *Let* $A(\zeta) \in \mathcal{O}(\Omega)$, *where* $A(\zeta) \neq 0$ *in* Ω. *Then for any* $h(z) \in \mathrm{Exp}_\Omega(\mathbf{C}_z^n)$ *there exists a unique solution* $u(z) \in \mathrm{Exp}_\Omega(\mathbf{C}_z^n)$ *of* (2.2.2), *which can be written in the form*

$$u(z) = \langle \tilde{h}(\zeta), A^{-1}(\zeta)e^{z\zeta}\rangle \tag{2.2.3}$$

($\tilde{h}(\zeta)$ *is the Fourier transform of* $h(z)$).

Proof. For the proof we use the isomorphism

$$F : \mathrm{Exp}_\Omega(\mathbf{C}_z^n) \to \mathcal{O}'(\Omega).$$

In view of this isomorphism, we obtain in the space $\mathcal{O}'(\Omega)$ the equation

$$A(\zeta)\tilde{u}(\zeta) = \tilde{h}(\zeta),$$

which is equivalent to the original equation (2.2.2). Hence

$$\tilde{u}(\zeta) = \frac{1}{A(\zeta)}\tilde{h}(\zeta)$$

and consequently, by the inversion formula

$$u(z) = \langle A^{-1}(\zeta)\tilde{h}(\zeta), e^{z\zeta}\rangle = \langle \tilde{h}(\zeta), A^{-1}(\zeta)e^{z\zeta}\rangle.$$

The theorem is proved.

Example 1. Let us find the solution of the equation

$$A(D)u(z) = P(z)e^{\lambda z}, \quad z \in \mathbf{C}^n,$$

where $A(\zeta)$ is analytic in a neighbourhood of $\lambda \in \mathbf{C}^n$ and $P(z)$ is a polynomial.

By formula (2.2.3) we have

$$u(z) = \langle \delta(\zeta), P(\partial)e^{\lambda \partial}[A^{-1}(\zeta)e^{z\zeta}]\rangle =$$

$$= \langle \delta(\zeta), P(\partial)[A^{-1}(\zeta + \lambda)e^{z(\zeta+\lambda)}]\rangle =$$

$$= \langle \delta(\zeta), e^{z(\zeta+\lambda)}P(\partial + z)A^{-1}(\zeta + \lambda)\rangle,$$

that is,

$$u(z) = e^{\lambda z}P(\partial + z)A^{-1}(\zeta + \lambda)\big|_{\zeta=0} = Q(z)e^{\lambda z},$$

where $Q(z)$ is a polynomial in z, the degree of which is not greater than that of $P(z)$.

Thus, the solution of an arbitrary PD-equation of (2.2.2) with quasi-polynomial right hand is also a quasi-polynomial (incidentally, we already know this fact, see §1.3).

One can state the dual result in the same way.

Theorem 2.2.2 *Let the functions $A(\zeta)$ and $A^{-1}(\zeta)$ be analytic in the same domain Ω. Then for any right hand $h(z) \in \mathrm{Exp}'_{\Omega^-}(\mathbf{C}^n_z)$ equation (2.2.2) has the unique solution*

$$u(z) = F[A^{-1}(-\zeta)\langle h(z), e^{z\zeta}\rangle](z). \tag{2.2.4}$$

Proof. Here we use the isomorphism

$$F : \mathcal{O}(\Omega^-) \to \mathrm{Exp}'_{\Omega^-}(\mathbf{C}^n_z)$$

(let us emphasize that, by comparison with the general theory, the variables z and ζ switch roles). Then, using the inversion formula from equation (2.2.2) we obtain the equivalent equation

$$A(-\zeta)\hat{u}(\zeta) = \hat{h}(\zeta), \quad \zeta \in \Omega^-, \tag{2.2.5}$$

where $\hat{u}(\zeta) = \langle u(z), \exp z\zeta \rangle \in \mathcal{O}(\Omega^-)$.

The desired formula (2.2.4) now follows immediately from (2.2.5). The theorem is proved.

Example 2. Let us find the fundamental solution of the PD-operator $A(D)$, where $A(\zeta) \in \mathcal{O}(\Omega)$, $\Omega \subset \mathbb{C}^n_\zeta$, is arbitrary. We have

$$A(D)u(z) = \delta(z) \in \text{Exp}'_{\Omega^-}(\mathbb{C}^n_z)$$

and in accordance with (2.2.4),

$$u(z) = F[A^{-1}(-\zeta)\langle \delta(z), e^{z\zeta} \rangle] = F[A^{-1}(-\zeta)],$$

that is,

$$u(z) = \frac{I}{A(D)}\delta(z).$$

Let us now turn to the Cauchy problem. Namely, we consider the Cauchy problem $(t \in \mathbb{C}^1, \ z \in \mathbb{C}^n)$

$$\frac{\partial^m u}{\partial t^m} + \sum_{k=0}^{m-1} A_k(t, D)\frac{\partial^k u}{\partial t^k} = h(t, z), \tag{2.2.6}$$

$$\frac{\partial^k u}{\partial t^k}(0, z) = \phi_k(z), \quad k = 0, 1, \ldots, m - 1, \tag{2.2.7}$$

where $A_k(t, D)$ are PD-operators with analytic symbols in some domain $\Omega \subset \mathbb{C}^n_\zeta$ for all t, $|t| < \delta$ $(\delta > 0)$.

Theorem 2.2.3. *Let the function* $h(t, \cdot) \in \text{Exp}_\Omega(\mathbb{C}^n_z)$ *be analytic in* t, $|t| < \delta$. *Further, let* $\phi_k(z) \in \text{Exp}_\Omega(\mathbb{C}^n_z)$, $k = 0, 1, \ldots, m - 1$. *Then the Cauchy problem* (2.2.6), (2.2.7) *has a unique solution* $u(t, \cdot) \in \text{Exp}_\Omega(\mathbb{C}^n_z)$ *which is analytic in* t, $|t| < \delta$.

Proof. After taking the Fourier transform, the problem (2.2.6), (2.2.7) gives the Cauchy problem for the ordinary differential equation with parameter $\zeta \in \Omega$:

$$\tilde{u}^{(m)}(t, \zeta) + \sum_{k=0}^{m-1} A_k(t, \zeta)\tilde{u}^{(k)}(t, \zeta) = \tilde{h}(t, \zeta) \tag{2.2.8}$$

$$\tilde{u}^{(k)}(0, \zeta) = \tilde{\phi}_k(\zeta), \quad k = 0, 1, \ldots, m - 1. \tag{2.2.9}$$

In view of the isomorphism

$$F : \mathrm{Exp}_\Omega(\mathbf{C}_z^n) \to \mathcal{O}'(\Omega),$$

the problem (2.2.8), (2.2.9) is the ordinary differential problem in the space of analytic functionals $\mathcal{O}'(\Omega)$. This problem has a unique solution which can be obtained in the usual way.

It is clear that the formula

$$u(t, z) = F_{\zeta \to z}\, \tilde{u}(t, \zeta)$$

gives the solution of the original problem (2.2.6), (2.2.7). The theorem is proved.

In conclusion we state a theorem on the solvability of the Cauchy problem in the space of functionals. Namely, the following theorem is true.

Theorem 2.2.4. *Let* $h(t, \cdot) \in \mathrm{Exp}'_{\Omega-}(\mathbf{C}_z^n)$, $\phi_k(z) \in \mathrm{Exp}'_{\Omega-}(\mathbf{C}_z^n)$, $k = 0, 1, \ldots$ $\ldots, m-1$, *be analytic in* t, $|t| < \delta$. *Then the Cauchy problem* (2.2.6), (2.2.7) *has a unique analytic solution* $u(t, \cdot) \in \mathrm{Exp}'_{\Omega-}(\mathbf{C}_z^n)$ *in* t, $|t| < \delta$.

The proof is the same.

Chapter 3. PD-Equations whose Symbols are Formal Series

3.1. Differential operators of infinite order
with constant coefficients

1. THE SPACE $E_{q,r}(\mathbf{C}_z^n)$ OF ENTIRE FUNCTIONS OF ORDER $q < 1$

Let $0 < q < 1$ and $r > 0$ be given numbers. We set

$$E_{q,r}(\mathbf{C}_z^n) = \{u(z) : \mathbf{C}^n \to \mathbf{C}^1 \text{ such that } \|u\|_{q,r} \stackrel{\text{def}}{=} \sup_{z \in \mathbf{C}^n} |u(z)| \exp(-r|z|^q) < \infty\}$$

(here $|z|^q = |z_1|^q + \ldots + |z_n|^q$).

It is easy to prove that for any $r_1 \le r_2$ we have the inclusions $E_{q,r_1}(\mathbf{C}_z^n) \subset E_{q,r_2}(\mathbf{C}_z^n)$ which are compact if $r_1 < r_2$.

We give two examples of functions belonging to $E_{q,r}(\mathbf{C}_z^n)$.

Example 1. Let $q = m/p < 1$ be a rational number with m and p coprime. We consider

$$f(z) = \sum_{k=0}^{p-1} \exp\left[\left(\exp \frac{2\pi i k}{p}\right)\nu z\right],$$

where $\nu \in \mathbf{C}^n$ and $\nu z = \nu_1 z_1 + \ldots + \nu_n z_n$. The Taylor expansion of this function is

$$f(z) = \sum_{|\alpha|=0}^{\infty} \frac{\nu^{p\alpha} z^{p\alpha}}{(p\alpha)!},$$

where $(p\alpha)! = (p\alpha_1)! \ldots (p\alpha_n)!$, $\nu^{p\alpha} = \nu_1^{p\alpha_1} \ldots \nu_n^{p\alpha_n}$, $z^{p\alpha} = z_1^{p\alpha_1} \ldots z_n^{p\alpha_n}$. From this it follows that the function

$$\phi(z) = f(z^q) \equiv f(z_1^q, \ldots, z_n^q)$$

is an entire function, belonging to $E_{q,r}(\mathbf{C}_z^n)$, where $r \ge \max(|\nu_1|, \ldots, |\nu_n|)$.

Example 2. Let $0 < q < 1$. We consider the function

$$\phi(z) = \sum_{|\alpha|=0}^{\infty} a_\alpha z^\alpha \equiv \sum_{|\alpha|=0}^{\infty} \frac{\nu^\alpha}{(\alpha!)^{1/q}} z^\alpha,$$

66

where $(\alpha!)^{1/q} = (\alpha_1!)^{1/q} \dots (\alpha_n!)^{1/q}$. Since for any $\gamma > 1$ and $b \geq 0, \dots, b_j \geq 0$

$$b_1^\gamma \dots b_j^\gamma \leq (b_1 + \dots + b_j)^\gamma, \quad j = 1, 2, \dots,$$

it follows that

$$\phi(z) \leq \sum_{|\alpha|=0}^\infty \frac{|\nu_1|^{\alpha_1} \dots |\nu_n|^{\alpha_n} |z_1|^{\alpha_1} \dots |z_n|^{\alpha_n}}{(\alpha!)^{1/q}} \leq$$

$$\leq \left(\sum_{|\alpha|=0}^\infty \frac{|\nu_1|^{q\alpha_1} \dots |\nu_n|^{q\alpha_n} \cdot |z_1|^{q\alpha_1} \dots |z_n|^{q\alpha_n}}{\alpha!} \right)^{1/q} \leq \exp\left(\frac{1}{q} |\nu|^q |z|^q \right).$$

Thus $\phi(z) \in E_{q,r}(\mathbf{C}_z^n)$, where $r = \max(|\nu_1|, \dots, |\nu_n|)$.

2. THE BASIC ESTIMATE

The following lemma plays an important role in the sequel.

Lemma 3.1.1. Let $\phi(z) \in E_{q,r}(\mathbf{C}_z^n)$ be an arbitrary function. Then for any $z \in \mathbf{C}^n$ and $|\alpha| = 0, 1, \dots$ the inequality

$$|D^\alpha u(z)| \leq \min \left\{ \frac{\|\phi\|_{q,r}}{\prod_{i=1}^n (1 + |z_i|)^{(1-q)\alpha_i}} e^{r|z|^q} (qr)^{|\alpha|} \prod_{i=1}^n \sqrt{\alpha_i} \; ; \right.$$

$$\left. \frac{\|\phi\|_{q,r}}{(\alpha!)^{\frac{1}{q}-1}} e^{r|z|^q} (qr)^{\frac{|\alpha|}{q}} \prod_{i=1}^n (\sqrt{\alpha_i})^{1/q} \right\}$$

holds.

Proof. In accordance with Cauchy's formula

$$D^\alpha \phi(z) = \frac{\alpha!}{(2\pi i)^n} \int_{|\zeta - z| = |a|} \frac{\phi(\zeta)}{(\zeta - z)^{\alpha+1}} d\zeta,$$

where $a = (a_1, \dots, a_n)$, $a_1 > 0, \dots, a_n > 0$ are arbitrary numbers; here $|\zeta - z| = |a|$ means $|\zeta_1 - z_1| = a_1, \dots, |\zeta_n - z_n| = a_n$.

By definition of the norm $\|\phi\|_{q,r}$ we have

$$|\phi(z)| \leq \|\phi\|_{q,r} \exp r|z|^q.$$

Hence

$$|D^\alpha \phi(z)| \leq \frac{\alpha!}{(2\pi)^n} \int_{|\zeta - z| = |a|} \frac{\|\phi\|_{q,r} e^{r|\zeta|^q} |d\zeta|}{|(\zeta - z)^{\alpha+1}|} \leq$$

$$\leq \frac{\alpha!}{(2\pi)^n} \|\phi\|_{q,r} \exp\left(r[(|z_1| + a_1)^q + \dots + (|z_n| + a_n)^q] \right) \frac{\prod_{i=1}^n 2\pi a_i}{\prod_{i=1}^n a_i^{\alpha_i+1}}.$$

Putting $(|z_i| + a_i)^q = |z_i|^q + \epsilon_i$, where $\epsilon_i > 0$, we obtain the inequality

$$|D^\alpha \phi(z)| \le \|\phi\|_{q,r} e^{r|z|^q} \alpha! \prod_{i=1}^{n} \frac{e^{r\epsilon_i}}{[(|z_i|^q + \epsilon_i)^{1/q} - |z_i|]^{\alpha_i}}. \qquad (3.1.1)$$

Let us consider the factors $\mu_i(\epsilon_i, z_i) = e^{r\epsilon_i}[(|z_i|^q + \epsilon_i)^{1/q} - |z_i|]^{-\alpha_i}$. Clearly each $\mu_i(\epsilon_i, z_i)$ tends to a finite limit as $|z_i| \to 0$ or $|z_i| \to \infty$. Therefore

$$\mu_i(\epsilon_i, z_i) \le \max_{z_i \in \mathbf{C}^1} \mu_i(\epsilon_i, z_i)$$

and

$$|D^\alpha \phi(z)| \le \|\phi\|_{q,r} e^{r|z|^q} \alpha! \prod_{i=1}^{n} \min_{\epsilon_i} \max_{z_i} \mu_i(\epsilon_i, z_i).$$

Setting (for simplicity) $|z_i| = \rho$, $\epsilon_i = \epsilon$, $\alpha_i = \alpha$, we obtain

$$\max_{\rho > 0} \frac{1}{(\rho^q + \epsilon)^{1/q} - \rho} = \frac{1}{\epsilon^{1/q}}.$$

Then in order to find $\min_{\epsilon} e^{r\epsilon} e^{-\frac{\alpha}{q}}$ we note that the derivative $(e^{r\epsilon} \epsilon^{-\frac{\alpha}{q}})' = 0$ for $\epsilon = \alpha/qr$ and, consequently,

$$\min_{\epsilon > 0} e^{r\epsilon} \epsilon^{-\frac{\alpha}{q}} = \left(\frac{e}{\alpha}\right)^\alpha (qr)^{\frac{\alpha}{q}}.$$

Substituting these factors for each $i = 1, \ldots, n$ in the inequality (3.1.1) and using Stirling's formula, we obtain

$$|D^\alpha \phi(z)| \le \|\phi\|_{q,r} e^{r|z|^q} \frac{(qr)^{|\alpha|/q}}{(\alpha!)^{1/q-1}} \prod_{i=1}^{n} (\sqrt{\alpha_i})^{1/q} \qquad (3.1.2)$$

(which is "one half" of the desired inequality).

Let us again turn to the inequality (3.1.1) and write it in the form

$$|D^\alpha \phi(z)| \le \|\phi\|_{q,r} e^{r|z|^q} \alpha! \prod_{i=1}^{n} (1 + |z_i|)^{-(1-q)\alpha_i} \times$$

$$\times \prod_{i=1}^{n} \frac{e^{r\epsilon_i}(1 + |z_i|)^{(1-q)\alpha_i}}{[(|z_i|^q + \epsilon_i)^{1/q} - |z_i|]^{\alpha_i}}. \qquad (3.1.3)$$

Let us study the function

$$f(\epsilon, \rho) = \frac{(1 + \rho)^{1-q}}{(\rho^q + \epsilon)^{1/q} - \rho} \qquad (\epsilon \ge 0, \ \rho \ge 0).$$

Since $f(\epsilon,\rho) \to \epsilon^{-1/q}$ $(\rho \to 0)$ and $f(\epsilon,\rho) \to \epsilon^{-1}$ $(\rho \to +\infty)$, the quantity $\sup_{\rho} f(\epsilon,\rho)$ is finite. To find it, we examine the derivative (with respect to ρ) of the function $f(\epsilon,\rho)$. (We note that the following calculations are elementary but (in our opinion) not all that trivial; for this reason we give them completely.) We have

$$f'(\epsilon,\rho) = \left\{(1-q)\rho\left[(1+\epsilon\rho^{-q})^{1/q} - 1\right] - (\rho+1)\left[(1+\epsilon\rho^{-q})^{\frac{1-q}{q}} - 1\right]\right\} \times$$

$$\times \frac{1}{(\rho+1)^q\left[(\rho^q+\epsilon)^{1/q} - \rho\right]^2}.$$

It is not difficult to see (using the Taylor expansion of $\{\ldots\}$) that

$$f'(\epsilon,\rho) \to -\infty \ (\rho \to 0) \text{ with asymptotic } -\rho^{q-1},$$
$$f'(\epsilon,\rho) \to 0 \ (\rho \to +\infty) \text{ with asymptotic } \rho^{-(1+q)}.$$

We note further that for all ρ $f'(\epsilon,\rho) > 0$.

We set

$$(1+\epsilon\rho^{-q})^{\frac{1-q}{q}} \equiv x > 1, \qquad \frac{1}{1-q} \equiv \nu > 1.$$

Then the condition $f'(\epsilon,\rho) = 0$ may be written in the form

$$\frac{1}{\nu}\frac{x^{\nu}-1}{x-1} - 1 = \left(\frac{x^{\nu-1}-1}{\epsilon}\right)^{\frac{\nu}{\nu-1}}. \tag{3.1.4}$$

It is easy to see that the the right hand side of (3.1.4) increases more rapidly than the left hand side. On the other hand, both sides tend to zero as $x \to 1$. Defining the left hand to be zero at $x = 1$, we find (again using the Taylor expansion) that the derivatives at the point $x = 1$ are respectively given by

$$\lim_{x\to 1}\left(\frac{x^{\nu-1}-1}{\nu(x-1)} - 1\right)/(x-1) = \frac{\nu-1}{2} > 0, \quad \frac{\nu}{\epsilon}\left(\frac{x^{\nu-1}-1}{\epsilon}\right)^{\frac{1}{\nu-1}}x^{\nu-2}\Big|_{x=1} = 0.$$

Thus, equation (3.1.4) has only one root for $x > 1$. Hence the derivative $f'(\epsilon,\rho)$ has only one zero. Taking into account the asymptotic behaviour of $f(\epsilon,\rho)$ (as $\rho \to 0$ and $\rho \to +\infty$) we find that this zero is the minimum point of $f(\epsilon,\rho)$. Therefore, for all $\rho > 0$ we have the inequality

$$f(\epsilon,\rho) \le \max(\epsilon^{-\frac{1}{q}}, q\epsilon^{-1}) \equiv M_q(\epsilon).$$

We now evaluate $\min_{\epsilon>0} e^{r\epsilon}M_q^{\alpha}(\epsilon)$.

1. Let $0 < \epsilon < q^{q-1)/q}$. Then $\epsilon^{1/q} \geq \rho\epsilon^{-1}$ and by definition $M_q(\epsilon) = \epsilon^{-1/q}$. On this half-interval the function $\epsilon^{-\alpha/q}e^{r\epsilon}$ decreases if α is sufficiently large. Consequently,

$$\min_{0<\epsilon<q^{(q-1)/q}} \epsilon^{-\alpha/q}e^{r\epsilon} = \left(\frac{1}{q}\right)^{\frac{\alpha}{q}(1-\frac{1}{q})} \exp r\left(\frac{1}{q}\right)^{\frac{1}{q}-1}.$$

2. Let $\epsilon \geq q^{(q-1)/q}$. Then $\epsilon^{-1/q} \leq q\epsilon^{-1}$; hence $M_q(\epsilon) = q\epsilon^{-1}$. In this case

$$\min_{\epsilon \geq q^{(q-1)/q}} \frac{\exp r\epsilon}{\epsilon^{\alpha}}q^{\alpha} = \frac{\exp \alpha}{\alpha^{\alpha}}(qr)^{\alpha}.$$

(This value is attained at the point $\epsilon = \alpha/r \geq q^{(q\epsilon-1)/q}$ for α sufficiently large.)

Since for $|\alpha| \gg 1$ the minimum for case 2 is less than the minimum for case 1, it follows that for all $\epsilon > 0$

$$\min_{\epsilon} M_q^{\alpha}(\epsilon) \exp r\epsilon = \frac{1}{\alpha!}(qr)^{\alpha} \exp \alpha.$$

Substituting this expression into (3.1.3) ($\alpha \leftrightarrow \alpha_i$, $\rho \leftrightarrow |z_i|$, $\epsilon \leftrightarrow \epsilon_i$) and using Stirling's formula, we obtain

$$|D^{\alpha}\phi(z)| \leq \|\phi\|_{q,r}e^{r|z|^q}\frac{1}{\prod_{i=1}^{n}(1+|z_i|)^{(1-q)\alpha_i}}(qr)^{|\alpha|}\prod_{i=1}^{n}\sqrt{\alpha_i}.$$

Combining this inequality with (3.1.2) we finally obtain the inequality of the lemma. Q.E.D.

3. THE NON-FORMAL ACTION OF D.O.I.O.'S

Again let $0 < q < 1$. We consider the infinitely differentiable function $A(\xi)$, $\xi \in \mathbf{R}^n$. For this function the formal Taylor series

$$A(\xi) \sim \sum_{|\alpha|=0}^{\infty} a_{\alpha}\xi^{\alpha}, \quad a_{\alpha} = \frac{1}{\alpha!}\partial^{\alpha}A(0),$$

may diverge everywhere except $\xi = 0$. However, a d.o.i.o. with such a formal symbol may act non-formally in the space $E_{q,r}(\mathbf{C}_z^n)$.

Namely, let us formulate the following condition.

Condition A. We say that the symbol $A(\xi)$ satisfies Condition A if there exists a number $\gamma \in]0,(1-q)/q[$ such that for any α

$$|a_{\alpha}| \leq P(\alpha!)^{\gamma}, \tag{A}$$

where $P > 0$ is a constant (depending, in general, on the function $A(\xi)$).

As usual, we define the action of the operator $A(D)$, $D = (\partial/\partial z_1, \ldots \ldots, \partial/\partial z_n)$, by the formula

$$A(D)\phi(z) = \sum_{|\alpha|=0}^{\infty} a_\alpha D^\alpha \phi(z).$$

Lemma 3.1.2. *If Condition A holds, then for any $r \in]0, +\infty[$ the map*

$$A(D) : E_{q,r}(\mathbf{C}_z^n) \to E_{q,r}(\mathbf{C}_z^n)$$

is well-defined; moreover, this map is continuous.

Proof. In fact, by Lemma 3.1.1, for any $\phi(z) \in E_{q,r}(\mathbf{C}_z^n)$ we have

$$|A(D)\phi(z)| \le \sum_{|\alpha|=0}^{\infty} |a_\alpha| \cdot |D^\alpha \phi(z)| \le$$

$$\le \sum_{|\alpha|=0}^{\infty} |a_\alpha| \cdot \|\phi\|_{q,r} e^{r|z|^q} \frac{1}{(\alpha!)^{(1-q)/q}} (qr)^{|\alpha|/q} \prod_{i=1}^{n} (\sqrt{\alpha_i})^{1/q}.$$

Hence, taking into account Condition A, we obtain

$$|A(D)\phi(z)| \le M \|\phi\|_{q,r} e^{r|z|^q},$$

where $M > 0$ is a constant not depending on $\phi(z)$. This means that

$$\|A(D)\phi(z)\|_{q,r} \le M \|\phi\|_{q,r}.$$

The lemma is proved.

Remark. Condition A is a sufficient condition for the well-definedness of the action of $A(D)$ in the space of entire functions $E_{q,r}(\mathbf{C}_z^n)$. The following proposition gives the "almost" necessity of this condition. Namely:

In order that the action of $A(D)\phi(z)$ be well-defined in $E_{q,r}(\mathbf{C}_z^n)$, it is necessary that

$$|a_\alpha|(\alpha!)^{(1-q)/q} \to 0 \text{ as } |\alpha| \to \infty.$$

In fact, let us consider the function (see Example 2 in subsection 1)

$$\phi(z) = \sum_{|\alpha|=0}^{\infty} \frac{z^\alpha}{(\alpha!)^{1/q}}.$$

Then

$$D^\beta \phi(z) = \sum_{\alpha - \beta \geq 0}^{\infty} \frac{\alpha(\alpha - 1)\ldots(\alpha - \beta + 1)z^{\alpha - \beta}}{(\alpha!)^{1/q}},$$

where $\alpha(\alpha - 1)\ldots(\alpha - \beta + 1) = \alpha_1 \ldots \alpha_n(\alpha_1 - 1)\ldots(\alpha_n - 1)\ldots(\alpha_1 - \beta_1 + 1)\ldots(\alpha_n - \beta_n + 1)$; here, the inequality $\alpha - \beta \geq 0$ means that $\alpha_1 - \beta_1 \geq 0, \ldots, \alpha_n - \beta_n \geq 0$. In other words,

$$D^\beta \phi(z) = \frac{1}{(\beta!)^{(1-q)/q}} + \sum_{\alpha - \beta > 0} \frac{\alpha(\alpha - 1)\ldots(\alpha - \beta + 1)z^{\alpha - \beta}}{(\alpha!)^{1/q}}.$$

Hence at the point $z = 0$ we have

$$A(D)\phi(z)\big|_{z=0} = \sum_{|\beta|=0}^{\infty} \frac{a_\beta}{(\beta!)^{(1-q)/q}}.$$

Clearly, the condition $|a_\beta|(\beta!)^{(q-1)/q} \to 0$ is necessary for the convergence of this series.

4. THE ALGEBRA OF D.O.I.O.'S

In this subsection we prove that the set of d.o.i.o.'s with domain of definition $E_{q,r}(\mathbf{C}_z^n)$ is an algebra. The operations in this algebra are addition and composition of d.o.i.o.'s satisfying Condition A.

We denote by Q the set of infinitely differentiable functions satisfying Condition A.

Lemma 3.1.3. *The set Q is an algebra, the ring operation for which is ordinary multiplication.*

Proof. Let $A(\xi) \in Q$ and $B(\xi) \in Q$. This means that the formal Taylor series

$$A(\xi) \sim \sum_{|\alpha|=0}^{\infty} a_\alpha \xi^\alpha, \quad B(\xi) \sim \sum_{|\beta|=0}^{\infty} b_\beta \xi^\beta$$

are such that

$$|a_\alpha| \leq P_A(\alpha!)^{\gamma_A}, \quad |b_\beta| \leq P_B(\beta!)^{\alpha_B},$$

where $P_A > 0$, $P_B > 0$ and $\gamma_A \in\,]0, (1 - q)/q[$, $\gamma_B \in\,]0, (1 - q)/q[$ are certain constants.

Formal multiplication gives

$$A(\xi)B(\xi) \sim \sum_{|\alpha'|=0}^{\infty} \Big(\sum_{\alpha+\beta=\alpha'} a_\alpha b_\beta \Big) \xi^{\alpha'},$$

where $\alpha + \beta = (\alpha_1 + \beta_1, \ldots, \alpha_n + \beta_n)$.

In accordance with Condition A on the coefficients a_α and b_β we have the inequality

$$\Big| \sum_{\alpha+\beta=\alpha'} a_\alpha b_\beta \Big| \leq \sum_{\alpha+\beta=\alpha'} |a_\alpha|\,|b_\beta| \leq P_A P_B \sum_{\alpha+\beta=\alpha'} (\alpha!)^{\gamma_A} (\beta!)^{\gamma_B}.$$

Let $\gamma = \max(\gamma_A, \gamma_B)$ and $P = P_A P_B$. Then in view of the inequality $\alpha!\beta! \leq (\alpha')!$ we immediately obtain

$$\Big| \sum_{\alpha+\beta=\alpha'} a_\alpha b_\beta \Big| \leq P \Big(\sum_{\alpha+\beta=\alpha'} 1 \Big) [(\alpha')!]^\gamma = P \cdot O(|\alpha'|^n)[(\alpha')!]^\gamma \leq [(\alpha')!]^{\gamma+\epsilon}$$

for any $\epsilon > 0$, provided that the multi-indices are sufficiently large. Clearly, this means that the function $A(\xi)B(\xi)$ satisfies Condition A. The lemma is proved.

Corollary. *It follows from Lemma 3.1.2 and Lemma 3.1.3 that the set of d.o.i.o.'s*

$$A(D) : E_{q,r}(\mathbf{C}_z^n) \to E_{q,r}(\mathbf{C}_z^n)$$

whose symbols satisfy Condition A (we denote this set of d.o.i.o.'s by $\mathcal{A}$) forms an algebra with composition as the ring operation. Further, we have the isomorphism

$$\mathcal{A} \leftrightarrow Q,$$

where

$$\lambda A(D) + \mu B(D) \leftrightarrow \lambda A(\xi) + \mu B(\xi),$$
$$A(D) \circ B(D) \leftrightarrow A(\xi)B(\xi).$$

5. THE CAUCHY PROBLEM

We consider the Cauchy problem for the equation

$$u' - A(t, D)u = 0, \qquad\qquad (3.1.5)$$
$$u(0, z) = \phi(z), \qquad\qquad (3.1.6)$$

where $\phi(z) \in E_{q,r}(\mathbf{C}_z^n)$.

The symbol $A(t,\xi)$ of the operator $A(t,D)$ is analytic in t for $|t| < T$ (for some number $T > 0$) and $A(t,\cdot) \in Q$, that is, for every t, $|t| < T$,

$$\frac{1}{\alpha!}|\partial^\alpha A(t,0)| \le P(\alpha!)^\gamma,$$

where $\partial = (\partial/\partial\xi_1, \ldots, \partial/\partial\xi_n)$, $P > 0$, $0 < \gamma < (1-q)/q$ are certain numbers. We shall denote this class of symbols by $\mathcal{O}(T;Q)$. Similarly we introduce the class $\mathcal{O}(T;E_{q,r}(\mathbf{C}_z^n))$ as the Banach space of functions $u(t,z)$ that are analytic in t for $|t| < T$ and take their values in the space $E_{q,r}(\mathbf{C}_z^n)$.

Theorem 3.1.1. *Let the symbol $A(t,\xi) \in \mathcal{O}(T;Q)$. Then for any initial function $\phi(z) \in E_{q,r}(\mathbf{C}_z^n)$ there exists a unique solution $u(t,z) \in \mathcal{O}(T;E_{q,r}(\mathbf{C}_z^n))$ of the Cauchy problem (3.1.5), (3.1.6).*

Proof. We find the solution by the operator method. Namely, let $E(t,\xi)$ be the solution of the ordinary differential equation with parameter $\xi \in \mathbf{R}^n$

$$E'(t,\xi) - A(t,\xi)E(t,\xi) = 0$$

with initial condition

$$E(0,\xi) = 1.$$

Clearly,

$$E(t,\xi) = \exp \int_0^t A(\tau,\xi)d\tau.$$

Lemma 3.1.4. *If $A(t,\xi) \in \mathcal{O}(T;Q)$, then the solution $E(t,\xi) \in \mathcal{O}(T;Q)$ as well.*

Proof. (To avoid cumbersome calculations we assume that $n = 1$.) It is clear that the inclusion $A(t,\xi) \in \mathcal{O}(T;Q)$ implies that also

$$B(t,\xi) \equiv \int_0^t A(\tau,\xi)d\tau \in \mathcal{O}(T;Q),$$

that is,

$$\frac{1}{\alpha!}|\partial^\alpha B(t,0)| \le P(\alpha!)^\gamma \tag{3.1.7}$$

for some numbers $P(= P(t)) > 0$ and $\gamma \in]0, (1-q)/q[$. Further, using the well known formula

$$\frac{1}{\alpha!}\partial^\alpha[f(g(\xi))] = \sum_{\beta_1 + 2\beta_2 + \ldots + \alpha\beta_\alpha = \alpha} \frac{1}{\beta_1! \ldots \beta_\alpha!} \times$$

$$\times f^{(\beta_1 + \ldots + \beta_\alpha)}(g(\xi)) \left(\frac{\partial g(\xi)}{1!}\right)^{\beta_1} \cdots \left(\frac{\partial^\alpha g(\xi)}{\alpha!}\right)^{\beta_\alpha},$$

we obtain

$$c_\alpha \equiv \frac{\partial^\alpha B(t,0)}{\alpha!} = \sum_{b_1 + 2\beta_2 + \ldots + \alpha\beta_\alpha = \alpha} \frac{b_1^{\beta_1} \ldots b_\alpha^{\beta_\alpha}}{\beta_1! \ldots \beta_\alpha!},$$

where $b_1, \ldots, b_\alpha$ are the Taylor coefficients of the function $B(t, \xi)$ at the point $\xi = 0$. From this and in view of (3.1.7) we have

$$|c_\alpha| \leq \sum_{\beta_1 + 2\beta_2 + \ldots + \beta_\alpha = \alpha} \frac{P^{\beta_1 + \ldots + \beta_\alpha}}{\beta_1! \ldots \beta_\alpha!} \left[(1!)^{\beta_1}(2!)^{\beta_2} \ldots (\alpha!)^{\beta_\alpha}\right]^\gamma.$$

Since $((1!)^{\beta_1}(2!)^{\beta_2} \ldots (\alpha!)^{\beta_\alpha} \leq (\beta_1 + 2\beta_2 + \ldots + \alpha\beta_\alpha)!$, it follows that

$$|c_\alpha| \leq \sum_{\beta_1 + 2\beta_2 + \ldots + \alpha\beta_\alpha = \alpha} \frac{P^{\beta_1 + \ldots + \beta_\alpha}}{\beta_1! \ldots \beta_\alpha!} (\alpha!)^\gamma \leq$$

$$\leq \exp(\alpha P)(\alpha!)^\gamma \leq P(\epsilon)(\alpha!)^{\gamma + \epsilon},$$

where $\epsilon > 0$ is arbitrarily small and $P(\xi) > 0$ is some constant. This means precisely that $E(t, \xi) \in \mathcal{O}(T; Q)$. Q.E.D.

We return to the Cauchy problem (3.1.5), (3.1.6). It is easy to see that the solution of this problem is given by the formula

$$u(t, z) = \exp\left(\int_0^t A(\tau, D)d\tau\right)\phi(z) \equiv E(t, D)\phi(z).$$

Indeed, taking into account Lemma 3.1.4, just proved, and the isomorphism

$$\mathcal{A} \leftrightarrow Q$$

(see subsection 4) we have $u(t, z) \in \mathcal{O}(T; E_{q,r}(\mathbf{C}_z^n))$ and (non-formally) for any $\phi(z) \in E_{q,r}(\mathbf{C}_z^n)$

$$u' - A(t, D)u = E'(t, D)\phi(z) - A(t, D)E(t, D)\phi(z) =$$

$$= [E'(t, D) - A(t, D)E(t, D)]\phi(z) = 0.$$

Moreover,

$$u(t, 0) = E(0, D)\phi(z) \equiv I\phi(z) = \phi(z),$$

where I is the identity operator. The theorem is proved.

3.2. Differential operators of infinite order with variable coefficients

1. DEFINITION OF A D.O.I.O. WITH VARIABLE COEFFICIENTS

Let $A(z,\xi)$ be a function that is entire in $z \in \mathbf{C}^n$ and infinitely differentiable in $\xi \in \mathbf{R}^n$. Then we can associate with this function its Taylor expansion

$$A(z,\xi) \sim \sum_{|\alpha|=0}^{\infty} a_\alpha(z)\xi^\alpha, \tag{3.2.1}$$

where the coefficients $a_\alpha(z)$ are entire functions. Further, according to this expansion we define the formal action of the d.o.i.o. $A(z,D)$ as

$$A(z,D)u(z) \overset{\text{def}}{=} \sum_{|\alpha|=0}^{\infty} a_\alpha(z)D^\alpha u(z). \tag{3.2.2}$$

Our first aim is to show that this definition has a non-formal sense in the space $E_{q,r}(\mathbf{C}_z^n)$. For this we need

Condition B. A function $A(z,\xi)$ is said to satisfy Condition B if the functions $a_\alpha(z)$ are polynomials and

$$|a_\alpha(z)| \leq c^{|\alpha|} \prod_{i=1}^{n}(1+|z_i|)^{(1-q)\alpha_i},$$

where c is a constant such that $0 < c < 1/qr$.

Lemma 3.2.1. *If the symbol $A(z,\xi)$ satisfies Condition B, then the mapping*

$$A(z,D) : E_{q,r}(\mathbf{C}_z^n) \to E_{q,r}(\mathbf{C}_z^n)$$

is well-defined.

Proof. First we recall that, in accordance with Lemma 3.1.1 (basic estimate), for any function $u(z) \in E_{q,r}(\mathbf{C}_z^n)$ the inequalities

$$|D^\alpha u(z)| \leq \|u\|_{q,r} e^{r|z|^q} \frac{1}{\prod_{i=1}^{n}(1+|z_i|)^{(1-q)\alpha_i}}(qr)^{|\alpha|}\prod_{i=1}^{n}\sqrt{\alpha_i},$$

$z \in \mathbf{C}^n$, $|\alpha| = 0, 1, \ldots$, hold. Then, taking into account Condition B, we find that

$$|A(z, D)u(z)| \leq \sum_{|\alpha|=0}^{\infty} |a_\alpha(z)| \cdot |D^\alpha u(z)| \leq$$

$$\leq \sum_{|\alpha|=0}^{\infty} c^{|\alpha|} \prod_{i=1}^{n} (1 + |z_i|)^{(1-q)\alpha_i} \frac{1}{\prod_{i=1}^{n}(1 + |z_i|)^{(1-q)\alpha_i}} (qr)^{|\alpha|} \times$$

$$\times \prod_{i=1}^{n} \sqrt{\alpha_i} \|u\|_{q,r} e^{r|z|^q}.$$

Since $(cqr) < 1$, it follows that for any $z \in \mathbf{C}^n$

$$|A(z, D)u(z)| \leq M \|u\|_{q,r} e^{r|z|^q},$$

where

$$M = \sum_{|\alpha|=0}^{\infty} (cqr)^{|\alpha|} \prod_{i=1}^{n} \sqrt{\alpha_i} < \infty.$$

This means that $A(z, D)u(z) \in E_{q,r}(\mathbf{C}_z^n)$ and, moreover, the map

$$A(z, D) : E_{q,r}(\mathbf{C}_z^n) \to E_{q,r}(\mathbf{C}_z^n)$$

is continuous. The lemma is proved.

Below we shall study the Cauchy problem for a d.o.i.o. with variable coefficients. For this we need to introduce the space $\mathcal{O}(\delta; E_{q,r}(\mathbf{C}_z^n))$ of analytic functions $u(t, z)$. Namely, let

$$\mathcal{O}(\delta; E_{q,r}(\mathbf{C}_z^n)) = \{u(t, z) : u(t, \cdot) \in E_{q,r}(\mathbf{C}_z^n),\ |t - t_0| < \delta\}$$

be the Banach space with the norm

$$\|u\|_{\delta;q,r} \stackrel{\text{def}}{=} \sup_{|t-t_0|<\delta} \sup_{z \in \mathbf{C}^n} |u(t, z)| \exp(-r|z|^q)$$

(the notation does not indicate its dependence on t_0).

Further, let

$$A(t, z, \xi) \sim \sum_{|\alpha|=0}^{\infty} a_\alpha(t, z)\xi^\alpha,$$

where the $a_\alpha(t,z)$ are analytic in t, $|t - t_0| \leq \delta$, and entire in $z \in \mathbf{C}^n$. Then we set

$$A(t,z,D)u(t,z) \stackrel{\text{def}}{=} \sum_{|\alpha|=0}^{\infty} a_\alpha(t,z)D^\alpha u(t,z).$$

Lemma 3.2.2. *Suppose that the symbol $A(t,z,\xi)$ satisfies Condition B, that is,*

$$|a_\alpha(t,z)| \leq c^{|\alpha|} \prod_{i=1}^{n}(1 + |z_i|)^{(1-q)\alpha_i}, \quad |\alpha| = 0,1,\dots,$$

for all t, $|t - t_0| \leq \delta$ and for all $z \in \mathbf{C}^n$. Then the map

$$A(t,z,D) : \mathcal{O}(\delta; E_{q,r}(\mathbf{C}_z^n)) \to \mathcal{O}(\delta; E_{q,r}(\mathbf{C}_z^n))$$

is non-formal and continuous.

The proof of this lemma is the same as that of Lemma 3.1.1.

2. THE CAUCHY PROBLEM IN THE SPACES $E_{q,r}(\mathbf{C}_z^n)$

In this subsection we consider the Cauchy problem

$$u' - A(t,z,D)u = h(t,z), \tag{3.2.3}$$

$$u(t_0,z) = \phi(z), \quad z \in \mathbf{C}^n. \tag{3.2.4}$$

Here $t \in G$, where $G \subset \mathbf{C}^1$ is a domain, $t_0 \in G$ is any fixed point and

$$A(t,z,D) = \sum_{|\alpha|=0}^{\infty} a_\alpha(t,z)D^\alpha$$

is a d.o.i.o. with variable coefficients $a_\alpha(t,z)$. These coefficients are analytic functions with respect to $t \in G$ and $z \in \mathbf{C}^n$.

We find the solution $u(t,z) \in \mathcal{O}(\delta; E_{q,r}(\mathbf{C}_z^n))$ of this problem, where $0 < q < 1$ and $r > 0$ are given.

Definition 3.2.1. We say that the Cauchy problem is well-posed in the spaces $E_{q,r}(\mathbf{C}_z^n)$ if for any $t_0 \in G$ there exists a number $\delta > 0$ such that for any $\phi(z) \in E_{q,r}(\mathbf{C}_z^n)$ and $h(t,z) \in \mathcal{O}(\delta; E_{q,r}(\mathbf{C}_z^n))$ the problem (3.2.3), (3.2.4) has a unique solution $u(t,z) \in \mathcal{O}(\delta; E_{q,r}(\mathbf{C}_z^n))$ and the estimate

$$\|u\|_{\delta;q,r} \leq M(\|\phi\|_{q,r} + \|h\|_{\delta;q,r}) \tag{3.2.5}$$

holds for some constant $M > 0$.

Theorem 3.2.1. *Suppose that Condition B holds, that is,*

$$|a_\alpha(t, z)| \le c^{|\alpha|} \prod_{i=1}^n (1 + |z_i|)^{(1-q)\alpha_i}, \quad z \in \mathbf{C}^n, \ |\alpha| = 0, 1, \dots,$$

where $0 < c < (qr)^{-1}$. Then the Cauchy problem is well-posed in the space $E_{q,r}(\mathbf{C}_z^n)$.

Proof. In fact, let $\phi(z) \in E_{q,r}(\mathbf{C}_z^n)$ and $h(t, z) \in \mathcal{O}(\delta; E_{q,r}(\mathbf{C}_z^n))$. Clearly, the problem (3.2.3), (3.2.4) is equivalent to the integral-differential equation

$$u(t, z) = \phi(z) + \int_{t_0}^t A(\tau, z, D)u(\tau, z)d\tau + \int_{t_0}^t h(\tau, z)d\tau. \tag{3.2.6}$$

To solve (3.2.6) we use the method of successive approximations:

$$u_0(t, z) = \phi(z) + \int_{t_0}^t h(\tau, z)d\tau,$$

$$u_{k+1}(t, z) = \phi(z) + \int_{t_0}^t A(\tau, z, D)u_k(\tau, z)d\tau + \int_{t_0}^t h(\tau, z)d\tau,$$

where $k = 0, 1, \dots$.

It is easy to see that for any k the inclusion $u_k(t, z) \in \mathcal{O}(\delta; E_{q,r}(\mathbf{C}_z^n))$ is valid. Let us obtain an estimate for the norm of the operator

$$Bu \equiv \int_{t_0}^t A(\tau, z, D)u(\tau, z)d\tau.$$

Taking into account Condition B and the basic estimate in Lemma 3.1.1, we have

$$|Bu(t, z)| = \left| \int_{t_0}^t A(\tau, z, D)u(\tau, z)d\tau \right| = \left| \int_{t_0}^t \sum_{|\alpha|=0}^\infty a_\alpha(\tau, z)D^\alpha u(\tau, z)d\tau \right| \le$$

$$\le \int_0^{|t-t_0|} \sum_{|\alpha|=0}^\infty c^{|\alpha|} \prod_{i=1}^n (1 + |z_i|)^{(1-q)\alpha_i} \|u(\tau - t_0, z)\|_{q,r} \times$$

$$\times \frac{e^{r|z|^q}}{\prod_{i=1}^n (1 + |z_i|)^{(1-q)\alpha_i}} (qr)^{|\alpha|} \prod_{i=1}^n \sqrt{\alpha_i} |d\tau| \le$$

$$\le M \sup_{|\tau - t_0| \le |t - t_0|} \|u(\tau - t_0, z)\|_{q,r} |t - t_0| e^{r|z|^q},$$

where

$$M = \sum_{|\alpha|=0}^{\infty} (cqr)^{|\alpha|} \prod_{i=1}^{n} \sqrt{\alpha_i} < \infty,$$

since $cqr < 1$.

This means that

$$\sup_{|t-t_0|<\delta} \sup_{z\in\mathbf{C}^n} |Bu(t,z)| \exp(-r|z|^q) \leq M\delta \|u(t,z)\|_{\delta;q,r}$$

or, what is the same,

$$\|Bu(t,z)\|_{\delta;q,r} \leq M\delta \|u(t,z)\|_{\delta;q,r}.$$

Clearly, if $\delta > 0$ is so small that $M\delta < 1$, then the sequence $u_k(t,z)$ converges in $\mathcal{O}(\delta; E_{q,r}(\mathbf{C}_z^n))$ to a function $u(t,z) \in \mathcal{O}(\delta; E_{q,r}(\mathbf{C}_z^n))$ which is the desired solution of the Cauchy problem (3.2.3), (3.2.4). Moreover, the estimate (3.2.5) is valid. The theorem is proved.

Remark. Under Definition 3.2.1. (in a slightly refined version) one can show that the well-posedness of the Cauchy problem in the spaces $E_{q,r}(\mathbf{C}_z^n)$ implies the inequalities

$$|a_\alpha(t,z)| \leq M(qr)^{-|\alpha|}(1+|z|)^{(1-q)|\alpha|},$$

where $M > 0$ is an absolute constant. This means that Condition B is an "almost" necessary condition for the well-posedness of the Cauchy problem in the spaces $E_{q,r}(\mathbf{C}_z^n)$.

In particular, for differential operators of finite order, the requirement $cqr < 1$ clearly plays no role and a necessary and sufficient condition for the well-posedness of the Cauchy problem in the spaces $E_{q,r}(\mathbf{C}_z^n)$ is the following:

The coefficients $a_\alpha(t,z)$ are polynomials in z with the property that $\deg a_\alpha(t,z) \leq (1-q)|\alpha|$.

3. THE CAUCHY PROBLEM IN THE SPACES $E_{q,r+\sigma|t|}(\mathbf{C}_z^n)$

Here we consider the Cauchy problem

$$u' - A(t,z,D)u = h(t,z), \tag{3.2.7}$$

$$u(t_0, z) = \phi(z) \tag{3.2.8}$$

in the spaces $E_{q,r+\sigma|t-t_0|}(\mathbf{C}_z^n)$.

We begin with two definitions. Namely, let $\delta > 0$, $r > 0$ and $\sigma > 0$ be given numbers.

Definition 3.2.2. We set

$$\mathcal{O}(\delta; E_{q,r+\sigma|t-t_0|}(\mathbf{C}_z^n)) = \{u(t,z) : u(t,\cdot) \in E_{q,r+\sigma|t-t_0|}(\mathbf{C}_z^n), \ |t-t_0| < \delta\}$$

with norm

$$\|u\|_{\delta;q,r,\sigma} = \sup_{|t-t_0|<\delta} \|u(t,\cdot)\|_{q,r+\sigma|t-t_0|} < \infty.$$

Every function $u(t,z) \in \mathcal{O}(\delta; E_{q,r+\sigma|t-t_0|}(\mathbf{C}_z^n))$ is an analytic function in t, $|t-t_0| < \delta$, and in $z \in \mathbf{C}^n$. Moreover, for every t the value $u(t,\cdot)$ belongs to the space $E_{q,r+\sigma|t-t_0|}(\mathbf{C}_z^n)$ which becomes larger as $|t-t_0|$ increases. It is in this space that we shall find a solution of our original problem.

Definition 3.2.3. We say that the Cauchy problem (3.2.7), (3.2.8) is well-posed in the space $E_{q,r+\sigma|t|}(\mathbf{C}_z^n)$ if for any $t_0 \in G$ there exist numbers $\sigma > 0$ and $\delta > 0$ such that for any initial data $\phi(z) \in E_{q,r}(\mathbf{C}_z^n)$ and for any right hand side $h(t,z) \in \mathcal{O}(\delta; E_{q,r+\sigma|t-t_0|}(\mathbf{C}_z^n))$ the problem (3.2.7), (3.2.8) has a unique solution $u(t,z) \in \mathcal{O}(\delta; E_{q,r+\sigma|t-t_0|}(\mathbf{C}_z^n))$; moreover,

$$\|u\|_{\delta;q,r,\sigma} \leq M\left(\|\phi\|_{q,r} + \|h\|_{\delta;q,r,\sigma}\right), \tag{3.2.9}$$

where $M > 0$ is a constant.

We now state the basic condition on the coefficients $a_\alpha(t,z)$ of the operator $A(t,z,D)$.

Condition C. We say that Condition C is satisfied if the functions $a_\alpha(t,z)$ are polynomials in z and

$$|a_\alpha(t,z)| \leq c^{|\alpha|}(1+|z|)^q \prod_{i=1}^{n}(1+|z_i|)^{(1-q)\alpha_i},$$

where, as before, $cqr < 1$.

The main result of this subsection is the following theorem.

Theorem 3.2.2. *Suppose that Condition C holds. Then the problem (3.2.7), (3.2.8) is well-posed in the spaces $E_{q,r+\sigma|t|}(\mathbf{C}_z^n)$ (in the sense of Definition 3.2.3).*

Proof. As before, we consider the sequence of successive approximations $(k = 0, 1, \ldots)$

$$u_0(t, z) = \phi(z) + \int_0^t h(\tau, z) d\tau$$

$$u_{k+1}(t, z) = \phi(z) + \int_0^t A(\tau, z, D) u_k(\tau, z) d\tau + \int_0^t h(\tau, z) d\tau$$

(without loss of generality one may set $t_0 = 0$).

First we note that, clearly, $\phi(z) \in \mathcal{O}(\delta; E_{q, r+\sigma|t|}(\mathbf{C}_z^n))$. Further, since $h(t, z) \in \mathcal{O}(\delta; E_{q, r+\sigma|t|}(\mathbf{C}_z^n))$, it follows that

$$\left| \int_0^t h(\tau, z) d\tau \right| \leq M \int_0^{|t|} e^{r|z|^q} e^{\sigma|\tau||z|^q} |d\tau| \leq$$

$$\leq M e^{r|z|^q} \frac{1}{\sigma|z|^q} \left(e^{\sigma|t||z|^q} - 1 \right) \leq M e^{(r+\sigma|t|)|z|^q}.$$

This means that

$$\int_0^t h(\tau, z) d\tau \in \mathcal{O}(\delta; E_{q, r+\sigma|t|}(\mathbf{C}_z^n))$$

and consequently, $u_0(t, z) \in \mathcal{O}(\delta; E_{q, r+\sigma|t|}(\mathbf{C}_z^n))$ also.

Let us now give the estimate of the norm of the operator

$$Bu(t, z) = \int_0^t A(\tau, z, D) u(\tau, z) d\tau$$

in the space $\mathcal{O}(\delta; E_{q, r+\sigma|t|}(\mathbf{C}_z^n))$. We have, in view of Condition C and the basic estimate (see Lemma 3.1.1),

$$|Bu(t, z)| = \left| \int_0^t A(\tau, z, D) u(\tau, z) d\tau \right| \leq \int_0^{|t|} \sum_{|\alpha|=0}^{\infty} |a_\alpha(t, z)| |D^\alpha u(\tau, z)| |d\tau| \leq$$

$$\leq \sum_{|\alpha|=0}^{\infty} \int_0^{|t|} c^{|\alpha|} q^{|\alpha|} (r + \sigma|\tau|)^{|\alpha|} (1 + |z|)^q \times$$

$$\times \prod_{i=1}^{n} \sqrt{\alpha_i} e^{(r+\sigma|\tau|)|z|^q} \|u(\tau, z)\|_{q, r+\sigma|\tau|} |d\tau| \leq$$

$$\leq M(1 + |z|)^q \int_0^{|t|} e^{(r+\sigma|\tau|)|z|^q} \|u(\tau, z)\|_{q, r+\sigma|\tau|} d|\tau|, \qquad (3.2.10)$$

where

$$M = \sum_{|\alpha|=0}^{\infty} (cq(r+\sigma\delta))^{|\alpha|} \prod_{i=1}^{n} \sqrt{\alpha_i} < \infty,$$

if $\sigma\delta$ is so small that $cq(r+\sigma\delta) < 1$.

It is then not difficult to verify that

$$(1+|z|)^q \int_0^{|t|} e^{(r+\sigma|\tau|)|z|^q} \|u(\tau,z)\|_{q,r+\sigma|\tau|} d|\tau| \le \sup_{|\tau|<|t|} \|u(\tau,\cdot)\|_{q,r+\sigma|\tau|} \times$$

$$\times \frac{1+|z|^q}{\sigma|z|^q} e^{r|z|^q} \left(e^{\sigma|t|\,|z|^q} - 1 \right) \le \frac{2}{\sigma} \max(\sigma|t|,1) e^{(r+\sigma|t|)|z|^q} \|u\|_{\delta;q,r,\sigma} \le$$

$$\le \frac{2}{\sigma} e^{(r+\sigma|t|)|z|^q} \|u(t,z)\|_{\delta;q,r,\sigma},$$

if $\sigma|t| \le \sigma\delta < 1$.

Taking into account this inequality, we see from (3.2.10) that for any t and z

$$|Bu(t,z)| \le \frac{2M}{\sigma} e^{(r+\sigma|t|)|z|^q} \|u(t,z)\|_{\delta;q,r,\sigma},$$

that is,

$$\|Bu\|_{\delta;q,r,\sigma} \le \frac{2M}{\sigma} \|u(t,z)\|_{\delta;q,r,\sigma}.$$

Let us now choose $\sigma > 0$ so large and, conversely, $\delta > 0$ so small that

1) $cq(r+\sigma\delta) \le q_0 < 1$, where $q_0 \in]cqr, 1[$ is a fixed number;

2) $\sigma\delta < 1$;

3) $2M/\sigma < 1$.

Clearly this is possible. Finally, this means that for every $k = 1, 2, \ldots$ the successive approximations $u_k(t,z) \in \mathcal{O}(\delta; E_{q,r+\sigma|t|}(\mathbf{C}_z^n))$ and, moreover, this process converges. Consequently, there exists a unique solution

$$u(t,z) \in \mathcal{O}(\delta; E_{q,r+\sigma|t|}(\mathbf{C}_z^n))$$

of the original Cauchy problem (3.2.7), (3.2.8). The estimate (3.2.10) is trivial. The theorem is proved.

Remark. We may make an assertion similar to the remark at the end of sub-section 2. Namely, if the Cauchy problem (3.2.7), (3.2.8) is well-defined in the scale $E_{q,r+\sigma|t|}(\mathbf{C}_z^n)$, then the inequalities

$$|a_\alpha(t,z)| \le M(qr)^{-|\alpha|}(1+|z|)^{(1-q)|\alpha|+q}$$

($M > 0$ is a constant) are valid. Hence, Condition C is an "almost" necessary condition for the well-posedness of the Cauchy problem in the scale $E_{q,r+\sigma|t|}(\mathbf{C}_z^n)$.

In conclusion we note that in the case of a differential operator of finite order we do not need the requirement $cqr < 1$; therefore the necessary and sufficient conditions for well-posedness in the scale $E_{q,r+\sigma|t|}(\mathbf{C}_z^n)$ are the following:

the coefficients $a_\alpha(t, z)$ are polynomials in z and

$$\deg a_\alpha(t, z) \le (1 - q)|\alpha| + q.$$

Part II. The Cauchy Problem in the Complex Domain

Introduction

Part II is devoted to the complex Cauchy problem, more precisely, to three questions concerning this problem:

1. local analytic solvability,

2. global exponential solvability,

3. the connection between these theories.

Let us briefly describe the more general results of this part.

As is well known, in 1842 Cauchy considered the system

$$\frac{\partial^{s_i} u_i}{\partial t^{s_i}} + \sum_{j,k} A_{ij}^k(t,z,\mathcal{D}) \frac{\partial^k u_j}{\partial t^k} = h_i(t,z) \tag{1}$$

$$(i,j = 1,\ldots,N; \quad k = 0,1,\ldots,s_i - 1)$$

and proved that if the coefficients of the differential operators $A_{ij}^k(t,z,\mathcal{D})$ are analytic in a domain $V \subset \mathbf{C}_{t,z}^{1+n}$ and, moreover,

$$\operatorname{ord} A_{ij}^k \le s_i - k, \tag{*}$$

then for any right hand sides $h_i(t,z)$ that are analytic in some neighbourhood $U(t_0,z_0) \subset V$ of any $(t_0,z_0) \in V$ and for any functions $\phi_{ik}(z)$ that are analytic in the intersection $U(t_0,z_0) \cap \{t = t_0\}$ there exists a unique solution $u(t,z) = (u_1(t,z),\ldots,u_N(t,z))$ of the system (1) satisfying the initial data

$$\frac{\partial^k u_i}{\partial t^k}(t_0,z) = \phi_{ik}(z), \quad k = 0,1,\ldots,s_i - 1. \tag{2}$$

It is important to observe that the solution of (1), (2) is defined in a neighbourhood $V(t_0,z_0) \subset U(t_0,z_0)$, which is (in general) smaller than $U(t_0,z_0)$.

However, this result remained unnoticed by Cauchy's contemporaries and (thirty years later) Weierstrass posed this problem again to his student Kovalevskaya. Kovalevskaya discovered the conditions (∗) once more; moreover she showed that these conditions are essential. After this, the conditions (∗) were called the Kovalevskaya conditions, and the results of Cauchy and Kovalevskaya the Cauchy-Kovalevskaya theory.

It was not until 1975 that Misohata proved that for one equation ($N = 1$) the conditions (∗) are not only sufficient but also necessary for the well-posedness of the Cauchy problem in the Cauchy-Kovalevskaya sense.

For the system (1) ($N > 1$) it is false. Indeed, in 1964 Leray, Gårding and Kotake proved that the solvability of the Cauchy problem (1), (2) is true for more general systems satisfying the Leray-Volevich conditions

$$\mathrm{ord} A_{ij}^k \le m_i - m_j + s_i - k \qquad (**)$$

where $m_1, \ldots, m_N$ are arbitrary non-negative integers (we put $A_{ij}^k \equiv 0$ if the right hand side is negative).

It turns out (we are now talking about the results of the present part) that the well-posedness of the Cauchy problem is connected with the way in which the Cauchy problem itself is posed, more precisely, the evolution in time t of the singularities of the solution. Namely, if the solution has singularities on the lateral surface of a cylinder, then the conditions

$$\mathrm{ord}\, A_{ij}^k \le m_i - m_j \qquad (3)$$

are necessary and sufficient conditions for the solvability of the Cauchy problem in such classes of analytic functions.

The class of systems satisfying (3) is not sufficiently rich. And if the solution has singularities on the lateral surface of a cone, then the more general Leray-Volevich conditions (∗∗) are necessary and sufficient for the solvability of the Cauchy problem in such classes of analytic functions.

The class of systems, satisfying (∗∗) is rich enough (clearly, for $m_1 = \ldots = m_N$ they are the Kovalevskaya systems).

We note in passing that the Leray-Volevich indices $m_1, \ldots, m_N$ now assume a simple geometrical meaning, namely, they are the orders of the singularities of the components $u_1(t, z), \ldots, u_N(t, z)$ of the solution.

On a level with the Cauchy problem in spaces of analytic functions with pole-type singularities there is the dual problem, namely, the Cauchy problem in the corresponding spaces of analytic functionals. Here the cases of cylindrical and conical evolution are also different.

Indeed, in the first case both the direct problem and the dual problem are easily included (as special cases) into the theory of the differential-operator equations in Banach spaces

$$u' + A(t)u = h(t) \tag{4}$$

with linear bounded operator $A(t)$. By contrast, in the second case there is no such inclusion, since the corresponding operator $A(t)$ has a domain of definition that essentially depends on t. Therefore, here new methods of investigation for the dual problem are required. As is known, one such method is the Ovsyannikov theorem on the solvability of the Cauchy problem for equation (4) in the scale of Banach spaces.

We note that the Ovsyannikov theorem (more precisely, a modification of it) was applied to the dual analytic Cauchy problem by F. Treves (see, for example, his lectures [1]).

Here we propose another approach to the investigation of the dual problem. Namely, we base our investigation on the fact of the existence of the solution of the original Cauchy problem and give an exact formula for the solution of the dual Cauchy problem.

Of course, the idea of using the original problem in order to investigate the dual problem is well known and goes back to the classical paper by Holmgren [1], where this idea was proposed for the proof of the uniqueness of the solution.

We show that this idea "works" well enough in the proof of the existence of the dual Cauchy problem also. Moreover, the dual result will be just as exact as the original result was.

The reader will find all the above material in Chapter 4.

Chapter 5 is devoted to the well-posedness of the Cauchy problem (1), (2) in spaces of exponential type functions. By contrast with the Cauchy-Kovalevskaya theory the exponential theory is less popular. This theme first arose in 1935 in the paper by Tikhonov [1], where the author proved that the correctness class of the global (in $x \in \mathbf{R}^1$) Cauchy problem for the heat equation is the class of exponential functions of order two precisely.

Later, questions of the correctness of the Cauchy problem in the exponential scale were extensively investigated by Petrovskii, Gel'fand, Shilov and their schools. A wide survey of the results in this direction is given in the article by Oleinik and Palamodov [1]. We emphasize, however, that all these investigations are devoted to the real Cauchy problem.

It turns out that for the complex Cauchy problem one can describe exactly the correctness classes of the Cauchy problem.

As in the analytic theory, the main results of the exponential theory also depend on the character of the evolution of the solution with respect to t. Namely, if there is the evolution $u(t, z) = (u_1(t, z), \ldots, u_N(t, z))$ in the class of initial data, that is, for all t

$$|u_j(t, z)| \leq M_j(1 + |z|)^{m_j} \exp r|z|^q, \quad z \in \mathbb{C}^n,$$

$(j = 1, \ldots, N; \; q \geq 1)$ then the following conditions are necessary and sufficient for the well-posedness of the Cauchy problem (1), (2):

1)   the coefficients $a_{ij}^{k\alpha}(t, z)$ of the operators $A_{ij}^k(t, z, D)$ are polynomials in z,

2)   the degrees of these polynomials satisfy the inequalities

$$\deg a_{ij}^{k\alpha} \leq m_i - m_j - |\alpha|(q - 1)$$

(as before, we set $a_{ij}^{k\alpha} \equiv 0$ if the right hand side is negative).

On the other hand, if we look for a solution in more extensive spaces, namely, in spaces of functions satisfying the inequalities

$$|u_j(t, z)| \leq M_j(1 + |z|)^{m_j} \exp(r + \sigma|t|)|z|^q, \quad z \in \mathbb{C}^n,$$

$(j = 1, \ldots, N)$, then the following conditions are necessary and sufficient for the well-posedness of the Cauchy problem (1), (2):

1)   the coefficients $a_{ij}^{k\alpha}(t, z)$ of the operators $A_{ij}^k(t, z, D)$ are polynomials in z (as before),

2)   the less stringent inequalities

$$\deg a_{ij}^{k\alpha} \leq m_i - m_j - |\alpha|(q - 1) + (s_i - k)q.$$

Thus, the Cauchy problem (1), (2) is well-posed in the scale of exponential type functions if and only if the system (1) is a system with polynomial coefficients.

We note especially the case $q = 1$, for which the last inequalities assume the form

$$\deg a_{ij}^{k\alpha} \leq m_i - m_j + s_i - k.$$

It is clear that these inequalities have a form which is the Fourier dual to the Leray-Volevich conditions. This is not mere chance. However, in order to understand this phenomenon we need the fact of the well-posedness of the Cauchy problem (1), (2) not only for differential equations but for analytic PD-equations as well.

The final chapter of Part II (Chapter 6) is devoted to just these problems.

Chapter 4. Cauchy-Kovalevskaya Theory in Spaces of Analytic Functions with Pole-type Singularities

4.1. The Cauchy problem in the spaces $D_{m,r}$
(Case of cylindrical evolution)

1. THE TEST-FUNCTION SPACE $D_{m,r}$

Let $z = (z_1, \ldots, z_n)$ be an arbitrary point of n-dimensional complex space $\mathbf{C}^n$ and $|z| = |z_1| + \ldots + |z_n|$. We denote by $U_r(z_0)$ the polycylinder

$$U_r(z_0) = \{z \in \mathbf{C}^n : |z - z_0| < r\},$$

where $r > 0$ is a real number and $z_0 = (z_{10}, \ldots, z_{n0})$ is a fixed point of $\mathbf{C}^n$.

Further, let $m = (m_1, \ldots, m_N)$ be a vector with non-negative integer components, where N is also a non-negative integer. We denote by $D_{m,r}(z_0)$ the space whose elements are vector-valued functions $u(z) = (u_1(z), \ldots, u_N(z))$ for which the inequalities

$$|u_j(z)| \leq M_j(r - |z - z_0|)^{-m_j} \quad (j = 1, \ldots, N)$$

hold. (Here the $M_j > 0$ are constants, depending on $u_j(z)$.) It is easy to see that $D_{m,r}(z_0)$ is a Banach space with norm

$$\|u(z)\|_{m,r} \overset{\text{def}}{=} \sum_{j=1}^{N} \|u_j(z)\|_{m_j,r},$$

where

$$\|u_j(z)\|_{m_j,r} \overset{\text{def}}{=} \sup_{z \in U_r(z_0)} |u_j(z)|(r - |z - z_0|)^{m_j}.$$

The space $D_{m,r}(z_0)$ will be used below as the space of initial data in the Cauchy problem. For brevity we omit the z_0, mereley denoting it by $D_{m,r}$.

90

We now introduce the space of vector-valued functions $u(t, z) = (u_1(t, z), \dots$
$\dots, u_N(t, z))$ which will be used as the space of solutions of the Cauchy problem.
Namely, let $\delta > 0$ be a number, $t_0 \in \mathbf{C}^1$ a fixed point. By definition

$$\mathcal{O}(\delta; D_{m,r}) = \{u(t, \cdot) \in D_{m,r} : \sup_t \|u_j(t, \cdot)\|_{m_j, r} < \infty,$$

$$|t - t_0| < \delta, \quad j = 1, \dots, N\}$$

(the dependence on t_0 in the notation of this space is also omitted). Thus in
the $(n+1)$-dimensional polycylinder $U_{\delta, r} = \{|t - t_0| < \delta\} \times U_r(z_0)$ the functions
$u(t, z) \in \mathcal{O}(\delta; D_{m,r})$ are analytic functions satisfying the inequalities

$$|u_j(t, z)| \le M_j (r - |z - z_0|)^{-m_j}, \quad j = 1, \dots, N,$$

where the $M_j > 0$ are constants.

This means that for $t \ne t_0$ the vector-valued function $u(t, z)$ may have
pole-type singularities on the lateral surface of the cylinder $U_{\delta, r}$, the character
of which is the same as for $t = t_0$, that is, for the initial function $u(t_0, z)$.

The space $\mathcal{O}(\delta; D_{m,r})$ becomes a Banach space if we set

$$\|u(t, z)\|_{\delta; m, r} \overset{\text{def}}{=} \sum_{j=1}^{N} \|u_j(t, z)\|_{\delta; m_j, r},$$

where

$$\|u_j(t, z)\|_{\delta; m_j, r} \overset{\text{def}}{=} \sup_{|t - t_0| < \delta} \|u_j(t, \cdot)\|_{m_j, r}.$$

2. CRITERION FOR THE WELL-POSEDNESS OF THE CAUCHY PROBLEM
IN THE SPACES $D_{m,r}$

We consider the Cauchy problem for the complex system

$$u' - A(t, z, \mathcal{D})u = h(t, z) \tag{4.1.1}$$

$$u(t_0, z) = \phi(z). \tag{4.1.2}$$

Here $u = (u_1, \dots, u_N)$, $h(t, z) = (h_1(t, z), \dots, h_N(t, z))$, $\phi(z) = (\phi_1(z), \dots$
$\dots, \phi_N(z))$ are vector-valued functions, $\mathcal{D} = (\partial/\partial z_1, \dots, \partial/\partial z_n)$ and $A(t, z, \mathcal{D})$
$\equiv (A_{ij}(t, z, \mathcal{D}))_{N \times N}$ is the matrix of differential operators of finite order

$$A_{ij}(t, z, \mathcal{D}) \overset{\text{def}}{=} \sum_{|\alpha|=0}^{m_{ij}} a_{ij}^{\alpha}(t, z) \mathcal{D}^{\alpha}$$

whose coefficients are analytic functions in a domain $V \subset \mathbf{C}^{n+1}_{t,z}$, where $\mathbf{C}^{n+1}_{t,z}$ denotes the complex space of independent variables $t \in \mathbf{C}^1$ and $z \in \mathbf{C}^n$.

The solution $u(t, z)$ of the problem (4.1.1), (4.1.2) is defined in a neighbourhood of each point $(t_0, z_0) \in V$.

In order to formulate the problem exactly, we first indicate the number $r > 0$, which is the radius of the initial polycylinder with centre z_0. Namely, let R be the distance from the point (t_0, z_0) to the boundary of the section $\{t = t_0\} \cap V$. Then we can take r to be an arbitrary number less than R (that is, $r < R$). It is clear that for such r there is always a number $\delta > 0$ such that the $(n + 1)$-dimensional polycylinder $U_{\delta,r}(t_0, z_0) \subset V$.

Definition 4.1.1. The Cauchy problem is said to be locally well-posed (in the domain V) in the space $D_{m,r}$ if for any point $(t_0, z_0) \in V$ there exists a number $\delta > 0$ such that for any initial functions $\phi(z) \in D_{m,r}$ and for any right hand side $h(t, z) \in \mathcal{O}(\delta; D_{m,r})$ the problem (4.1.1), (4.1.2) has a unique solution $u(t, z) \in \mathcal{O}(\delta; D_{m,r})$. Moreover, the estimate

$$\|u(t, z)\|_{\delta;m,r} \leq M(\|\phi(z)\|_{m,r} + \|h(t, z)\|_{\delta;m,r}) \qquad (4.1.3)$$

holds, where $M > 0$ is constant.

The main result of the present subsection is

Theorem 4.1.1. *The Cauchy problem* (4.1.1), (4.1.2) *is locally well-posed in the space* $D_{m,r}$ *if and only if the inequalities*

$$\operatorname{ord} A_{ij}(t, z, \mathcal{D}) \leq m_i - m_j \ (\forall i, j) \qquad (4.1.4)$$

hold (we set $\operatorname{ord} A_{ij}(t, z, \mathcal{D}) = -\infty$ *if and only if* $A_{ij}(t, z, \mathcal{D}) \equiv 0$).

Proof. Sufficiency. Suppose that the conditions (4.1.4) hold. We prove that the Cauchy problem (4.1.1), (4.1.2) is locally well posed in the space $D_{m,r}$.

For the proof we use the method of contraction maps. Clearly we have

$$u(t, z) = \int_0^t A(\tau, z, \mathcal{D})u(\tau, z)d\tau + \phi(z) + \int_0^t h(\tau, z)d\tau, \qquad (4.1.5)$$

where the integration is taken along the straight line segment joining 0 and t. (Without loss of generality one can put $t_0 = 0, z_0 = 0$.) We show that the operator

$$Bu \equiv \int_0^t A(\tau, z, \mathcal{D})u(\tau, z)d\tau$$

is a contraction operator in the space $\mathcal{O}(\delta; D_{m,r})$ if the number $\delta > 0$ is sufficiently small. For this purpose we need

Lemma 4.1.1. *Let $v(z)$ be an analytic function in the polycylinder $U_r = \{z \in \mathbf{C}^n : |z| < r\}$, such that*

$$|v(z)| \le M(r - |z|)^{-q},$$

where $M > 0$, $q > 0$ and $r > 0$. Then for all $z \in U_r$ and $k = 1, \ldots, n$

$$\left| \frac{\partial v(z)}{\partial z_k} \right| \le \frac{M(q+1)^{q+1}}{q^q} \frac{1}{(r - |z|)^{q+1}},$$

where $M > 0$ is the same.

Proof. First let $v(z)$ be a function of one complex variable. Then from Cauchy's formula we have $(\mathcal{D} \equiv d/dz)$

$$|\mathcal{D}v(z)| \le \frac{1}{2\pi} \int_{|\zeta|=a} \frac{|v(\zeta)|}{|\zeta - z|^2} |d\zeta| \le$$

$$\le \frac{M}{(r-a)^q} \frac{1}{2\pi} \int_0^{2\pi} \frac{a\,d\phi}{a^2 + |z|^2 - 2a|z|\cos\phi} = \frac{M}{(r-a)^q} \frac{a}{a^2 - |z|^2} \le$$

$$\le \frac{M}{(r-a)^q(a - |z|)},$$

where a is an arbitrary number satisfying $|z| < a < r$. Setting $a = (r + q|z|) \div (q+1)$, we obtain the inequality of the lemma for $n = 1$.

It only remains to note that for $n > 1$ finding an estimate of $\partial v(z)/\partial z_k$ reduces to the case $n = 1$ after making the substitution $r \leftrightarrow r - |z'|$, where $z' = (z_1, \ldots, z_{k-1}, z_{k+1}, \ldots, z_n)$. The lemma is proved.

Let us turn to the direct investigation of the operator Bu. In fact, for any $u(t, z) \in \mathcal{O}(\delta; D_{m,r})$ we have, in accordance with Lemma 4.1.1, the inequality

$$\left| \int_0^t a_{ij}^\alpha(\tau, z) \mathcal{D}^\alpha u_j(\tau, z) d\tau \right| \le M \int_0^{|t|} \|u_j(\tau, z)\|_{m_j, r} (r - |z|)^{-m_j - |\alpha|} d|\tau| \le$$

$$\le M\delta \sup_{|t| < \delta} \|u_j(t, \cdot)\|_{m_j, r} (r - |z|)^{-m_j - |\alpha|} \equiv$$

$$\equiv M\delta \|u_j(t, z)\|_{\delta; m_j, r} (r - |z|)^{-m_j - |\alpha|},$$

where $M > 0$ is constant, depending on $\max |a_{ij}^\alpha(t,z)|$ in the polycylinder $\overline{U}_{\delta,r}(0,0)$. Hence for each component $(Bu)_i$, $i = 1, \ldots, N$, we have the inequality

$$|(Bu)_i(t,z)| \leq M\delta \sum_{j=1}^{N} \|u_j(t,z)\|_{\delta;m_j,r}(r - |z|)^{-m_j - m_{ij}},$$

where m_{ij} denotes the order of the differential operator $A_{ij}(t, z, \mathcal{D})$. Since, in view of (4.1.4), $m_{ij} \leq m_i - m_j$, the last estimate gives us the inequality

$$|(Bu)_i(t,z)| \leq M\delta \sum_{j=1}^{N} \|u_j(t,z)\|_{\delta;m_j,r}(r - |z|)^{-m_i}.$$

This means that $Bu \in \mathcal{O}(\delta; D_{m,r})$ and

$$\|Bu\|_{\delta;m,r} \leq M\delta\|u\|_{\delta;m,r} \leq \theta\|u\|_{\delta;m,r},$$

where $\theta < 1$, if $\delta > 0$ is sufficiently small.

It is clear that the last estimates guarantee the existence and uniqueness of a fixed point of the operator

$$Bu + \phi(z) + \int_0^t h(\tau,z)d\tau : \mathcal{O}(\delta; D_{m,r}) \to \mathcal{O}(\delta; D_{m,r}).$$

Thus the existence and uniqueness of the solution of the integral equation (4.1.5) are proved and we have the estimate

$$\|u\|_{\delta;m,r} \leq \|Bu\|_{\delta;m,r} + \|\phi(z)\|_{\delta;m,r} + \left\|\int_0^t h(\tau,z)d\tau\right\|_{\delta;m,r} \leq$$

$$\leq \theta\|u\|_{\delta;m,r} + \|\phi(z)\|_{m,r} + \delta\|h(t,z)\|_{\delta;m,r},$$

from which the estimate (4.1.5) immediately follows. Thus the well-posedness of the Cauchy problem (4.1.1), (4.1.2) is finally proved.

Necessity. Suppose that the Cauchy problem (4.1.1), (4.1.2) is locally well posed in the space $D_{m,r}$. We show that the restrictions (4.1.5) on the orders of the differential operators $A_{ij}(t, z, \mathcal{D})$ are valid.

To prove these inequalities we choose $h(t,z) \equiv 0$, $\phi(z) = (0, \ldots, 0, \phi_j(z), 0, \ldots, 0)$, where j is a fixed number and

$$\phi_j(z) = (w - \tau_1 z_1 - \ldots - \tau_n z_n)^{-m_j} \equiv (w - \tau z)^{m_j};$$

$w \in \mathbf{C}^1$, $\tau_k \in \mathbf{C}^1$ ($k = 1, \ldots, n$) are parameters such that $|w| = r$, $|\tau_k| \leq 1$. Clearly, $\phi(z) \in D_{m,r}$ and $\|\phi(z)\|_{m,r} \leq 1$ for all w and τ.

Let $u(t, z)$ be the corresponding solution of the problem (4.1.1), (4.1.2) (we omit the indices j, w, τ in the notation).

Let us consider the system (4.1.1) for this solution for $t = 0$. We have, trivially,

$$A_{ij}(0, z, \mathcal{D})(w - \tau z)^{-m_j} = u_i'(0, z) \quad (i = 1, \ldots, N)$$

or, what is the same,

$$\sum_{q=0}^{m_{ij}} \sum_{|\alpha|=q} c_{\alpha j} a_{ij}^{\alpha}(0, z) \tau^{\alpha} (w - \tau z)^{-m_j - q} = u_i'(0, z), \tag{4.1.6}$$

where

$$c_{\alpha j} = \prod_{k=0}^{n} m_j(m_j + 1) \ldots (m_j + \alpha_k).$$

Further, since $u(t, z) \in \mathcal{O}(\delta; D_{m,r})$, it follows from Cauchy's formula that the inequality

$$|u_i'(0, z)| \leq M_i(r - |z|)^{-m_i}$$

holds (M_i is a constant). It is important to emphasize that the constants M_i do not depend on the parameters w and τ since, as has been pointed out, $\|\phi_i(z)\|_{m_i, r} \leq 1$ for all w, τ. Thus in view of (4.1.6) we have

$$\sum_{q=0}^{m_{ij}} \sum_{|\alpha|=q} c_{\alpha j} a_{ij}^{\alpha}(0, z) \tau^{\alpha} (w - \tau z)^{m_{ij} - q} = (w - \tau z)^{m_{ij} + m_j} O((r - |z|)^{-m_i}).$$

We set $z_2 = 0, \ldots, z_n = 0$. Then for all z_1, $|z_1| < r$, and for all (fixed) $\tau_1, \ldots, \tau_n$ ($|\tau_j| \leq 1$, $j = 1, \ldots, n$) we have

$$\sum_{q=0}^{m_{ij}} \sum_{|\alpha|=q} c_{\alpha j} a_{ij}^{\alpha}(0, z_1, 0, \ldots, 0) \tau^{\alpha} (w - \tau_1 z_1)^{m_{ij} - q}$$

$$= (w - \tau_1 z_1)^{m_{ij} + m_j} O((r - |z|)^{-m_i}). \tag{4.1.7}$$

Now let $|\tau_1| = 1$ and $w = s\tau$, where $s \in \mathbf{C}^1$, $|s| = 1$, is arbitrary. Moreover, we choose $z_1 = \lambda s$, where $\lambda \in [0, 1]$ is arbitrary as well. Then (4.1.7) yields the equality

$$\sum_{q=0}^{m_{ij}} \sum_{|\alpha|=q} c_{\alpha j} a_{ij}^{\alpha}(0, \lambda s, 0, \ldots, 0) \tau^{\alpha} (s\tau_1)^{m_{ij} - q} (1 - \lambda)^{m_{ij} - q} =$$

$$= (s\tau_1)^{m_{ij} + m_j} (1 - \lambda)^{m_{ij} + m_j} O((1 - \lambda)^{-m_i}).$$

Assuming that $m_{ij} > m_i - m_j$ and letting $\lambda \to 1$, we immediately obtain

$$\mathcal{F}(s, \tau) \equiv \sum_{|\alpha|=m_{ij}} c_{\alpha j} a_{ij}^{\alpha}(0, s, 0, \ldots, 0)\tau^{\alpha} = 0,$$

where $s \in \mathbb{C}^1, |s| = r$, and $\tau \in \mathbb{C}^n$, $|\tau_1| = 1, |\tau_2| \leq 1, \ldots, |\tau_n| \leq 1$ are arbitrary. It follows from this that $\mathcal{F}(s, \tau) \equiv 0$ for all $s \in \mathbb{C}^1$, $|s| \leq 1$ and $\tau \in \mathbb{C}^n$. In particular, $\mathcal{F}(0, \tau) \equiv 0$ for all $\tau \in \mathbb{C}^n$. It is clear that this is possible if and only if $a_{ij}^{\alpha}(0, 0) = 0$ for all α such that $|\alpha| = m_{ij}$.

Since the investigation of the Cauchy problem in a neighbourhood of $(t_0, z_0) \in V$ clearly reduces to the case $t_0 = 0, z_0 = 0$ (by the substitution $t \leftrightarrow t - t_0$, $z \leftrightarrow z - z_0$), we find by analogy with the previous result that $a_{ij}^{\alpha}(t, z) = 0$ for all $(t, z) \in V$ if $m_{ij} > m_i - m_j$. This means that $\operatorname{ord} A_{ij}(t, z, \mathcal{D}) \leq m_i - m_j$. The necessity is proved. The theorem is also completely proved.

3. THE STRUCTURE OF SYSTEMS WITH $\operatorname{ord} A_{ij} \leq m_i - m_j$

Let us elucidate what are the systems

$$u' - A(t, z, \mathcal{D})u = h(t, z), \tag{4.1.8}$$

for which the restrictions

$$\operatorname{ord} A_{ij}(t, z, \mathcal{D}) \leq m_i - m_j \quad (i, j = 1, \ldots, N)$$

are satisfied. First we consider the case of different m_j, $j = 1, \ldots, N$. Without loss of generality we can assume that $m_1 < \ldots < m_N$. In this case the matrix of operators $A_{ij}(t, z, \mathcal{D})$ has the following triangular form (recall that $A_{ij}(t, z, \mathcal{D}) \equiv 0$ if $m_i - m_j < 0$)

$$A = \begin{bmatrix} a_{11} & & & & \\ A_{21} & a_{22} & & \mathbf{0} & \\ \vdots & & \ddots & & \\ A_{N1} & A_{N2} & \ldots \ldots & & a_{NN} \end{bmatrix},$$

where the $A_{ii}(t, z, \mathcal{D}) \equiv a_{ii}(t, z)$ are certain analytic functions in the domain V and the $A_{ij}(t, z, \mathcal{D})$ $(i > j)$ are (arbitrary) partial differential operators of orders $\leq m_i - m_j$.

Thus, in this case the system (4.1.8) also has the triangular structure

$$u_1' - a_{11}(t,z)u_1 \qquad\qquad\qquad\qquad\qquad\qquad = h_1(t,z)$$

$$u_2' - A_{21}(t,z,\mathcal{D})u_1 - a_{22}(t,z)u_2 \qquad\qquad\qquad\quad = h_2(t,z)$$

. . . .

$$u_N' - A_{N,1}(t,z,\mathcal{D})u_1 - \ldots - A_{N,N-1}(t,z,\mathcal{D})u_{N-1} - a_{NN}(t,z)u_N \;\; = h_N(t,z)$$

Essentially, this system is a sequence of separate ordinary differential equations for the unknown function $u_1(t,z)$, then for $u_2(t,z)$ etc.

Let us now consider the other extremal case $m_1 = \ldots = m_N$. Then for all i,j we have the inequalities ord $A_{ij}(t,z,\mathcal{D}) \leq 0$. Consequently, the system (4.1.8) is the system of ordinary differential equations

$$u' - A(t,z)u = h(t,z),$$

where $A(t,z) \equiv (A_{ij}(t,z))_{N\times N}$ is a matrix of arbitrary functions.

Looking at these two cases, it is easy to understand the structure of system (4.1.8) in the general case. Namely, the matrix of such a system (maybe after renumbering) will have a stepped structure containing several blocks:

$$
\begin{bmatrix}
a_{11} & & & & & & & & \\
 & a_{22} & 0 & & & & & & 0 \\
 & & \ddots & & & & & & \\
 & & & a_{kk} & 0 & \cdots & 0 & 0 & \\
 & & & & a_{k+1,k+1} & \cdots & a_{k+1,k+l} & 0 & \\
 & & & & \vdots & \ddots & \vdots & \vdots & 0 \\
 & & & & a_{k+l,k+1} & \cdots & a_{k+l,k+l} & 0 & \\
 & A_{ij} & & & & & & a_{k+l+1,k+l+1} & \\
 & & & & & & & & \ddots \\
 & A_{ij} & & & A_{ij} & & & & a_{ss}
\end{bmatrix}
$$

These blocks correspond to the set $m_1 < \ldots < m_k < m_{k+1} = m_{k+2} = \ldots = m_{k+l} < m_{k+l+1} < \ldots < m_s$, where $1 \leq k \leq N$, $k+l \leq s \leq N$.

Finally, we see that the systems with ord $A_{ij}(t, z, \mathcal{D}) \leq m_i - m_j$ are systems with block-triangular form, the solution of which reduces to the successive solution of ordinary differential equations or systems of ordinary differential equations.

Thus, we have a description of the class of systems (4.1.8) for which the Cauchy problem is well-posed in the space $D_{m,r}$. It is clear that the conditions ord $A_{ij}(t, z, \mathcal{D}) \leq m_i - m_j$ are very strong and, consequently, the class of such systems is somewhat special.

4. THE CAUCHY PROBLEM IN THE DUAL SPACES $D'_{m,r}$

Before describing the main results on the solvability of the Cauchy problem in the spaces of functionals $D'_{m,r}$, we return for the time being to the Cauchy problem

$$u' - A(t, z, \mathcal{D})u = h(t, z) \tag{4.1.9}$$

$$u(t_0, z) = \phi(z) \tag{4.1.10}$$

in the space $D_{m,r}$. Namely, let us note that the solvability of the problem (4.1.9), (4.1.10) is a corollary of a general theorem on the solvability of the abstract Cauchy problem for the differential-operator equation

$$u'(t) - A(t)u(t) = h(t) \tag{4.1.11}$$

$$u(t_0) = \phi, \tag{4.1.12}$$

where $A(t)$ is a bounded operator in Banach space.

More precisely, let X be a Banach space and $A(t) : X \to X$ a bounded linear operator depending on t, $|t - t_0| < \delta$, and which is analytic in t. We seek the solution of the Cauchy problem (4.1.11), (4.1.12) in the space of functions $u(t) \in X$ (for all t such that $|t - t_0| < \delta$). Namely, let $u(t)$ be an analytic function with values in X such that

$$\|u(t)\|_{\delta;X} = \sup_{|t - t_0| < \delta} \|u(t)\|_X < \infty.$$

We denote this space by $\mathcal{O}(\delta; X)$.

Thus, the Cauchy problem (4.1.11), (4.1.12) is the abstract Cauchy problem in the Banach space $\mathcal{O}(\delta; X)$, where $u'(t)$ is the strong derivative, that is,

$$u'(t) = \lim_{\Delta t \to 0} (\Delta t)^{-1}[u(t + \Delta t) - u(t)]$$

in the norm of X.

There is the well known theorem

Theorem 4.1.2. *Let $A(t)$ be a bounded linear operator in X. Then for any $\phi \in X$ and $h(t) \in \mathcal{O}(\delta; X)$ the problem (4.1.11), (4.1.12) has a unique solution $u(t) \in \mathcal{O}(\delta; X)$. Moreover,*

$$\|u(t)\|_{\delta;X} \leq M(\|\phi\|_X + \|h(t)\|_{\delta;X},$$

where $M > 0$ is a constant.

We give no proof of this theorem, which can be obtained, in particular, by an elementary contraction method.

We further note that if the operator $A(t) : X \to X$ is a bounded linear operator, then the adjoint operator $A^*(t) : X^* \to X^*$ is also bounded and linear. Consequently, the dual problem

$$u'(t) + A^*(t, z, \mathcal{D})u(t) = h(t), \quad |t - t_0| < \delta,$$
$$u(t_0) = \phi \in X^*$$

in the space $\mathcal{O}(\delta; X^*)$ is well-posed also.

Let us return to the Cauchy problem (4.1.9), (4.1.10) in the spaces $D_{m,r}$. It is easy to see that the calculations performed during the course of the proof of Theorem 4.1.1 show precisely that under the conditions

$$\text{ord}\, A_{ij}(t, z, \mathcal{D}) \leq m_i - m_j \quad (i, j = 1, \ldots, N)$$

the operator $A(t, z, \mathcal{D}) \equiv (A_{ij}(t, z, \mathcal{D}))_{N \times N}$ is a bounded linear operator in $D_{m,r}$. Moreover, these conditions are also necessary conditions for the boundedness of the map

$$A(t, z, \mathcal{D}) : D_{m,r} \to D_{m,r} \tag{4.1.13}$$

(see the estimates obtained in the course of the proof of Theorem 4.1.1).

Thus, if $\text{ord}\, A_{ij} \leq m_i - m_j$, then the map (4.1.13) and hence the adjoint map

$$A^*(t, z, \mathcal{D}) : D'_{m,r} \to D'_{m,r}$$

are bounded linear maps. Here $A^*(t, z, \mathcal{D}) \equiv (A^*_{ij}(t, z, \mathcal{D}))_{N \times N}$, where

$$A^*_{ij}(t, z, \mathcal{D}) \cdot \equiv \sum_\alpha (-1)^{|\alpha|} \mathcal{D}^\alpha (a^\alpha_{ij}(t, z) \cdot)$$

are the adjoint operators.

Hence we have the following theorem.

Theorem 4.1.3. *Suppose that the conditions*

$$\text{ord } A_{ij}(t, z, \mathcal{D}) \leq m_i - m_j \quad (i, j = 1, \ldots, N)$$

hold (as usual, $A_{ij} \equiv 0 \Leftrightarrow \text{ord } A_{ij} = -\infty$). Then for any $\phi(z) \in D'_{m,r}$ and any $h(t, z) \in \mathcal{O}(\delta; D'_{m,r})$ (here $r > 0$ and $\sigma > 0$ are the same as in Theorem 4.1.1) there exists a unique solution $u(t, z) \in \mathcal{O}(\delta; D'_{m,r})$ of the problem

$$u' + A^*(t, z, \mathcal{D})u = h(t, z) \tag{4.1.14}$$

$$u(t_0, z) = \phi(z). \tag{4.1.15}$$

Moreover, the estimate

$$\|u\|_{\delta; X^*} \leq M(\|h\|_{\delta; X^*} + \|\phi\|_{X^*})$$

holds, where $M > 0$ is a constant and $X^* = D'_{m,r}$.

Remark. We end this subsection with a remark concerning the form of the representation for the dual problem (4.1.14), (4.1.15). Namely, it is obvious that every operator

$$A^*_{ij}(t, z, \mathcal{D})u \equiv \sum_{\alpha}(-1)^{|\alpha|}\mathcal{D}^{\alpha}[a^{\alpha}_{ij}(t, z)u]$$

may be written in the form

$$A^*_{ij}(t, z, \mathcal{D})u = \sum_{\beta} b_{\beta}(t, z)\mathcal{D}^{\beta}u$$

which is also an operator with analytic coefficients $b_{\beta}(t, z)$ of the same order as that of $A_{ij}(t, z, \mathcal{D})u$. Moreover, if $\text{ord } A_{ij} \leq m_i - m_j$, then also $\text{ord } A^*_{ij} \leq m_i - m_j$ and conversely. This means that the operators $A_{ij}(t, z, \mathcal{D})$ and $A^*_{ij}(t, z, \mathcal{D})$ satisfy the conditions of the Theorem 4.1.3 simultaneously. Thus, the solvability of the dual Cauchy problem (4.1.14), (4.1.15) may also be stated for the original system (4.1.9), (4.1.10).

4.2. The Cauchy problem in the spaces $D_{m,r-\sigma|t|}$.
(Case of conical evolution)

1. THE TEST-FUNCTION SPACE $D_{m,r-\sigma|t|}$

As usual, $U_r(z_0)$ is the polycylinder of radius $r > 0$ centred at the point $z_0 \in \mathbf{C}^n$. Further, let $\sigma > 0$ be a number. We denote by

$$V_{\sigma,r}(t_0, z_0) = \{(t, z) \in \mathbf{C}^{n+1} : r - |z - z_0| - \sigma|t - t_0| > 0\}$$

the cone centred at the point (t_0, z_0) with the height r/σ.

We consider the space $\mathcal{O}(r/\sigma; D_{m,r-\sigma|t-t_0|})$ defined as the set of all $\{u(t, z) \in \mathcal{O}(V_{\sigma,r}(t_0, z_0))$ such that

1) $u(t, \cdot) \in D_{m,r-\sigma|t-t_0|}$, $|t - t_0| < r/\sigma$;

2) $\displaystyle\sup_{|t-t_0|<r/\sigma} \|u(t, \cdot)\|_{m,r-\sigma|t-t_0|} < \infty\}$.

It is the space of functions $u(t, z) = (u_1(t, z), \ldots, u_N(t, z))$ that are analytic in the cone $V_{\sigma,r}(t_0, z_0)$ and for which the inequalities

$$|u_j(t, z)| \leq \frac{M_j}{(r - |z - z_0| - \sigma|t - t_0|)^{m_j}}, \quad (t, z) \in V_{\sigma,r}(t_0, z_0)$$

hold. Here the $M_j > 0$ are constants depending on $u_j(t, z)$; the m_j are natural numbers, $j = 1, \ldots, N$. In other words, the functions $u(t, z) \in \mathcal{O}(r/\sigma; D_{m,r-\sigma|t-t_0|})$ have (or may have) pole-type singularities on the lateral boundary of the cone $V_{\sigma,r}(t_0, z_0)$.

The norm in $\mathcal{O}(r/\sigma; D_{m,r-\sigma|t-t_0|})$ is defined by the formula

$$\|u(t, z)\|_{r/\sigma;m,r} \stackrel{\text{def}}{=} \sum_{j=1}^{N} \|u_j(t, z)\|_{r/\sigma;m_j,r},$$

where

$$\|u_j(t, z)\|_{r/\sigma;m_j,r} \stackrel{\text{def}}{=} \sup_{|t-t_0|<r/\sigma} \|u_j(t, \cdot)\|_{m_j,r-\sigma|t-t_0|}$$

or, what is the same,

$$\|u_j(t, z)\|_{r/\sigma;m_j,r} = \sup_{(t,z)\in V_{\sigma,r}(t_0,z_0)} |u_j(t, z)|(r - |z - z_0| - \sigma|t - t_0|)^{m_j}.$$

We shall write $u_j(t, z) \in \mathcal{O}(r/\sigma; D_{m_j,r-\sigma|t-t_0|})$ if $\|u_j(t, z)\|_{r/\sigma;m_j,r} < \infty$. In other words, $u(t, z) \in \mathcal{O}(r/\sigma; D_{m,r-\sigma|t-t_0|})$ if and only if $u_j(t, z) \in \mathcal{O}(r/\sigma; D_{m_j,r-\sigma|t-t_0|})$, $j = 1, \ldots, N$.

It is worth noting that the scale $D_{m,r-\sigma|t-t_0|}$ tends to zero "linearly" as $|t - t_0| \to 0$. It seems that this scale is the most natural one for the evolution of the analytic Cauchy problem.

The main property of the scale $D_{m,r-\sigma|t-t_0|}$ is contained in the following lemma.

Lemma 4.2.1. *The maps*

$$\frac{\partial}{\partial z_k} : \mathcal{O}(r/\sigma; D_{m,r-\sigma|t-t_0|}) \to \mathcal{O}(r/\sigma; D_{m+1,r-\sigma|t-t_0|}),$$

$$\frac{\partial}{\partial t} : \mathcal{O}(r/\sigma; D_{m,r-\sigma|t-t_0|}) \to \mathcal{O}(r/\sigma; D_{m+1,r-\sigma|t-t_0|})$$

are continuous (here $m + 1 = (m_1 + 1, \ldots, m_N + 1)$, $k = 1, \ldots, n$).

Proof. In fact, according to the definition of the norm $\|u(t, z)\|_{r/\sigma;m,r}$, we have the following inequality for all $(t, z) \in V_{\sigma,r}(t_0, z_0)$:

$$|u_j(t, z)| \le \|u_j(t, z)\|_{r/\sigma;m_j,r}(r - |z - z_0| - \sigma|t - t_0|)^{-m_j},$$

where $m_j > 0$, $j = 1, \ldots, N$.

Then in view of Lemma 4.2.1, we obtain the inequalities

$$\left|\frac{\partial u_j}{\partial z_k}\right| \le M_j \|u_j\|_{r/\sigma;m_j,r}(r - |z - z_0| - \sigma|t - t_0|)^{-m_j-1},$$

$$\left|\frac{\partial u_j}{\partial t}\right| \le M_j \sigma \|u_j\|_{r/\sigma;m_j,r}(r - |z - z_0| - \sigma|t - t_0|)^{-m_j-1},$$

where $M_j = (m_j + 1)^{m_j+1}/m_j^{m_j}$. The assertion of the lemma follows immediately from these inequalities.

2. CRITERION FOR THE WELL-POSEDNESS OF THE CAUCHY PROBLEM IN THE SPACES $D_{m,r-\sigma|t|}$. NECESSITY OF KOVALEVSKAYA CONDITIONS AND LERAY-VOLEVICH CONDITIONS

We consider the Cauchy problem

$$u' - A(t, z, \mathcal{D})u = h(t, z), \qquad (4.2.1)$$

$$u(t_0, z) = \phi(z) \qquad (4.2.2)$$

in the spaces $\mathcal{O}(r/\sigma; D_{m,r-\sigma|t-t_0|})$. Here, as usual, $A(t,z,\mathcal{D}) \equiv (A_{ij}(t,z,\mathcal{D}))_{N \times N}$ is a matrix of differential operators

$$A_{ij}(t,z,\mathcal{D}) \equiv \sum_{|\alpha|=0}^{m_{ij}} a_{ij}^{\alpha}(t,z)\mathcal{D}^{\alpha}$$

with analytic coefficients in the domain $V \subset \mathbf{C}_{t,z}^{n+1}$.

Before stating the correctness of the Cauchy problem in the scale $D_{m,r-\sigma|t|}$ we note that there is the inclusion $u'(t,z) \in \mathcal{O}(r/\sigma; D_{m+1,r-\sigma|t-t_0|})$ (see Lemma 4.2.1) if $u(t,z) \in \mathcal{O}(r/\sigma; D_{m,r-\sigma|t-t_0|})$. Therefore, using the system (4.2.1) it is natural to assume that $h(t,z) \in \mathcal{O}(r/\sigma; D_{m+1,r-\sigma|t-t_0|})$.

Definition 4.2.1. The Cauchy problem is said to be locally well-posed in the scale $D_{m,r-\sigma|t|}$ if for any point $(t_0, z_0) \in V$ there exists $\sigma > 0$ (recall that numbers $r > 0$ were chosen earlier) such that for any right hand side $h(t,z) \in \mathcal{O}(r/\sigma; D_{m+1,r-\sigma|t-t_0|})$ and any initial data $\phi(z) \in D_{m,r}$ the Cauchy problem (4.2.1) (4.2.2) has a unique solution $u(t,z) \in \mathcal{O}(r/\sigma; D_{m,r-\sigma|t-t_0|})$ and, in addition,

$$\|u\|_{r/\sigma;m,r} \le M(\|\phi\|_{m,r} + \|h\|_{r/\sigma;m+1,r}),$$

where $M > 0$ is a constant.

Theorem 4.2.1. *The Cauchy problem* (4.2.1), (4.2.2) *is locally well-posed in the scale* $D_{m,r-\sigma|t|}$ *if and only if the following inequalities hold:*

$$\operatorname{ord} A_{ij}(t,z,\mathcal{D}) \le m_i - m_j + 1 \tag{4.2.3}$$

(by definition $\operatorname{ord}A_{ij} = -\infty$ *if* $A_{ij} = 0$).

Proof. In general, the proof of this theorem is parallel to that of Theorem 4.1.1.

Sufficiency. We reduce the problem (4.2.1), (4.2.2) to the integral equation

$$u(t,z) = \int_0^t A(t,z,\mathcal{D})u(\tau,z)d\tau + \phi(z) + \int_0^t h(\tau,z)d\tau \tag{4.2.4}$$

(we may suppose without loss of generality that $z_0 = 0$, $t_0 = 0$) and prove that the operator on the right hand side of (4.2.4) acts in the Banach space $\mathcal{O}(r/\sigma; D_{m,r-\sigma|t|})$; moreover it is a contraction, if $\sigma > 0$ is sufficiently large.

First we note that vector-valued functions $h(t, z) = (h_1(t, z), \ldots, h_N(t, z))$ and $\phi(z) = (\phi_1(z), \ldots, \phi_N(z))$ belong to $\mathcal{O}(r/\sigma; D_{m, r-\sigma|t|})$. Indeed,

$$\left| \int_0^t h_j(\tau, z) d\tau \right| \leq M \int_0^{|t|} (r - |z| - \sigma|\tau|)^{-m_j - 1} d|\tau| \leq \frac{M_1}{(r - |z| - \sigma|t|)^{m_j}},$$

where $M > 0$, $M_1 > 0$ are constants. The inclusion $\phi(z) \in \mathcal{O}(r/\sigma; D_{m, r-\sigma|t|})$ is obvious.

Next we consider the operator

$$Bu \equiv \int_0^t A(\tau, z, \mathcal{D}) u(\tau, z) d\tau.$$

For each component $(Bu)_i$, $i = 1, \ldots, N$, we have

$$(Bu)_i(t, z) \equiv \sum_{j=1}^N \int_0^t A_{ij}(\tau, z, \mathcal{D}) u_j(\tau, z) d\tau =$$

$$= \sum_{j=1}^N \sum_{|\alpha| \leq m_{ij}} \int_0^t a_{ij}^\alpha(\tau, z) \mathcal{D}^\alpha u_j(\tau, z) d\tau,$$

where the m_{ij} are the orders of the operators $A_{ij}(t, z, \mathcal{D})$. Since the coefficients $a_{ij}^\alpha(t, z)$ are analytic functions in V, then they are continuous functions in the closure of the cone $V_{\sigma, r}(0, 0)$. Consequently,

$$\left| \int_0^t a_{ij}^\alpha(\tau, z) \mathcal{D}^\alpha u_j(\tau, z) d\tau \right| \leq M \int_0^{|t|} |\mathcal{D}^\alpha u_j(\tau, z)| d|\tau|.$$

Furthermore, by Lemma 4.2.1, each function $u_j(\tau, z) \in \mathcal{O}(r/\sigma; D_{m_j, r-\sigma|t|})$ satisfies the inequalities

$$\left| \int_0^t a_{ij}^\alpha(\tau, z) \mathcal{D}^\alpha u_j(\tau, z) d\tau \right| \leq M \|u_j(t, z)\|_{r/\sigma; m_j, r} \times$$

$$\times \int_0^{|t|} (r - |z| - \sigma|\tau|)^{-m_j - |\alpha|} d|\tau|$$

$$\leq \frac{M}{\sigma} \|u_j(t, z)\|_{r/\sigma; m_j, r} (r - |z| - \sigma|t|)^{-m_j - |\alpha| + 1},$$

where $M > 0$ is a constant. Thus for each component $(Bu)_i(t, z)$ we obtain the following estimate

$$|Bu_i(t, z)| \leq M\sigma^{-1} \sum_{j=1}^N \|u_j(t, z)\|_{r/\sigma; m, r} (r - |z| - \sigma|t|)^{-m_j - m_{ij} + 1}.$$

From these estimates we immediately have

$$|(Bu)_i(t,z)| \leq M\sigma^{-1}\|u(t,z)\|_{r/\sigma;m,r}(r - |z| - \sigma|t|)^{-m_i},$$

since, by condition (4.2.4), $m_{ij} \leq m_i - m_j + 1$.

These estimates show that for any $u(t,z) \in \mathcal{O}(r/\sigma; D_{m,r-\sigma|t|})$ the function $Bu(t,z)$ is also an element of this space and

$$\|Bu\|_{r/\sigma;m,r} \leq M\sigma^{-1}\|u\|_{r/\sigma;m,r}.$$

It is clear that the operator Bu is a contraction if $\sigma > 0$ is sufficiently large, namely $\sigma > M$. Thus the solution of the Cauchy problem exists and (as is easy to see) the estimate

$$\|u\|_{r/\sigma;m,r} \leq M(\|\phi\|_{m,r} + \|h\|_{r/\sigma;m+1,r})$$

holds.

Necessity. The proof of the necessity of the inequalities ord $A_{ij} \leq m_i - m_j + 1$ is analogous to the case of the inequalities ord $A_{ij} \leq m_i - m_j$. The only difference is that for $u(t,z) \in \mathcal{O}(r/\sigma; D_{m,r-\sigma|t|})$ the derivative $u'(t,z) \in \mathcal{O}(r/\sigma; D_{m+1,r-\sigma|t|})$ and, in particular, for $t = 0$

$$|u'(0,z)| = O((r - |z|)^{-m_i-1}), \quad i = 1,\ldots,N.$$

The rest is a verbatim repetition. The theorem is proved.

Example. Consider the Cauchy problem

$$u' - z^2 \frac{\partial u}{\partial z} + u = 0 \tag{4.2.5}$$

$$u(0,z) = \phi(z), \quad z \in \mathbb{C}^1. \tag{4.2.6}$$

According to Theorem 4.2.1, for any $r > 0$ there exists $\sigma > 0$ such that for any $\phi(z) \in D_{m,r}$ the problem (4.2.5), (4.2.6) has the unique solution $u(t,z) \in \mathcal{O}(r/\sigma; D_{m,r-\sigma|t|})$. On the other hand, well-known calculations show that the solution of this problem is

$$u(t,z) = \phi\left(\frac{z}{1+tz}\right)e^{-t}.$$

It follows from this formula that the Cauchy problem (4.2.5), (4.2.6) is solved locally both for t and for z, in spite of the fact that the coefficients and $\phi(z)$ are entire functions. Namely, the solution exists if $|tz| < 1$ so that there is no $\delta > 0$ such that $u(t, z)$ exists for all z and $|t| < \delta$. On the other hand, if the initial function $\phi(z)$ is analytic for $|z| < r$, then there exists an analytic solution $u(t, z)$ for $|t| < 1/r$ and, in particular, in the cone $r - |z| - \sigma|t| > 0$, where $\sigma = r^2$.

3. THE STRUCTURE OF SYSTEMS WITH ord $A_{ij} \leq m_i - m_j + 1$

Let us describe the structure of the systems

$$u' - A(t, z, \mathcal{D})u = h(t, z), \tag{4.2.7}$$

whose operators $A_{ij}(t, z, \mathcal{D})$ satisfy the conditions

$$\text{ord } A_{ij}(t, z, \mathcal{D}) \leq m_i - m_j + 1. \tag{4.2.8}$$

First we consider several special cases.

1) Kovalevskaya systems. Let $m_1 = m_2 = \ldots = m_N$. Then the conditions (4.2.8) imply that ord $A_{ij} \leq 1$, which are the well-known Kovalevskaya conditions. Hence the Cauchy problem is well-posed in the isotropic scale $D_{m,r-\sigma|t|}$ if and only if the sytem (4.2.7) is a Kovalevskaya system.

2) Let $m_1 < m_2 < \ldots < m_N$ with $m_{i+1} - m_i = 1$ for all $i = 1, \ldots, N$. In this case, as it is easy to see, the operator-matrix $A(t, z, \mathcal{D})$ has the form

$$\begin{bmatrix} A_{11} & a_{12} & & & \\ A_{21} & A_{22} & a_{23} & & \\ \vdots & & \ddots & & \ddots \\ \vdots & A_{ij} \ (i > j) & & \ddots & & a_{N-1,N} \\ \vdots & & & & \ddots & \\ A_{N1} & A_{N2} & \ldots & \ldots & & A_{NN} \end{bmatrix}, \tag{4.2.9}$$

where the $A_{ij} \equiv A_{ij}(t, z, \mathcal{D})$ are differential operators, the degrees of which are at most $i - j + 1$, and the $a_{i,i+1} \equiv a_{i,i+1}(t, z)$ are analytic functions in the domain V. The corresponding systems are

$$u_1' - A_{11}(t, z, \mathcal{D})u_1 - a_{12}(t, z)u_2 = h_1(t, z)$$

$$u_2' - A_{21}(t, z, \mathcal{D})u_1 - A_{22}(t, z, \mathcal{D})u_2 - a_{23}(t, z)u_3 = h_2(t, z)$$

$$\cdots \cdots \cdots \cdots \cdots \cdots \cdots \cdots \cdots \cdots \cdots \cdots$$

$$u_N' - A_{N1}(t, z, \mathcal{D})u_1 - A_{N2}(t, z, \mathcal{D})u_2 - \ldots - A_{NN}(t, z, \mathcal{D})u_N = h_N(t, z)$$

It is clear that by eliminating $u_2(t, z)$ from the first equation and substituting $u_2(t, z)$ into the second equation, then eliminating $u_3(t, z)$ from the second equation and so on, we finally obtain a single equation of order N (with respect to t) in $u_1(t, z)$:

$$u_1^{(N)} - \sum_{j=0}^{N-1} A_j(t, z, \mathcal{D})u_1^{(j)} = h(t, z). \tag{4.2.10}$$

Calculations show that $\operatorname{ord} A_j(t, z, \mathcal{D}) \leq N - j$, that is, equation (4.2.10) is of Kovalevskaya type.

On the other hand, if equation (4.2.10) is considered, then the standard substitution

$$u_1 \equiv u_1(t, z), \quad u_2 \equiv u_1'(t, z), \ldots, \quad u_N \equiv u_1^{(N-1)}(t, z)$$

gives the first order system with matrix

$$\begin{bmatrix} 0 & 1 & 0 & \cdots & 0 \\ 0 & 0 & 1 & \cdots & 0 \\ \vdots & & & & \\ 0 & 0 & 0 & \cdots & 1 \\ A_0 & A_1 & A_2 & \cdots & A_{N-1} \end{bmatrix},$$

which is a matrix of type (4.2.9). Moreover, if $u_1(t, z) \in \mathcal{O}(\delta; D_{m_1, r - \sigma|t|})$, then

$$u_2(t, z) \equiv u_1'(t, z) \in \mathcal{O}(\delta_1; D_{m_1+1, r-\sigma|t|}),$$

$$\cdots \cdots \cdots \cdots \cdots \cdots \cdots \cdots \cdots \cdots$$

$$\cdots \cdots \cdots \cdots \cdots \cdots \cdots \cdots \cdots \cdots$$

$$u_N(t, z) \equiv u_1^{(N-1)}(t, z) \in \mathcal{O}(\delta_1; D_{m_1+N-1, r-\sigma|t|}),$$

where $\delta_1 < \delta$ is arbitrary. Thus, the question of the well-posedness of the Cauchy problem in $D_{m_1, r - \sigma|t|}$ for one equation of order N is reduced to the same question for the system of first order in the scale $D_{m, r - \sigma|t|}$, where $m = (m_1, m_1 + 1, \ldots, m_1 + N - 1)$, that is, the previous case $m_{i+1} - m_i = 1$. Consequently,

$$\operatorname{ord} A_j(t, z, \mathcal{D}) \leq N - j, \quad j = 0, 1, \ldots, N - 1, N.$$

Thus, if $a_{i,i+1}(t, z) \neq 0$ in V, then the system (4.2.7) with operator-valued matrix of type (4.2.9) is equivalent to a single equation of order N of Kovalevskaya type.

3) Suppose that $m_1 \leq \ldots \leq m_k < m_{k+1} \leq \ldots \leq m_N$, where $m_{k+1} - m_k > 1$ (k is a fixed number). Then the system (4.2.7) has the following matrix

$$
\begin{bmatrix}
\begin{array}{c} A_{ij} \\ 1 \leq i, j \leq k \end{array} & 0 \\
\begin{array}{c} A_{ij} \\ k + 1 \leq i \leq N \\ 1 \leq j \leq k \end{array} & \begin{array}{c} A_{ij} \\ k + 1 \leq i, j \leq N \end{array}
\end{bmatrix}
$$

It is clear that the solution of the corresponding system

$$u_i' - \sum_{j=1}^{k} A_{ij}(t, z, \mathcal{D})u_j = h_i(t, z), \quad (1 \leq i \leq k) \tag{4.2.11}$$

$$u_i' - \sum_{j=1}^{k} A_{ij}(t, z, \mathcal{D})u_j - \sum_{j=k+1}^{N} A_{ij}(t, z, \mathcal{D})u_j = h_i(t, z) \tag{4.2.12}$$

$$(k + 1 \leq j \leq N)$$

can be obtained in two stages. First the unknown functions $u_1(t, z), \ldots, u_k(t, z)$ are independently defined from the subsystem (4.2.11). Then the remaining functions $u_{k+1}(t, z), \ldots, u_N(t, z)$ are defined from the second subsystem (4.2.12). Thus if $m_{k+1} - m_k > 1$, then the solution of the basic system (4.2.7) reduces to the successive solution of two Cauchy problems for systems whose dimensions are equal to $k < N$ and $N - k < N$ respectively. Essentially, such systems are disintegrated.

$$
\left[
\begin{array}{cccccccccc}
A_{11} & a_{12} & 0 & & & & & & & \\
A_{22} & a_{23} & & & & & & & & \\
 & \ddots & \ddots & & & & 0 & & & \\
 & & A_{kk} & a_{k-1,k} & & & & & & \\
 & & A_{k+1,k+1} & a_{k,k+1} & 0 & & & & & \\
 & & A_{k+2,k+2} & A_{k+1,k+2} & a_{k,k+2} & 0 & & & & \\
 & & \ddots & \vdots & \cdots & \ddots & & & & \\
 & & A_{k+l,k+l} & A_{k+1,k+l} & A_{k+2,k+l} & \cdots & a_{k,k+l} & 0 & \cdots & 0 \\
 & & A_{k+l+1,k+l+1} & A_{k+1,k+l+1} & A_{k+l,k+l+1} & \cdots & a_{k,k+l+1} & a_{k+l+1,k+l+2} & 0 & 0 \\
 & & & & & \cdots & \cdots & 0 & 0 & \vdots \\
 & & A_{ij}\ (i>j) & & & & & & & 0 \\
 & & & & a_{N-1,N} & \cdots & 0 & 0 & 0 & 0 \\
 & & & & A_{N,N} & & & & &
\end{array}
\right]
$$

Table 1

4) Finally we turn to the general case. First we note that (as the last example shows) one can suppose without loss of generality that $m_1 \leq m_2 \leq \ldots \leq m_N$, moreover, if $m_i < m_{i+1}$, then $m_{i+1} - m_i = 1$. Suppose, for definiteness, that

$$m_1 < \ldots < m_{k+1} = m_{k+2} = \ldots = m_{k+l} < m_{k+l+1} < \ldots < m_N,$$

where k and l are certain positive integers $(k + l \leq N)$. The corresponding matrix has the form indicated in Table 1.

The system of differential equations with such a matrix is

$$u_i' - \sum_{j=1}^{i} A_{ij}(t, z, \mathcal{D})u_j - a_{i,i+1}(t, z)u_{i+1} = h_i(t, z), \quad (i = 1, \ldots, k-1)$$

$$u_k' - \sum_{j=1}^{k} A_{kj}(t, z, \mathcal{D})u_j - \sum_{j=k+1}^{k+l} a_{kj}(t, z)u_j = h_k(t, z),$$

$$u_i' - \sum_{j=1}^{i+l} A_{ij}(t, z, \mathcal{D})u_j - a_{i,k+l+1}(t, z)u_{k+l+1} = h_i(t, z), \quad (i = k+1, \ldots, k+l)$$

$$u_i' - \sum_{j=1}^{i} A_{ij}(t, z, \mathcal{D})u_j - a_{i,i+1}(t, z)u_{i+1} = h_i(t, z), \quad (i = k+l+1, \ldots, N-1)$$

$$u_N' - \sum_{j=1}^{N} A_{Nj}(t, z, \mathcal{D})u_j = h_N(t, z).$$

As in the case of systems with $\operatorname{ord} A_{ij} \leq m_i - m_j$ this system also has block-triangular form. However, in this case, when $\operatorname{ord} A_{ij} \leq m_i - m_j + 1$, the solution of the Cauchy problem does not reduce to the solution of the Cauchy problem for the partial systems corresponding to the separate blocks.

4. THE CAUCHY PROBLEM IN THE DUAL SCALE $D'_{m+1,r+\sigma|t|}$

In this subsection we solve the Cauchy problem dual to the Cauchy problem

$$u' A(t, z, \mathcal{D})u = h(t, z) \tag{4.2.13}$$

$$u(t_0, z) = \phi(z), \quad |z - z_0| < R, \tag{4.2.14}$$

in the scale $D_{m,r-\sigma|t|}$, $r \leq R$.

Before talking about the Cauchy problem in spaces of analytic functionals we make two remarks which are essential for the well posedness of the dual problem.

First, as was proved in subsection 2, the problem (4.2.13), (4.2.14) is well-posed in the scale $D_{m,r-\sigma|t|}$ if and only if

$$\operatorname{ord} A_{ij}(t, z, \mathcal{D}) \le m_i - m_j + 1. \tag{4.2.15}$$

It is easy to verify that at the same time these conditions are necessary and sufficient for the continuity of the map

$$A(t, z, \mathcal{D}) : D_{m,r-\sigma|t|} \to D_{m+1,r-\sigma|t|},$$

where t is a fixed number and r is any number such that $r \le R$. (This is established by a repetition of the argument given in Theorem 4.1.1.) It is natural, therefore, to suppose these inequalities to be satisfied.

The second remark is the following. We note that for fixed z the "life time" for the solution of the problem (4.2.13), (4.2.14) is defined by the inequality $|t - t_0| < (R - |z - z_0|)/\sigma$ and hence by the choice of the numbers $R > 0$ and $\sigma > 0$. In turn, the choice of $\sigma > 0$ is defined by the inequality $M\sigma < 1$, where $M > 0$ is a constant which (as is easily seen from the calculations carried out in the proof of Theorem 4.1.1) depends on $\max |a_{ij}^{\alpha}(t, z)|$ in the closure of the cone $V_{\sigma,R}(t_0, z_0)$.

This means, in particular, that for any compact set $K \subset V$ these numbers may be the same for all the points $(t_0, z_0) \in K$. In other words, for any compact $K \subset V$ there exist numbers $R > 0$ and $\sigma > 0$ such that for any $(t_0, z_0) \in K$ the solution $u(t, z) \in \mathcal{O}(V_{\sigma,R}(t_0, z_0))$, moreover,

$$u(t, \cdot) \in D_{m,R-\sigma|t-t_0|}, \quad |t - t_0| < R/\sigma.$$

From this follows the next observation, which as will be shown below, is essential for the organization of the dual Cauchy problem. Namely, let

$$W_{\sigma,R}^r(t_0, z_0) = \left\{ (t, z) : |z - z_0| < r + \sigma|t - t_0|, \ |t - t_0| < \frac{R - r}{\sigma} \right\}$$

where $r < R$ is arbitrary. Then for any point $(t_1, z_0) \in W_{\sigma,R}^r(t_0, z_0)$, $|t_1 - t_0| \le (R - r)/\sigma$, the solution of the problem

$$u' - A(t, z, \mathcal{D})u = h(t, z),$$
$$u(t_1, z) = \phi(z)$$

exists and is an analytic function in the cone

$$V_{\sigma, r+\sigma|t_1-t_0|}(t_1, z_0) = \{(t, z) : r + \sigma|t_1 - t_0| - |z - z_0| - \sigma|t - t_1| > 0\}.$$

In particular, this solution is defined for all points $t = t_0$, $|z-z_0| < r$. Moreover, if $\phi(z) \in D_{m+\sigma|t_1-t_0|}$, then $u(t, z) \in D_{m, r+\sigma|t-t_0|}$ for every $t \in [t_1, t_0]$. These $R > 0$ and $\sigma > 0$ are precisely chosen below.

This concludes our second remark.

Let us turn to the organization of the dual Cauchy problem. Namely, suppose that conditions (4.2.15) hold. Then, in accordance with the first remark, the adjoint map

$$A^*(t, z, \mathcal{D}) : D'_{m+1, r-\sigma|t-t_0|} \to D'_{m, r-\sigma|t-t_0|}$$

is a bounded linear map.

It turns out that the natural scale for the investigation of the dual Cauchy problem

$$u' + A^*(t, z, \mathcal{D})u = h(t, z) \tag{4.2.16}$$

$$u(t_0, z) = \phi(z) \tag{4.2.17}$$

is the scale of the spaces of functionals $D'_{m+1, r+\sigma|t-t_0|}$. This scale is precisely the well-posedness scale for the dual Cauchy problem.

We give the precise statements. Let $|t - t_0| < (R - r)/\sigma$, where $r < R$ and $\sigma > 0$ are given numbers, and let $u(t, \cdot) \in D'_{m+1, r+\sigma|t-t_0|}$ be a functional depending analytically on t, $|t - t_0| < (R - r)/\sigma$.

We set

$$\mathcal{O}\left(\frac{R - r}{\sigma}; D'_{m+1, r+\sigma|t-t_0|}\right) = \Big\{u(t, z) : u(t, \cdot) \in D'_{m+1, r+\sigma|t-t_0|},$$

$$|t - t_0| < \frac{R - r}{\sigma}\Big\}.$$

It is the Banach space of analytic functionals $u(t, z)$ with norm

$$\|u(t, z)\|'_{\frac{R-r}{\sigma}; m+1, r, \sigma} \overset{\text{def}}{=} \sup_{|t-t_0| < \frac{R-r}{\sigma}} \|u(t, \cdot)\|'_{m+1, r+\sigma|t-t_0|},$$

where

$$\|u(t, \cdot)\|'_{m+1, r+\sigma|t-t_0|} = \sup_v \frac{\langle u(t, \cdot), v(z)\rangle}{\|v(z)\|_{m+1, r+\sigma|t-t_0|}}$$

is the norm in $D'_{m+1,r+\sigma|t-t_0|}$.

Definition 4.2.2. The functional

$$u'(t,z) \in \mathcal{O}\left(\frac{R-r}{\sigma}; D'_{m,r+\sigma|t-t_0|}\right)$$

is called the derivative of the functional $u(t,z) \in \mathcal{O}\left(\frac{R-r}{\sigma}; D_{m,r+\sigma|t-t_0|}\right)$ if for any function $v(t,z) \in \mathcal{O}\left(\frac{R-r}{\sigma}; D_{m,r+\sigma|t-t_0|}\right)$ the following identity is satisfied:

$$\langle u'(t,z), v(t,z)\rangle \overset{\text{def}}{=} \frac{d}{dt}\langle u(t,z), v(t,z)\rangle - \langle u(t,z), v'(t,z)\rangle$$

or, what is the same,

$$\int_{t_0}^{t} \langle u'(\tau,z), v(\tau,z)\rangle d\tau = \langle u(t,z), v(t,z)\rangle -$$

$$-\langle u(t_0,z), v(t_0,z)\rangle - \int_{t_0}^{t} \langle u(\tau,z), v'(\tau,z)\rangle d\tau,$$

where, as before, the integration is carried out along the straight line segment joining the points t_0 and t.

By virtue of this definition, the equality

$$\int_{t_0}^{t} \langle u'(\tau,z) + A^*(\tau,z,\mathcal{D})u, \ v(\tau,z)\rangle d\tau \overset{\text{def}}{=} \langle u(t,z), \ v(t,z)\rangle -$$

$$-\langle u(t_0,z), \ v(t_0,z)\rangle - \int_{t_0}^{t} \langle u(\tau,z), \ v'(\tau,z) - A(\tau,z,\mathcal{D})v(\tau,z)\rangle d\tau$$

is well defined.

This equality is the foundation of the definition of the dual Cauchy problem. Namely,

Definition 4.2.3. Let $\phi(z) \in D'_{m+1,r}$ and $h(t,z) \in \mathcal{O}\left(\frac{R-r}{\sigma}; D'_{m+1,r+\sigma|t-t_0|}\right)$, where $0 \le r < R$. The functional $u(t,z) \in \mathcal{O}\left(\frac{R-r}{\sigma}; D'_{m+1,r+\sigma|t-t_0|}\right)$ is said to be a solution of the Cauchy problem (4.2.16), (4.2.17) if the following identity holds:

$$\langle u(t,z), v(t,z)\rangle = \int_{t_0}^{t} \langle u(\tau,z), v'(\tau,z) - A(\tau,z,\mathcal{D})v(\tau,z)\rangle d\tau +$$

$$+\langle \phi(z), v(t_0,z)\rangle + \int_{t_0}^{t} \langle h(\tau,z), v(\tau,z)\rangle d\tau, \ |t-t_0| < \frac{R-r}{\sigma}, \quad (4.2.18)$$

where $v(t, z) \in \mathcal{O}\left(\frac{R-r}{\sigma}; D_{m+1,r+\sigma|t-t_0|}\right)$ is an arbitrary function such that $v' - A(t, z, \mathcal{D})v \in \mathcal{O}\left(\frac{R-r}{\sigma}; D_{m+1,r+\sigma|t-t_0|}\right)$ (we call such functions test functions).

Theorem 4.2.2. *Let* ord $A_{ij}(t, z, \mathcal{D}) \leq m_i - m_j + 1$ *and* $0 \leq r < R$. *Then for any* $\phi(z) \in D'_{m+1,r}$ *and* $h(t, z) \in \mathcal{O}\left(\frac{R-r}{\sigma}; D'_{m+1,r+\sigma|t-t_0|}\right)$ *there exists a unique solution* $u(t, z) \in \mathcal{O}\left(\frac{R-r}{\sigma}; D'_{m+1,r+\sigma|t-t_0|}\right)$ *(in the sense of Definition 4.2.3) of the problem* (4.2.16), (4.2.17). *Moreover, the estimate*

$$\|u\|'_{\frac{R-r}{\sigma};m+1,r,\sigma} \leq M\left(\|\phi\|'_{m+1,r} + \|h\|'_{\frac{R-r}{\sigma};m+1,r,\sigma}\right)$$

holds, where $M > 0$ *is some constant.*

Proof. The proof will be carried out in two steps:

1) first we shall give an explicit formula for the solution of the dual problem based on the solution of the "direct" problem (4.2.7), (4.2.8);

2) secondly, we shall prove that the formula so obtained is indeed the solution of the dual problem.

Step 1 (Representation formula). Let $u(t, z) \in \mathcal{O}\left(\frac{R-r}{\sigma}; D'_{m+1,r+\sigma|t-t_0|}\right)$ be a solution of the Cauchy problem (4.2.16), (4.2.17) (without loss of generality we may suppose that $t_0 = 0, z_0 = 0$). This means that for any test function $v(t, z)$

$$\langle u(t, z), v(t, z)\rangle = \int_0^t \langle u(\tau, z), v'(\tau, z) - A(\tau, z, \mathcal{D})v(\tau, z)\rangle d\tau +$$

$$+ \langle \phi(z), v(0, z)\rangle + \int_0^t \langle h(\tau, z), v(\tau, z)\rangle d\tau, \quad |t| < \frac{R-r}{\sigma}. \tag{4.2.19}$$

We note that the left hand side of (4.2.19) does not depend on the value of $v(\tau, z)$ for $\tau \in]t, 0]$, that is, for $\tau \neq t$. Consequently, for $\tau \in]t, 0]$ on the right hand side we may take the extension $v(t; \tau, z)$ of the value of $v(t, z)$ such that

$$v'_\tau(t; \tau, z) - A(\tau, z, \mathcal{D})v(t; \tau, z) = 0 \tag{4.2.20}$$

$$v(t; \tau, z)\Big|_{\tau=t} = v(t, z), \tag{4.2.21}$$

where $v(t, z) \in D_{m+1,r+\sigma|t|}$ is the value of the test function at the point t.

Taking into account the second remark at the beginning of the present subsection, we see that the solution of (4.2.20), (4.2.21) is defined in the cone

$r + \sigma|t| - |z| - \sigma|\tau - t| > 0$ of the variables τ, z. In particular, this solution is defined for $\tau = 0$, $r - |z| > 0$. Moreover, for all $\tau \in [t, 0]$ we have the inclusion $v(t; \tau, \cdot) \in D_{m+1, r + \sigma|\tau|}$ and in particular, $v(t; 0, \cdot) \in D_{m+1, r}$.

Hence this extension is a test function and, consequently, we have

$$\langle u(t, z), v(t, z) \rangle = \langle \phi(z), v(t; 0, z) \rangle + \int_0^t \langle h(\tau, z), v(t; \tau, z) \rangle d\tau. \qquad (4.2.22)$$

This formula gives the desired representation of the solution $u(t, z)$.

Corollary. *The solution of the dual Cauchy problem is unique.*

Step 2. We show that for any $\phi(z) \in D'_{m+1, r}$ and $h(t, z) \in \mathcal{O}\left(\frac{R-r}{\sigma}; D'_{m+1, r + \sigma|t|}\right)$ the functional (4.2.22) is precisely the solution of the problem (4.2.20), (4.2.21). We must show that the identity (4.2.19) is satisfied.

Lemma 4.2.2. *Let* $w(\tau; \eta, z) \in \mathcal{O}\left(\frac{R-r}{\sigma}; D_{m+1, r + \sigma|\eta|}\right)$ *be the solution of the problem*

$$w'_\eta(\tau; \eta, z) - A(\eta, z, \mathcal{D})w(\tau; \eta, z) = 0, \quad \eta \in [\tau, 0],$$
$$w(\tau; \eta, z)\big|_{\eta = \tau} = v'(\tau, z) - A(\tau, z, \mathcal{D})v(\tau, z).$$

Then $w(\tau; \eta, z) \equiv v'_\tau(\tau; \eta, z)$ *for all* $\eta \in [\tau, 0]$.

Proof. Indeed, by definition of the function $v(\tau; \eta, z)$ we have

$$v'_\eta(\tau; \eta, z) - A(\eta, z, \mathcal{D})v(\tau; \eta, z) = 0, \quad \eta \in [\tau, 0],$$
$$v(\tau; \eta, z)\big|_{\eta = \tau} = v(\tau, z).$$

Hence, taking the derivative with respect to τ, we obtain

$$\frac{\partial}{\partial \eta} v'_\tau(\tau; \eta, z) - A(\eta, z, \mathcal{D})v'_\tau(\tau; \eta, z) = 0,$$

$$v'_\tau(\tau; \eta, z)\big|_{\eta = \tau} = v'(\tau, z) - v'_\eta(\tau; \tau, z) = v'(\tau, z) - A(\tau, z, \mathcal{D})v(\tau, z).$$

Thus, both functions $v'_\tau(\tau; \eta, z)$ and $w(\tau; \eta, z)$ are the solutions of the same problem. Consequently, $w(\tau; \eta, z) \equiv v'_\tau(\tau; \eta, z)$. The lemma is proved.

We now turn to the proof of the identity (4.2.22). By virtue of Lemma 4.2.2 we have

1) $\qquad \langle u(t,z), v(t,z) \rangle \overset{\text{def}}{=} \langle \phi(z), v(t;0,z) \rangle + \int_0^t \langle h(\tau,z), v(\tau,z) \rangle d\tau,$

2) $\qquad I(\tau) \equiv \langle u(\tau,z), v'(\tau,z) - A(\tau,z,\mathcal{D})v(\tau,z) \rangle \overset{\text{def}}{=}$

$$= \langle \phi(z), w(\tau;0,z) \rangle + \int_0^\tau \langle h(\eta,z), w(\tau;\eta,z) \rangle d\eta =$$

$$= \langle \phi(z), v'_\tau(\tau;\eta,z) \rangle + \int_0^\tau \langle h(\eta,z), v'_\tau(\tau;\eta,z) \rangle d\eta =$$

$$= \frac{\partial}{\partial \tau} \langle \phi(z), v(\tau;\eta,z) \rangle + \frac{\partial}{\partial \tau} \int_0^\tau \langle h(\eta,z), v(\tau;\eta,z) \rangle d\eta -$$

$$-\langle h(\tau,z), v(\tau;\tau,z) \rangle = \frac{\partial}{\partial \tau} \langle \phi(z), v(\tau;\eta,z) \rangle +$$

$$+ \frac{\partial}{\partial \tau} \int_0^\tau \langle h(\eta,z), v(\tau;\eta,z) \rangle d\eta - \langle h(\tau,z), v(\tau,z) \rangle,$$

since $v(\tau;\tau,z) = v(t,z)$.

Taking into account the equality $v(0;0,z) = v(0,z)$, from 1) and 2) we obtain

$$\langle u(t,z), v(t,z) \rangle - \int_0^t I(\tau) d\tau = \langle \phi(z), v(0;0,z) \rangle +$$

$$+ \int_0^t \langle h(\tau,z), v(\tau,z) \rangle d\tau = \langle \phi(z), v(0,z) \rangle + \int_0^t \langle h(\tau,z), v(\tau,z) \rangle d\tau,$$

which is the identity (4.2.22), as required.

The estimate

$$\|u\|'_{\frac{R-r}{\sigma};m+1,r,\sigma} \leq M \left(\|\phi\|'_{m+1,r} + \|h\|'_{\frac{R-r}{\sigma};m+1,r,\sigma} \right)$$

is trivial. The theorem is completely proved.

Example. Consider in the scale of functionals $D'_{m+1,r+\sigma|t|}$ the Cauchy problem

$$\frac{\partial u}{\partial t} + \frac{\partial}{\partial z}(z^2 u) - u = 0 \qquad (4.2.23)$$

$$u(0,z) = \phi(z). \qquad (4.2.24)$$

In accordance with (4.2.22) we have

$$\langle u(t,z), v(t,z) \rangle = \langle \phi(z), v(t;0,z) \rangle,$$

where the function $v(t; \tau, z)$ is the solution of the problem

$$\frac{\partial v}{\partial t} + z^2 \frac{\partial v}{\partial z} + v = 0, \quad \tau \in [t, 0],$$

$$v(t; \tau, z)\big|_{\tau=t} = v(t, z)$$

($v(t, z)$ is some arbitrary test function). We have (see the example in subsection 2)

$$v(t; \tau, z) = \exp(t - \tau) v \left(t, \frac{z}{1 - (t - \tau)z} \right);$$

hence the solution of the problem (4.2.23), (4.2.24) is the functional

$$\langle u(t,z), v(t,z) \rangle = \exp t \left\langle \phi(z), v \left(t, \frac{z}{1 - tz} \right) \right\rangle.$$

In conclusion, we note that the coefficient in equation (4.2.23) is, of course, an entire function, therefore, according to the general theory, the numbers $R > 0$ and $r > 0$ may be chosen arbitrarily, provided that $r < R$. Thus, if $\phi(z) \in D'_{m+1,r}$, then Theorem 4.2.2 guarantees the existence of the solution $u(t, z) \in \mathcal{O}\left(\frac{R-r}{\sigma}; D'_{m+1, r+\sigma|t|} \right)$, where $\sigma > 0$ may taken to be r^2.

4.3. Formulation of the basic results for arbitrary systems in normal form

1. In this subsection we reformulate the previous results for higher order systems. Namely, we consider the following Cauchy problem:

$$\frac{\partial^{s_i} u_i}{\partial t^{s_i}} + \sum_{k,j} A_{ij}^k(t, z, \mathcal{D}) \frac{\partial^k u_j}{\partial t^k} = h_i(t, z) \tag{4.3.1}$$

$$\frac{\partial^k u_i}{\partial t^k}(t_0, z) = \phi_{ik}(z), \quad k = 0, 1, \ldots, s_i - 1. \tag{4.3.2}$$

Here $i, j = 1, \ldots, N$; s_i are natural numbers. The differential operators

$$A_{ij}^k(t, z, \mathcal{D}) \equiv \sum_{\alpha} a_{ij}^{k\alpha}(t, z) \mathcal{D}^{\alpha}$$

are differential operators of arbitrary (finite) order with analytic coefficients in some domain $V \subset \mathbf{C}_{t,z}^{n+1}$.

For the problem (4.3.1), (4.3.2) we have results that are similar to those of §§4.1, 4.2.

Definition 4.3.1. The Cauchy problem (4.3.1), (4.3.2) is said to be well-posed in the spaces $D_{m,r}$ if for any point $(t_0, z_0) \in V$ there exists a number $\delta > 0$ such that for every $\phi(z) \in D_{m,r}$ and every $h(t, z) \in \mathcal{O}(\delta; D_{m,r})$ there exists a unique solution $u(t, z) \in \mathcal{O}(\delta; D_{m,r})$ of this problem and, in addition, the estimate

$$\|u\|_{\delta;m,r} \leq M\left(\|\phi\|_{m,r} + \|h\|_{\delta;m,r}\right)$$

holds, where $M > 0$ is a constant.

Theorem 4.3.1. *The Cauchy problem* (4.3.1), (4.3.2) *is well-posed in the spaces* $D_{m,r}$ *if and only if*

$$\operatorname{ord} A_{ij}(t, z, \mathcal{D}) \leq m_i - m_j \ (i, j = 1, \dots, N)$$

(as usual, if $A_{ij}(t, z, \mathcal{D}) \equiv 0$, *then* $\operatorname{ord} A_{ij}(t, z, \mathcal{D}) = -\infty$).

Let us now turn to the Cauchy problem in the spaces $D_{m,r-\sigma|t|}$. As was noted earlier (see Lemma 4.2.1) if $u(t, z) \in \mathcal{O}(r/\sigma; D_{m,r-\sigma|t|})$, then $\frac{\partial^k u}{\partial t^k}(t, z) \in \mathcal{O}(r/\sigma; D_{m+k,r-\sigma|t|})$, where $m + k = (m_1 + k, \dots, m_N + k)$. In particular, $\frac{\partial^k u_i}{\partial t^k}(t_0, z) \in D_{m_i+k,r}$, where $i = 1, \dots, N$; $k = 0, 1, \dots, s_i - 1$. We take this into consideration in posing the Cauchy problem of higher order with respect to t.

Definition 4.3.2. The Cauchy problem is said to be well-posed in the spaces $D_{m,r-\sigma|t|}$ if for any point $(t_0, z_0) \in V$ there exists a number $\sigma > 0$ such that for all initial functions $\phi_{ik}(z) \in D_{m_i+k,r}$ and for all right hand sides $h_i(t, z) \in \mathcal{O}(r/\sigma; D_{m_i+s_i,r-\sigma|t-t_0|})$ there exists a unique solution $u(t, z) \in \mathcal{O}(r/\sigma; D_{m,r-\sigma|t-t_0|})$; moreover,

$$\|u\|_{r/\sigma;m,r} \leq M\left(\sum_{i,k} \|\phi_{ik}\|_{m_i+k,r} + \sum_i \|h_i\|_{r/\sigma;m_i+s_i,r}\right),$$

where $M > 0$ is a constant.

Theorem 4.3.2. *The Cauchy problem* (4.3.1), (4.3.2) *is well-posed in the spaces* $D_{m,r-\sigma|t|}$ *if and only if*

$$\operatorname{ord} A_{ij}^k(t, z, \mathcal{D}) \leq m_i - m_j + s_i - k \ (i, j = 1, \dots, N; \ k = 0, 1, \dots, s_i - 1)$$

(*as before,* $A_{ij}^k(t, z, \mathcal{D}) \equiv 0 \Leftrightarrow \operatorname{ord} A_{ij}^k(t, z, \mathcal{D}) = -\infty$).

The proof of both theorems may be carried out (after the standard passage to the systems of integro-differential equations) by analogy with the case of first order systems.

One can proceed in another way. Namely, one can replace the system (4.3.1) by an equivalent first order system with respect to t. Moreover, in accordance with Lemma 4.1.1, after each differentiation with respect to t the degrees of the singularities of $u(t, z)$ increase by one precisely. In view of this, the necessary and sufficient conditions on the orders of the corresponding first order system provide the inequalities as given in in Theorem 4.3.1. and Theorem 4.3.2 for the original higher order system.

Chapter 5. Exponential theory of the Cauchy problem

5.1. The Cauchy problem in the scale of spaces of initial data

1. BANACH SPACES OF ENTIRE FUNCTIONS OF FINITE ORDER

Let q be a natural number and let $m = (m_1, \ldots, m_N)$ and $r = (r_1, \ldots, r_n)$ be vectors with non-negative components; moreover, let $m_1 \geq 0, \ldots, m_N \geq 0$ be integers. Further, let $u(z) = (u_1(z), \ldots, u_N(z))$ be an entire vector-valued function.

We set

$$\mathrm{Exp}_{m,r,q}(\mathbf{C}^n_z) = \Big\{ u(z) : |u_j(z)| \leq M_j (1 + |z|)^{m_j} \exp(r_1 |z_1|^q + \ldots + r_n |z_n|^q),$$
$$z \in \mathbf{C}^n, \; j = 1, \ldots, N \Big\},$$

where the $M_j > 0$ are constants (depending, in general, on $u(z)$). This space is a Banach space with norm

$$\|u\|_{m,r,q} \equiv \sum_{j=1}^{N} \|u_j\|_{m_j,r,q} \overset{\text{def}}{=} \sum_{j=1}^{N} \sup_{z \in \mathbf{C}^n} |u_j(z)| (1 + |z|)^{-m_j} \exp(-rz_+^q),$$

where we set (for brevity) $rz_+^q \equiv r_1 |z_1|^q + \ldots + r_n |z_n|^q$.

It is clear that each component $u_j(z)$ of the vector-valued function $u(z) \in \mathrm{Exp}_{m,r,q}(\mathbf{C}^n_z)$ is an entire function of order $q \geq 1$ and of vector type $r = (r_1, \ldots, r_n)$ for the growth in $z \in \mathbf{C}^n$ and has finite norm $\|u_j\|_{m_j,r,q}$. We shall denote this by writing $u_j(z) \in \mathrm{Exp}_{m_j,r,q}(\mathbf{C}^n_z)$.

In what follows, the spaces $\mathrm{Exp}_{m,r,q}(\mathbf{C}^n_z)$ will be taken as the spaces of initial data for the Cauchy problem. Moreover, in considering the Cauchy problem we shall fix the vector $m = (m_1, \ldots, m_N)$ and the number $q \geq 1$. On the other hand, we shall take $r = (r_1, \ldots, r_n)$ as a variable vector parameter (parameter of the scale $\mathrm{Exp}_{m,r,q}(\mathbf{C}^n_z)$).

Proposition 5.1.1. *Let $r \leq \bar{r}$, that is, $r_1 \leq \bar{r}_1, \ldots, r_n \leq \bar{r}_n$. Then*

$$\mathrm{Exp}_{m,r,q}(\mathbf{C}_z^n) \subset \mathrm{Exp}_{m,\bar{r},q}(\mathbf{C}_z^n). \tag{5.1.1}$$

Moreover if $r < \bar{r}$, then the inclusion (5.1.1) is compact.

Proof. Indeed, the inclusion (5.1.1) is evident, since for $r \leq \bar{r}$

$$\|u\|_{m,\bar{r},q} \equiv \sum_{j=1}^{N} \sup_{z \in \mathbf{C}^n} |u_j(z)|(1+|z|)^{-m_j} \exp(-\bar{r}z_+^q) \leq$$

$$\leq \sum_{j=1}^{N} \sup_{z \in \mathbf{C}^n} |u_j(z)|(1+|z|)^{-m_j} \exp(-rz_+^q) \equiv \|u\|_{m,r,q}.$$

We now prove the compactness of the inclusion (5.1.1) when $r < \bar{r}$. In fact, the inequality $\|u\|_{m,r,q} \leq 1$ implies, in particular, that for each compact set $K \subset \mathbf{C}_z^n$ there is the inequality

$$\max_{z \in K} |u(z)| \leq M,$$

where $M = M(K)$ is a constant. Hence, using the well-known diagonal process, we can find a sequence $u_\nu(z)$, $\nu = 1, 2, \ldots$, and an entire function $u(z)$ such that $u_\nu(z) \to u(z)$ locally uniformly in $\mathbf{C}_z^n$, that is, $u_\nu(z) \to u(z)$ uniformly on any compact set $K \subset \mathbf{C}_z^n$. Further, if $\|u_\nu\|_{m,r,q} \leq 1$, then for each component $u_j(z)$

$$\sup_{z \in \mathbf{C}^n} |u_{\nu j}(z)|(1+|z|)^{-m_j} \exp(-rz_+^q) \leq 1$$

as well. Hence, letting ν tend to $+\infty$, we obtain the inequalities

$$\sup_{z \in \mathbf{C}^n} |u_j(z)|(1+|z|)^{-m_j} \exp(-rz_+^q) \leq 1.$$

This means that $u(z) \in \mathrm{Exp}_{m,r,q}(\mathbf{C}_z^n)$ and $\|u_j\|_{m_j,r,q} \leq 1$ $(j = 1, \ldots, N)$. It is clear that $u(z) \in \mathrm{Exp}_{m,\bar{r},q}(\mathbf{C}_z^n)$.

We now show that $u_\nu(z) \to u(z)$ in the norm of the space $\mathrm{Exp}_{m,\bar{r},q}(\mathbf{C}_z^n)$. Indeed, for each compact set $K \subset \mathbf{C}_z^n$ we have

$$\|u - u_\nu\|_{m,\bar{r},q} \equiv \sum_{j=1}^{N} \sup_{z \in \mathbf{C}^n} |u_j(z) - u_{\nu j}(z)|(1+|z|)^{-m_j} \times$$

$$\times \exp(-\bar{r}z_+^q) \leq \sum_{j=1}^{N} \sup_{z \in K} |u_j(z) - u_{\nu j}(z)|(1+|z|)^{-m_j} \times$$

$$\times \exp(-\bar{r}z_+^q) + \sum_{j=1}^{N} \sup_{z \in \mathbf{C}^n \setminus K} |u_j(z) - u_{\nu j}(z)|(1 + |z|)^{-m_j} \times$$

$$\times \exp(-rz_+^q)\exp(r-\bar{r})z_+^q \leq \sum_{j=1}^{N} \sup_{z \in K} |u_j(z) - u_{\nu j}(z)| +$$

$$+ \sum_{j=1}^{N} \|u_j(z) - u_{\nu j}(z)\|_{m_j, r, q} \cdot \max_{z \in \mathbf{C}^n \setminus K} \exp(r-\bar{r})z_+^q \leq$$

$$\leq \sum_{j=1}^{N} \sup_{z \in K} |u_j(z) - u_{\nu j}(z)| + 2N \max_{z \in \mathbf{C}^n \setminus K} \exp(r-\bar{r})z_+^q,$$

since, as we have just seen, $\|u_j(z)\|_{m_j, r, q} \leq 1$.

We now observe that the second term in the last inequality is small if the compact set $K \subset \mathbf{C}^n$ is sufficiently large, while the first term is small if $\nu \gg 1$, since $u_{\nu j}(z) \to u_j(z)$ uniformly on each fixed compact set. This means that $u_\nu(z) \to u(z)$ in the space $\mathrm{Exp}_{m,\bar{r},q}(\mathbf{C}_z^n)$. The assertion is proved.

We now introduce the space of the solutions of the Cauchy problem. Namely, let $\delta > 0$ be a positive number. We set

$$\mathcal{O}(\delta; \mathrm{Exp}_{m,r,q}(\mathbf{C}_z^n)) = \{u(t,z) : u(t,\cdot) \in \mathrm{Exp}_{m,r,q}(\mathbf{C}_z^n), \ |t - t_0| < \delta\}$$

with norm

$$\|u\|_{\delta;m,r,q} \equiv \sup_{|t-t_0|<\delta} \|u(t,\cdot)\|_{m,r,q} < \infty$$

(for brevity we do not indicate the dependence of this norm on $t_0 \in G \subset \mathbf{C}^1$).

It follows from the above definition that the functions $u(t,z) \in \mathcal{O}(\delta; \mathrm{Exp}_{m,r,q}(\mathbf{C}_z^n))$ are analytic in t, $|t - t_0| < \delta$, and $z \in \mathbf{C}^n$. Moreover, the norms

$$\|u_j(t,z)_{\delta;m_j,r,q} \equiv \sup_{|t-t_0|<\delta} \sup_{z \in \mathbf{C}^n} |u_j(t,z)|(1 + |z|)^{-m_j} \exp(-rz_+^q)$$

are finite.

Remark. Instead of the spaces $\mathcal{O}(\delta; \mathrm{Exp}_{m,r,q}(\mathbf{C}_z^n))$ one may consider spaces of vector-valued functions $u(t,z) = (u_1(t,z), \ldots, u_N(t,z))$ such that each $u_j(t,z)$ $(j = 1, \ldots, N)$ is analytic in t, $|t - t_0| < \delta$, and continuous for $|t - t_0| \leq \delta$. Moreover, the quantity

$$\sum_{j=1}^{N} \max_{|t-t_0|\leq\delta} \sup_{z \in \mathbf{C}^n} |u_j(t,z)|(1 + |z|)^{-m_j} \exp(-rz_+^q)$$

may be chosen as the norm.

All the results below will be the same for such spaces.

2. CRITERION FOR THE WELL-POSEDNESS OF THE CAUCHY PROBLEM IN THE SCALE $\mathrm{Exp}_{m,r,q}(\mathbf{C}_z^n)$

Let us consider the Cauchy problem

$$u' - A(t, z, \mathcal{D})u = h(t, z), \tag{5.1.2}$$

$$u(t_0, z) = \phi(z), \ z \in \mathbf{C}^n, \tag{5.1.3}$$

where, as before, $u(t, z) = (u_1(t, z), \ldots, u_N(t, z))$, $\phi(z) = (\phi_1(z), \ldots, \phi_N(z))$ are vector-valued functions, $A(t, z, \mathcal{D}) = (A_{ij}(t, z, \mathcal{D}))_{N \times N}$ is a matrix of differential operators of finite order

$$A_{ij}(t, z, \mathcal{D}) \equiv \sum_{|\alpha| \le n_{ij}} a_{ij}^\alpha(t, z) \mathcal{D}^\alpha \tag{5.1.4}$$

with coefficients $a_{ij}^\alpha(t, z)$ that are analytic functions in $t \in G \subset \mathbf{C}^1$ and in $z \in \mathbf{C}^n$. Here G is a domain in the complex plane.

Definition 5.1.1. The Cauchy problem (5.1.2), (5.1.3) is said to be well-posed in the scale $\mathrm{Exp}_{m,r,q}(\mathbf{C}_z^n)$ (the scale parameter is the vector r) if for each point $t_0 \in G$ and each vector r there is a number $\delta > 0$ such that for any initial function $\phi(z) \in \mathrm{Exp}_{m,r,q}(\mathbf{C}_z^n)$ and any right hand side $h(t, z) \in \mathcal{O}(\delta; \mathrm{Exp}_{m,r,q}(\mathbf{C}_z^n))$ there exists a unique solution $u(t, z) \in \mathcal{O}(\delta; \mathrm{Exp}_{m,r,q}(\mathbf{C}_z^n))$. In addition, the estimate

$$\|u\|_{\delta;m,r,q} \le M(r)(\|h\|_{\delta;m,r,q} + \|\phi\|_{m,r,q})$$

holds, where $M(r) > 0$ is a constant which is bounded for bounded values of r.

The main result of the present subsection is

Theorem 5.1.1. *The Cauchy problem*(5.1.2), (5.1.3) *is well-posed in the scale* $\mathrm{Exp}_{m,r,q}(\mathbf{C}_z^n)$ *if and only if the following conditions hold:*
   1)   *the coefficients* $a_{ij}^\alpha(t, z)$ *are polynomials in* z,
   2) $\deg a_{ij}^\alpha(t, z) \le m_i - m_j - |\alpha|(q - 1)$ $(i, j = 1, \ldots, N)$
(*if* $a_{ij}^\alpha(t, z) \equiv 0$, *then* $\deg a_{ij}^\alpha(t, z)$ *is defined to be* $-\infty$).

Proof. Sufficiency. Suppose that conditions 1) and 2) hold. Then we must show that the Cauchy problem is well-posed in the scale $\mathrm{Exp}_{m,r,q}(\mathbf{C}_z^n)$.

Clearly we can set $t_0 = 0$ (without loss of generality). Then the problem (5.1.2), (5.1.3) is equivalent to the integro-differential equation

$$u(t,z) = \int_0^t A(\tau, z, \mathcal{D})u(\tau, z)d\tau + \phi(z) + \int_0^t h(\tau, z)d\tau,$$

where, as usual, $\tau \in [0, t]$.

As before, we note that the inclusions $\phi(z) \in \text{Exp}_{m,r,q}(\mathbf{C}_z^n)$ and $h(t, z) \in \mathcal{O}(\delta; \text{Exp}_{m,r,q}(\mathbf{C}_z^n))$ imply that

$$\phi(z) \in \mathcal{O}(\delta; \text{Exp}_{m,r,q}(\mathbf{C}_z^n)) \text{ and } \int_0^t h(\tau, z)d\tau \in \mathcal{O}(\delta; \text{Exp}_{m,r,q}(\mathbf{C}_z^n)).$$

We now show that the operator

$$Bu(t,z) \equiv \int_0^t A(\tau, z, \mathcal{D})u(\tau, z)d\tau, \ |t| < \delta,$$

acts in the space $\mathcal{O}(\delta; \text{Exp}_{m,r,q}(\mathbf{C}_z^n))$ and is a contraction if $\delta > 0$ is sufficiently small.

For this we need

Lemma 5.1.1. *Let $v(z) : \mathbf{C}^n \to \mathbf{C}^1$ be an entire function such that*

$$|v(z)| \le M(1 + |z|)^m \exp rz_+^q, \ z \in \mathbf{C}^n, \tag{5.1.5}$$

where $M > 0$, $m \ge 0$, $q \ge 1$ are constants, $r = (r_1, \ldots, r_n)$ is a vector with non-negative components. Then for every multi-index $\alpha = (\alpha_1, \ldots, \alpha_n)$ and any $z \in \mathbf{C}^n$ the inequality

$$|\mathcal{D}^\alpha v(z)| \le M(r, \alpha)(1 + |z|)^{m+|\alpha|(q-1)} \exp rz_+^q$$

holds. Here $M(r, \alpha)$ is a constant which is bounded if r and α are bounded.

Proof. In fact, by Cauchy's formula, we have for any $z \in \mathbf{C}^n$

$$\mathcal{D}^\alpha v(z) = \frac{\alpha!}{(2\pi i)^n} \int_\Gamma \frac{v(\zeta)d\zeta}{(\zeta - z)^{\alpha+1}},$$

where Γ is the skeleton of the polycylinder of radius $a = (a_1, \ldots, a_n)$ centred at z and $\alpha! = \alpha_1! \ldots \alpha_n!$. Hence we obtain the inequality

$$|\mathcal{D}^\alpha v(z)| \le M\alpha!(1 + |z| + |a|)^m a_1^{-\alpha_1} \ldots a_n^{-\alpha_n} \times$$

$$\times \exp[r_1(|z_1| + a_1)^q + \ldots + r_n(|z_n| + a_n)^q].$$

Choosing $a_j > 0$ such that $(|z_j| + a_j)^q = |z_j|^q + 1$ and making some trivial calculations we immediately obtain the inequality of the lemma. The lemma is proved.

In view of this lemma, for any function $u(t, z) \in \mathcal{O}(\delta; \mathrm{Exp}_{m,r,q}(\mathbf{C}_z^n))$ we have

$$\left| \int_0^t a_{ij}^\alpha(\tau, z) \mathcal{D}^\alpha u_j(\tau, z) d\tau \right| \leq M(r, \alpha)(1 + |z|)^{m_j + |\alpha|(q-1)} \times$$

$$\times \exp r z_+^q \int_0^{|t|} |a_{ij}^\alpha(\tau, z)| \, \|u_j(\tau, z)\|_{m_j, r, q} d\tau \leq$$

$$\leq M_1(r, \alpha)\delta \sup_{|\tau| \leq |t|} \|u_j(\tau, z)\|_{m_j, r, q} \cdot (1 + |z|)^{m_j + |\alpha|(q-1) + n_{ij}^\alpha} \exp r z_+^q,$$

where the n_{ij}^α are the degrees of the polynomials $a_{ij}^\alpha(t, z)$ and $M_1(r, \alpha)$ depends on $M(r, \alpha)$ and the coefficients $a_{ij}^\alpha(t, z)$. Hence taking into account the fact that $n_{ij}^\alpha \leq m_i - m_j - |\alpha|(q - 1)$, we obtain the inequality

$$\left| \int_0^t a_{ij}^\alpha(\tau, z) \mathcal{D}^\alpha u_j(\tau, z) d\tau \right| \leq \delta M_1(r, \alpha) \|u_j(t, z)\|_{\delta; m_j, r, q}(1 + |z|)^{m_i} \exp r z_+^q.$$

Consequently, for each component

$$(Bu)_i(t, z) \equiv \sum_{j=1}^N \int_0^t A_{ij}(\tau, z, D) u_j(\tau, z) d\tau =$$

$$= \sum_{j=1}^N \sum_{|\alpha| \leq n_{ij}} \int_0^t a_{ij}^\alpha(\tau, z) \mathcal{D}^\alpha u_j(\tau, z) d\tau$$

we have the inclusions $(Bu)_i(t, z) \in \mathcal{O}(\delta; \mathrm{Exp}_{m_i, r, q}(\mathbf{C}_z^n))$.

Moreover, it is clear that

$$\|Bu\|_{\delta; m, r, q} \leq \delta M(r) \|u\|_{\delta; m, r, q},$$

where $M(r) > 0$ is a constant which is bounded for bounded r. Thus, the operator Bu acts in the space $\mathcal{O}(\delta; \mathrm{Exp}_{m,r,q}(\mathbf{C}_z^n))$ and is a contraction when $\delta > 0$ is sufficiently small. In other words, for $\delta \ll 1$ the operator

$$Bu : \mathcal{O}(\delta; \mathrm{Exp}_{m,r,q}(\mathbf{C}_z^n)) \to \mathcal{O}(\delta; \mathrm{Exp}_{m,r,q}(\mathbf{C}_z^n))$$

is a contraction. Consequently, the integro-differential equation (5.1.4) and the original Cauchy problem have a unique solution $u(t, z) \in \mathcal{O}(\delta; \mathrm{Exp}_{m,r,q}(\mathbf{C}_z^n))$. The estimate in Theorem 5.1.1 is trivial. The sufficiency of conditions 1) and 2) of our theorem is proved.

Necessity. Suppose that the Cauchy problem (5.1.2), (5.1.3) is well posed in the scale $\mathrm{Exp}_{m,r,q}(\mathbf{C}_z^n)$ (see Definition 5.1.1). Let us show that the entire functions $a_{ij}^\alpha(t, z)$ are polynomials in z with

$$\deg a_{ij}^\alpha(t, z) \le m_i - m_j - |\alpha|(q - 1).$$

The proof will be carried out in two steps. First, we prove merely that coefficients $a_{ij}^\alpha(t, z)$ are polynomials in z. Then we obtain the values of the degrees of these polynomials precisely.

Step 1. We fix a number j $(1 \le j \le N)$ and choose the initial functions $\phi_j(z)$ as

$$u_j(0, z) \equiv \phi_j(z) = \exp(\zeta_1 z_1^q + \ldots + \zeta_n z_n^q),$$
$$u_i(0, z) \equiv \phi_i(z) = 0, \quad i \ne j \ (i = 1, \ldots, N).$$

Here $\zeta = (\zeta_1, \ldots, \zeta_n)$ is a complex vector parameter, the values of which will be chosen below. Clearly, $|\phi_j(z)| \le \exp r z_+^q$ and moreover,

$$|\phi_j(z)| \le (1 + |z|)^{m_j} \exp r z_+^q,$$

where $r = (|\zeta_1|, \ldots, |\zeta_n|)$.

Further, let $u(t, z) = (u_1(t, z), \ldots, u_N(t, z))$ be the corresponding solution of the Cauchy problem (5.1.2), (5.1.3) which is an element of the space $\mathcal{O}(\delta; \mathrm{Exp}_{m,r,q}(\mathbf{C}_z^n))$, where $\delta > 0$ (we omit the dependence of $u(t, z)$ on the index j). Then in accordance with Cauchy's formula we obtain the standard estimates

$$|u_i'(0, z)| \le M(1 + |z|)^{m_i} \exp r z_+^q, \qquad (5.1.6)$$

where $M = M(|\zeta|)$ is bounded for bounded $|\zeta|$.

Further, it is clear that

$$\mathcal{D}^\alpha \phi_j(z) = [q^{|\alpha|} \zeta^\alpha z^{(q-1)\alpha} + \ldots] \exp(\zeta_1 z_1^q + \ldots + \zeta_n z_n^q),$$

where the dots denote the terms of degree less than $|\alpha|$ in ζ. Hence, taking into account (5.1.6), we obtain for the system (5.1.2) the following inequalities for $t = 0$:

$$\left| \sum_{|\alpha| \leq n_{ij}} a_{ij}^{\alpha}(0, z)[q^{|\alpha|} \zeta^{\alpha} z^{(q-1)\alpha} + \ldots] \right| \leq$$

$$\leq M(1 + |z|)^{m_i} |\exp(r z_+^q - \zeta_1 z_1^q - \ldots - \zeta_n z_n^q)| \quad (i = 1, \ldots, N). \tag{5.1.7}$$

Next we prove that the estimates (5.1.7) imply that the entire functions $a_{ij}^{\alpha}(0, z)$ are polynomials in each variable separately. From this, as is well known, it follows that the $a_{ij}^{\alpha}(0, z)$ are polynomials in $z_1, \ldots, z_n$ jointly.

In fact, we fix the variables $z_2, \ldots, z_n$, in (5.1.7), that is, we set $z_2 = z_{20}, \ldots, z_n = z_{n0}$, where $z_{20}, \ldots, z_{n0}$ are arbitrary complex numbers. Then for the variable z_1 we obtain the inequality

$$\left| \sum_{|\alpha| \leq n_{ij}} a_{ij}^{\alpha}(0, z_1; z')[q^{|\alpha|} \zeta^{\alpha} z_1^{(q-1)\alpha_1} (z_0')^{(q-1)\alpha'} + \ldots] \right| \leq$$

$$\leq M(z_0') |e^{r_1 |z_1|^q - \zeta_1 z_1^q}| (1 + |z_1|)^{m_i},$$

where $M(z_0')$ is a constant, $z_0' = (z_{20}, \ldots, z_{n0})$, $\alpha' = (\alpha_2, \ldots, \alpha_n)$, $r_1 = |\zeta_1|$. We now set $\zeta_j = \tau_j$, where the τ_j ($0 \leq \tau_j \leq 1$, $j = 2, \ldots, n$) are arbitrary, and $\zeta_1 = \tau_1 / z_{10}^q$, where τ_1, $0 \leq \tau_1 \leq 1$ is also arbitrary; $z_{10} \in \mathbf{C}^1$ is such that $|z_{10}| \geq 1$. Then, clearly, $r_1 = \tau_1 / |z_{10}|^q$ and, consequently, putting $z_1 = z_{10}$ and multiplying the last inequality by $z_{10}^{n_{ij}}$, we obtain the inequality

$$\left| \sum_{|\alpha| \leq n_{ij}} a_{ij}^{\alpha}(0, z_{10}; z_0')[q^{|\alpha|} \tau^{\alpha} z_{10}^{n_{ij} - \alpha_1} (z_0')^{(q-1)\alpha'} + \ldots] \right| \leq$$

$$\leq M(z_0')(1 + |z_{10}|)^{m_i + n_{ij}}.$$

(We emphasize that, in view of the fact that $|\zeta_j| \leq 1$, the constant $M(z_0')$ may be considered to be a constant that does not depend on ζ_j, that is, τ_j, $j = 1, \ldots, n$.)

Since z_{10}, $|z_{10}| > 1$, is arbitrary, this means that the entire function (in z_1)

$$P(z_1; z_0', \tau) \equiv \sum_{|\alpha| \leq n_{ij}} a_{ij}^{\alpha}(0, z_1; z_0')[q^{|\alpha|} \tau^{\alpha} z_1^{n_{ij} - \alpha_1} (z_0')^{(q-1)\alpha'} + \ldots]$$

is a polynomial for all τ_j, $0 \leq \tau_j \leq 1$ ($j = 1, \ldots, n$). Moreover, the degree of this polynomial does not depend on z_0'.

Since for $|\alpha| = n_{ij}$ the derivatives

$$\mathcal{D}_\tau^\alpha P(z_1; z_0', \tau) \equiv a_{ij}^\alpha(0, z_1; z_0') q^{n_{ij}} \alpha! z_1^{n_{ij} - \alpha_1} (z_0')^{(q-1)\alpha'},$$

these functions are also polynomials in z_1 for all fixed $(z_{20}, \ldots, z_{n0}) \equiv z_0'$. Hence it immediately follows that functions $a_{ij}^\alpha(0, z_1; z_0')$ are also polynomials in z_1 for all fixed $(z_{20}, \ldots, z_{n0}) \equiv z_0'$.

The proof that the functions $a_{ij}^\alpha(0, z)$ are polynomials in $z_2, \ldots, z_n$ separately can be obtained in the same way. As was noted before, these facts imply that coefficients $a_{ij}^\alpha(0, z)$ are polynomials in $z_1, \ldots, z_n$ jointly.

It remains to remark that, after proving that the $a_{ij}^\alpha(0, z)$ are polynomials for $|\alpha| = n_{ij}$, a similar argument then shows that the $a_{ij}^\alpha(0, z)$ are polynomials for $|\alpha| = n_{ij} - 1$, $|\alpha| = n_{ij} - 2$ etc.

Finally, it can be shown that all the coefficients $a_{ij}^\alpha(0, z)$ are polynomials in z.

Step 2. We now give the exact estimate for the degrees of the polynomials $a_{ij}^\alpha(0, z)$. Namely, we prove that the inequalities

$$\deg a_{ij}^\alpha(0, z) \le m_i - m_j - |\alpha|(q - 1) \tag{5.1.8}$$

hold.

For this we must again turn to the basic system (5.1.2). We set

$$u_j(0, z) = (z_1 + \ldots + z_n)^{m_j} \exp(\zeta_1 z_1^q + \ldots + \zeta_n z_n^q),$$

$$u_i(0, z) = 0 \ (i \ne j \text{ a fixed number}).$$

Then (by analogy with (5.1.7)) we have the inequality

$$\left| \sum_{|\alpha| \le n_{ij}} a_{ij}^\alpha(0, z) [q^{|\alpha|} \zeta^\alpha (z_1 + \ldots + z_n)^{m_j} z^{(q-1)\alpha} + \ldots] \right| \le$$

$$\le M(r)(1 + |z|)^{m_i} |\exp(rz_+^q - \zeta_1 z_1^q - \ldots - \zeta_n z_n^q)|,$$

where $r = (|\zeta_1|, \ldots, |\zeta_n|)$ and the dots denote the terms of degree less than $|\alpha|$ in z and less than $m_j + (q - 1)|\alpha|$ in ζ.

Putting $\zeta_k = \tau_k$, $0 \le \tau_k \le 1$, and $z_k = x_k$, $x_k \ge 0$ $(k = 1, \ldots, N)$ we have

$$\left| \sum_{|\alpha| \le n_{ij}} a_{ij}^\alpha(0, x) [q^{|\alpha|} \tau^\alpha (x_1 + \ldots + x_n)^{m_j} x^{(q-1)\alpha} + \ldots] \right| \le M(r)(1 + |x|)^{m_i}.$$

These inequalities show that the polynomial (as was proved in Step 1)

$$P(\tau; x) \equiv \sum_{|\alpha| \le n_{ij}} a_{ij}^{\alpha}(0, x) \left[q^{|\alpha|} \tau^{\alpha}(x_1 + \ldots + x_n)^{m_j} x^{(q-1)\alpha} \right]$$

is of degree at most m_i in x. Then, on differentiating $P(\tau; x)$ with respect to τ, we obtain for $|\alpha| = n_{ij}$ the inequality

$$\deg \left[a_{ij}^{\alpha}(0, x)(x_1 + \ldots + x_n)^{m_j} x^{(q-1)\alpha} \right] \le m_i$$

and consequently

$$\deg a_{ij}^{\alpha}(0, x) \le m_i - m_j - |\alpha|(q - 1),$$

which are the inequalities (5.1.8) for $|\alpha| = n_{ij}$.

After this, a similar argument gives the desired inequalities for $|\alpha| = n_{ij} - 1,\ n_{ij} - 2, \ldots,\ 1,\ 0$.

In conclusion, we note that the proof of the necessary conditions for $t = t_0 \ne 0$ clearly reduces to the case $t = 0$ by means of the substitution $t \leftrightarrow t - t_0$. Hence, the necessity is proved. Theorem 5.1.1 is completely proved.

3. THE STRUCTURE OF SYSTEMS SATISFYING THE CONDITIONS

$$\deg a_{ij}^{\alpha} \le m_i - m_j - |\alpha|(q - 1)$$

1. *The case of isotropic growth* $(m_1 = \ldots = m_N)$. In this case the inequalities indicated in the title have the form

$$\deg a_{ij}^{\alpha}(t, z) \le -|\alpha|(q - 1)$$

and are clearly non-trivial either for $\alpha = 0$ (and arbitrary $q \ge 1$) or for $q = 1$ (and arbitrary α).

In the first case system (5.1.2) is any system of ordinary differential equations, the right hand side and initial data of which depend on $z \in \mathbb{C}^n$, as a parameter. Theorem 5.1.1 asserts that for all $h(t, z)$ and $\phi(z)$, of z-growth of finite order $q \ge 1$, the solution $u(t, z)$ has the same growth with respect to z, that is, growth of order q exactly.

In the second case, system (5.1.2) is

$$u_i' - \sum_{j=1}^{N} A_{ij}(t, \mathcal{D}) u_j = h_i(t, z), \tag{5.1.9}$$

where the $A_{ij}(t, \mathcal{D})$ are arbitrary differential operators of finite order with coefficients not depending on z. Then Theorem 5.1.1 asserts that in the spaces of exponential functions the Cauchy problem is well-posed for an arbitrary system (5.1.9).

   2. *A single higher order equation.* We consider the Cauchy problem

$$u^{(N)} - \sum_{j=0}^{N-1} A_j(t, z, \mathcal{D})u^{(j)} = h(t, z), \qquad (5.1.10)$$

$$u^{(j)}(0, z) = \phi_j(z), \quad j = 0, 1, \ldots, N - 1, \qquad (5.1.11)$$

where

$$A_j(t, z, \mathcal{D}) \equiv \sum_\alpha a_j^\alpha(t, z)\mathcal{D}^\alpha, \quad a_j^\alpha(t, z) \in \mathcal{O}(V).$$

Let $u(t, z) \in \mathcal{O}(\delta; \mathrm{Exp}_{m_1,r,q}(\mathbf{C}_z^n))$ be a solution of the problem (5.1.10), (5.1.11) where $m_1 \geq 0$ is some integer. Then, as follows from Cauchy's formula,

$$u'(t, z) \in \mathcal{O}(\delta_1; \mathrm{Exp}_{m_1,r,q}(\mathbf{C}_z^n)),$$

$$\cdot \quad \cdot \quad \cdot \quad \cdot \quad \cdot \quad \cdot \quad \cdot \quad \cdot \quad \cdot \quad \cdot \quad \cdot \quad \cdot \quad \cdot$$

$$\cdot \quad \cdot \quad \cdot \quad \cdot \quad \cdot \quad \cdot \quad \cdot \quad \cdot \quad \cdot \quad \cdot \quad \cdot \quad \cdot$$

$$u^{(N-1)}(t, z) \in \mathcal{O}(\delta_1; \mathrm{Exp}_{m_1,r,q}(\mathbf{C}_z^n)),$$

where $\delta_1 < \delta$ is arbitrary. This means that the standard substitution

$$u_1 = u, \ u_2 = u', \ldots, \ u_N = u^{(N-1)}$$

leads to the question of the well-posedness of the Cauchy problem for the system

$$u' - A(t, z, \mathcal{D})u = H(t, z),$$

$$u_1(0, z) = \phi_0(z), \ldots, \ u_N(0, z) = \phi_{N-1}(z),$$

where

$$A = \begin{pmatrix} 0 & 1 & 0 & \cdots & 0 \\ 0 & 0 & 1 & \cdots & 0 \\ \vdots & \vdots & \vdots & & \vdots \\ 0 & 0 & 0 & \cdots & 1 \\ A_0 & A_1 & A_2 & \cdots & A_{N-1} \end{pmatrix}, \quad H = \begin{pmatrix} 0 \\ 0 \\ \vdots \\ 0 \\ h \end{pmatrix}$$

in the scale $\mathrm{Exp}_{m,r,q}(\mathbf{C}_z^n)$, $m = (m_1, \ldots, m_1)$, that is, the isotropic scale.

Hence, in accordance with Theorem 5.1.1 for $q = 1$ (that is, for the class of functions of exponential type), the Cauchy problem (5.1.10), (5.1.11) is well-posed for equation (5.1.10), where $A_j(t, z, \mathcal{D}) \equiv A_j(t, \mathcal{D})$ are arbitrary differential operators not depending on z. If $q > 1$ then in the scale $\operatorname{Exp}_{m_1, r, q}(\mathbf{C}_z^n)$ the Cauchy problem (5.1.10), (5.1.11) is well-posed for the ordinary differential equations

$$u^{(N)} - \sum_{j=0}^{N-1} A_j(t) u^{(j)} = h(t, z),$$

$$u(0, z) = \phi_1(z), \ldots, \quad u^{(N-1)}(0, z) = \phi_{N-1}(z),$$

where $z \in \mathbf{C}^n$ is a parameter.

3. Let $m_1 \leq m_2 \leq \ldots \leq m_N$. Then for any α the following matrix of coefficients $a_{ij}^\alpha(t, z)$ is a triangular one:

$$\begin{pmatrix} a_{11}^\alpha(t) & & & \\ & & & 0 \\ & a_{22}^\alpha(t) & & \\ a_{ij}^\alpha(t) & & \ddots & \\ & & & a_{NN}^\alpha(t) \end{pmatrix}$$

where for $i > j$ the $a_{ij}^\alpha(t, z)$ are polynomials of degree less than $m_i - m_j - |\alpha|(q - 1)$. (We note that, if $q = 1$, then the principal diagonal is non-zero for any α, while if $q > 1$, then the principal diagonal is non-zero only for $\alpha = 0$.)

The corresponding systems are

$$u_1' - \sum_\alpha a_{11}^\alpha(t) \mathcal{D}^\alpha u_1 \qquad\qquad\qquad\qquad = h_1(t, z)$$

$$u_2' - \sum_\alpha a_{21}^\alpha(t, z) \mathcal{D}^\alpha u_1 - \sum_\alpha a_{22}^\alpha(t) \mathcal{D}^\alpha u_2 \qquad\qquad = h_2(t, z)$$

$$\cdots \cdots \cdots \cdots \cdots \cdots \cdots \cdots \cdots \cdots$$

$$\cdots \cdots \cdots \cdots \cdots \cdots \cdots \cdots \cdots \cdots$$

$$u_N' - \sum_\alpha a_{N1}^\alpha(t, z) \mathcal{D}^\alpha u_1 - \quad \cdots \quad \cdots \quad - \sum_\alpha a_{NN}^\alpha(t) \mathcal{D}^\alpha u_N = h_N(t, z).$$

Thus the Cauchy problem for these systems may be reduced to a sequence of Cauchy problems for the components $u_1(t, z), \ldots, u_N(t, z)$. Moreover, in the case $q = 1$ the corresponding equations are arbitrary partial differential equations with coefficients not depending on z. If $q > 1$ then the corresponding equations are systems of ordinary differential equations.

4. *The case* $m_1 < \ldots < m_{k+1} = m_{k+2} = \ldots = m_{k+l} < m_{k+l+1} < \ldots < m_N$. In this case the conditions

$$\deg a_{ij}^\alpha(t, z) \le m_i - m_j - |\alpha|(q - 1)$$

give (for each α) the following coefficient matrix:

$$
\begin{bmatrix}
a_{11}^\alpha(t) & & & & & & \\
& \ddots & & & 0 & & \\
& & a_{kk}^\alpha(t) & & & & 0 \\
& & & a_{k+1,k+1}^\alpha(t) & \cdots & a_{k+1,k+l}^\alpha(t) & \\
& & & & \ddots & \vdots & \\
a_{ij}^\alpha(t, z) & & & & & a_{k+l,k+l}^\alpha(t) & \\
(i > j) & & & & & & a_{k+l+1,k+l+1}^\alpha(t) \\
& & & & & & 0 \\
& & & & & & \ddots \\
& & & & & & a_{NN}^\alpha(t)
\end{bmatrix}
$$

The corresponding systems are

$$u_i' - \sum_\alpha \sum_{1 \le j < i} a_{ij}^\alpha(t, z) \mathcal{D}^\alpha u_j - \sum_\alpha a_{ii}^\alpha(t) \mathcal{D}^\alpha u_i = h_i(t, z) \quad (i = 1, \ldots, k)$$

$$u_i' - \sum_\alpha \sum_{j=1}^{k} a_{ij}^\alpha(t, z) \mathcal{D}^\alpha u_j - \sum_\alpha \sum_{j=k+1}^{k+l} a_{ij}^\alpha(t) \mathcal{D}^\alpha u_j = h_i(t, z) \; (i = k+1, \ldots, k+l)$$

$$u_i' - \sum_\alpha \sum_{1 \le j < i} a_{ij}^\alpha(t, z) \mathcal{D}^\alpha u_j - \sum_\alpha a_{ii}^\alpha(t) \mathcal{D}^\alpha u_i = h_i(t, z) \quad (i = k + l + 1, \ldots, N).$$

It is clear that the solution of these systems reduces to finding the components:

a) $u_1(t, z), \ldots, u_k(t, z)$; in this case it is necessary to solve one corresponding ordinary or partial differential equation,

b) $u_{k+1}(t, z), \ldots, u_{k+l}(t, z)$; in this case it is necessary to solve the corresponding system of ordinary or partial differential equations.

c) $u_{k+l+1}(t, z), \ldots, u_N(t, z)$; in this case it is again necessary to solve one differential equation.

It is clear that in the general case the matrix of coefficients $a_{ij}^{\alpha}(t, z)$ is some combination of matrices corresponding to the groups of different values of m_j and to the groups of equal values of m_j, that is, the combination of matrices described above.

5.2. The Cauchy problem in the scale of "linearly increasing" spaces of initial data

1. THE SCALE $\text{Exp}_{m,r+\sigma|t|,q}(\mathbf{C}_z^n)$ OF ENTIRE FUNCTIONS OF FINITE ORDER

Let $q \geq 1$, $m = (m_1, \ldots, m_N)$, $r = (r_1, \ldots, r_n)$ be as before and let $\sigma = (\sigma_1, \ldots, \sigma_n)$ be a vector with non-negative components. Further, let $u(t, z) = (u_1(t, z), \ldots, u_N(t, z))$ be a vector-valued function that is entire in z and analytic in t in the circle $|t| < \delta$, where $\delta > 0$.

We say that the vector-valued function takes its values in the scale $\text{Exp}_{m,r+\sigma|t|,q}(\mathbf{C}_z^n)$ if for any t, $|t| < \delta$,

$$u(t, \cdot) \in \text{Exp}_{m,r+\sigma|t|,q}(\mathbf{C}_z^n),$$

where, by definition, $r + \sigma|t| = (r_1 + \sigma_1|t|, \ldots, r_n + \sigma_n|t|)$.

Therefore the growth with respect to z of such a function with the evolution of t is of "linearly increasing" type $r + \sigma|t|$; more precisely, for each component $u_j(t, z)$ of the function $u(t, z)$ the following inequality holds:

$$|u_j(t, z)| \leq M_j(1 + |z|)^{m_j} \exp(r + \sigma|t|)z_+^q \quad (j = 1, \ldots, N),$$

where $M_j = M_j(t)$ is a constant and (we recall)

$$(r + \sigma|t|)z_+^q \equiv (r_1 + \sigma_1|t|)|z_1|^q + \ldots + (r_n + \sigma_n|t|)|z_n|^q.$$

We note this by writing $u_j(t, \cdot) \in \text{Exp}_{m_j,r+\sigma|t|,q}(\mathbf{C}_z^n)$.

Let $t_0 \in G \subset \mathbf{C}_t^1$ be a fixed point. We introduce the Banach space

$$\mathcal{O}(\delta; \text{Exp}_{m,r+\sigma|t-t_0|,q}(\mathbf{C}_z^n)) = \{u(t, z) : u(t, \cdot) \in \text{Exp}_{m,r+\sigma|t-t_0|,q}(\mathbf{C}_z^n)\}$$

with norm

$$\|u\|_{\delta;m,r,\sigma,q} = \sup_{|t-t_0|<\delta} \|u(t,\cdot)\|_{m,r+\sigma|t|,q} < \infty$$

(we do not indicate its dependence on t_0) or, what is the same,

$$\|u\|_{\delta;m,r,\sigma,q} = \sum_{j=1}^{N} \|u_j\|_{\delta;m_j,r,\sigma,q} =$$

$$= \sup_{|t-t_0|<\delta} \sup_{z\in\mathbb{C}^n} |u_j(t,z)|(1+|z|)^{-m_j} \exp[-(r+\sigma|t-t_0|)z_+^q].$$

Remark. As in the case of the scale $\mathrm{Exp}_{m,r,q}(\mathbb{C}^n)$, one can consider the Banach space, the elements of which are analytic functions in t in the circle $|t-t_0| < \delta$ and continuous in the closed circle $|t - t_0| \leq \delta$, where

$$\|u\|_{\delta;m,r,\sigma,q} = \max_{|t-t_0|\leq\delta} \|u(t,\cdot)\|_{m,r+\sigma|t-t_0|,q}.$$

2. CRITERION FOR THE WELL-POSEDNESS OF THE CAUCHY PROBLEM IN THE SCALE $\mathrm{Exp}_{m,r+\sigma|t|,q}(\mathbb{C}_z^n)$

Let us consider the Cauchy problem

$$u' - A(t,z,\mathcal{D})u = h(t,z), \quad t \in G, \ z \in \mathbb{C}^n, \tag{5.2.1}$$

$$u(t_0,z) = \phi(z), \quad z \in \mathbb{C}^n, \tag{5.2.2}$$

where $t_0 \in G$ is an arbitrary fixed point. Recall that the elements of the matrix $A(t,z,\mathcal{D})$ are differential operators

$$A_{ij}(t,z,\mathcal{D}) \equiv \sum_{|\alpha|\leq m_{ij}} a_{ij}^{\alpha}(t,z)\mathcal{D}^{\alpha}$$

with analytic coefficients in $G \times \mathbb{C}_z^n$.

Before giving the definition of the well-posedness of the problem (5.2.1), (5.2.2) in the scale $\mathrm{Exp}_{m,r+\sigma|t|,q}(\mathbb{C}_z^n)$ we shall prove the following lemma.

Lemma 5.2.1. *Let $u(t,z) \in \mathcal{O}(\delta;\mathrm{Exp}_{m,r+\sigma|t|,q}(\mathbb{C}_z^n))$. Then for each component $u_j(t,z)$ the inequality*

$$|u_j'(t,z)| \leq \frac{M_j}{\delta-|t|}(1+|z|)^{m_j+q} \exp(r+\sigma|t|)z_+^q$$

holds, where $M_j > 0$ is a constant.

Proof. The inequality of this lemma is, in essence, a simple corollary of Cauchy's formula. Indeed, by Cauchy's formula we have for any t, $|t| < \delta$,

$$u'_j(t, z) = \frac{1}{2\pi i} \int_{|\tau - t| = a} \frac{u_j(\tau, z)}{(\tau - t)^2} d\tau,$$

where $a < \delta - |t|$ is arbitrary. Hence, taking into account the equality $(r + \sigma|t|)z^q_+ = rz^q_+ + \sigma|t|z^q_+$, we have the inequality

$$|u'_j(t, z)| \leq M_j(2\pi)^{-1}(1 + |z|)^{m_j} a^{-2} \int_{|\tau - t| = a} \exp(r + \sigma|\tau|)z^q_+ |d\tau| \leq$$

$$\leq M_j(2\pi)^{-1}(1 + |z|)^{m_j} a^{-2} \int_{|\tau - t| = a} \exp(r + \sigma|t| + |a|)z^q_+ |d\tau| \leq$$

$$\leq M_j(2\pi)^{-1}(1 + |z|)^{m_j} a^{-1} \exp(r + \sigma|t|)z^q_+ \cdot \exp a z^q_+.$$

Setting $a = (\delta - |t|)(1 + \sigma z^q_+)^{-1}$, we immediately obtain the desired inequality. The lemma is proved.

In view of this lemma, it is natural to assume that the right hand side $h(t, z) \in \mathcal{O}(\delta; \mathrm{Exp}_{m+q, r+\sigma|t-t_0|, q}(\mathbf{C}^n_z))$.

Definition 5.2.1. We say that the Cauchy problem is well-posed in the scale $\mathrm{Exp}_{m, r+\sigma|t|, q}(\mathbf{C}^n_z)$ if for each point $t_0 \in G$ and for each vector type r there exist a number $\delta > 0$ and a vector σ such that for any right hand side $h(t, z) \in \mathcal{O}(\delta; \mathrm{Exp}_{m+q, r+\sigma|t-t_0|, q}(\mathbf{C}^n_z))$ problem (5.2.1), (5.2.2) has a unique solution $u(t, z) \in \mathcal{O}(\delta; \mathrm{Exp}_{m, r+\sigma|t-t_0|, q}(\mathbf{C}^n_z))$ and, in addition,

$$\|u\|_{\delta; m, r, \sigma, q} \leq M(\|\phi\|_{m, r, q} + \|h\|_{\delta; m, r, \sigma, q}),$$

where $M = M(r)$ is a constant, which is the bound for bounded values of r.

The main result of this subsection is the following.

Theorem 5.2.1. *The Cauchy problem (5.2.1), (5.2.2) is well-posed in the scale* $\mathrm{Exp}_{m, r+\sigma|t|, q}(\mathbf{C}^n_z)$ *if and only if*

    1)   *the coefficients $a^\alpha_{ij}(t, z)$ are polynomials in z,*

    2)   *the degrees $\deg a^\alpha_{ij}(t, z)$ of these polynomials satisfy the inequalities*

$$\deg a^\alpha_{ij}(t, z) \leq m_i - m_j - |\alpha|(q - 1) + q.$$

(*As before,* $\deg a_{ij}^{\alpha}(t,z) = -\infty$ *if* $a_{ij}^{\alpha}(t,z) \equiv 0$.)

Proof. Since the proof of this theorem is parallel to that of the analogous theorem in the case of the scale $\mathrm{Exp}_{m,r,q}(\mathbf{C}_z^n)$, we shall occasionally repeat ourselves. However, in view of the great importance of these results, we give the proof in its entirety.

Sufficiency. Suppose that conditions 1 and 2 hold. We prove that the problem (5.2.1), (5.2.2) is well-posed in the scale $\mathrm{Exp}_{m,r+\sigma|t|,q}(\mathbf{C}_z^n)$.

It is claear that we can set $t_0 = 0$. Then the Cauchy problem (5.2.1), (5.2.2) is equivalent to the integral equation

$$u(t,z) = \int_0^t A(\tau,z,\mathcal{D})u(\tau,z)d\tau + \phi(z) + \int_0^t h(\tau,z)d\tau \qquad (5.2.3)$$

where, as usual, the integral is calculated on the interval $(0,t)$.

First we note that the inclusion $h(t,z) \in \mathcal{O}(\delta; \mathrm{Exp}_{m+q,r+\sigma|t|,q}(\mathbf{C}_z^n))$ implies that

$$\int_0^t h(\tau,z)d\tau \in \mathcal{O}(\delta; \mathrm{Exp}_{m,r+\sigma|t|,q}(\mathbf{C}_z^n)). \qquad (5.2.4)$$

It is clear that $\phi(z) \in \mathcal{O}(\delta; \mathrm{Exp}_{m,r+\sigma|t|,q}(\mathbf{C}_z^n))$ also.

We now show that the operator

$$Bu \equiv \int_0^t A(\tau,z,\mathcal{D})u(\tau,z)d\tau, \quad |t| < \delta,$$

is defined in the space $\mathcal{O}(\delta; \mathrm{Exp}_{m,r+\sigma|t|,q}(\mathbf{C}_z^n))$; moreover, it is a contraction if $\delta > 0$ is sufficiently small and the vector $\sigma = (\sigma_1, \ldots, \sigma_n)$ is sufficiently large, that is, $\sigma_j \gg 1$, $j = 1, \ldots, n$. The proof of this fact is based on the estimates of the derivatives of the function $u(t,z) \in \mathcal{O}(\delta; \mathrm{Exp}_{m,r+\sigma|t|,q}(\mathbf{C}_z^n))$, which were obtained in Lemma 5.2.1 (see subsection 2 of §5.1). Namely, if $u(t,z) \in \mathcal{O}(\delta; \mathrm{Exp}_{m,r+\sigma|t|,q}(\mathbf{C}_z^n))$, then for each component $u_j(t,z)$

$$|\mathcal{D}^\alpha u_j(t,z)| \le M_j(r,|\alpha|)\|u_j(t,z)\|_{m_j,r+\sigma|t|,q} \cdot (1+|z|)^{m_j+|\alpha|(q-1)} \exp(r+\sigma|t|)z_+^q,$$

where $M(r,|\alpha|) > 0$ is a constant which is bounded for bounded r and $|\alpha|$. (In what follows we denote the various constants, which do not differ essentially, by the same letter.)

We have for $i = 1, \ldots, N$

$$(Bu)_i(t,z) \equiv \sum_{j=1}^{N} \int_0^t A_{ij}(\tau,z,\mathcal{D})u_j(\tau,z)d\tau = \sum_{j=1}^{N} \sum_{|\alpha| \le m_{ij}} \int_0^t a_{ij}^\alpha(\tau,z)\mathcal{D}^\alpha u_j(\tau,z)d\tau.$$

$$(5.2.5)$$

Taking into account the inclusion $u(t, z) \in \mathcal{O}(\delta; \mathrm{Exp}_{m, r+\sigma|t|, q}(\mathbb{C}_z^n))$, we obtain from (5.2.5) the inequality

$$\left| \int_0^t a_{ij}^\alpha(\tau, z) \mathcal{D}^\alpha u_j(\tau, z) d\tau \right| \leq M (1 + |z|)^{m_j + |\alpha|(q-1)} \times$$

$$\times \int_0^{|t|} |a_{ij}^\alpha(\tau, z)| \cdot \|u_j(\tau, \cdot)\|_{m_j, r+\sigma|\tau|, q} \exp(r + \sigma|\tau|) z_+^q |d\tau| \leq$$

$$\leq M \sup_{|\tau| \leq |t|} \|u_j(\tau, \cdot)\|_{m_j, r+\sigma|\tau|, q} (1 + |z|)^{m_j + |\alpha|(q-1) + \deg a_{ij}^\alpha} \times$$

$$\times \frac{1}{\sigma z_+^q} \left[\exp(r + \sigma|t|) z_+^q - 1 \right],$$

where the value of M depends continuously on $r + \sigma|t|$ and on the maximum modulus of the coefficients of the polynomials $a_{ij}^\alpha(t, z)$ for $|t| \leq \delta$.

Since, in accordance with condition 2, $\deg a_{ij}^\alpha(t, z) \leq m_i - m_j - |\alpha| \times \times (q - 1) + q$, we immediately obtain

$$\left| \int_0^t a_{ij}^\alpha(\tau, z) \mathcal{D}^\alpha u_j(\tau, z) d\tau \right| \leq \frac{M}{\sigma_0} \max_{|\tau| \leq |t|} \|u_j(\tau, \cdot)\|_{m_j, r+\sigma|\tau|, q} \times$$

$$\times (1 + |z|)^{m_i} \exp(r + \sigma|t|) z_+^q,$$

where $\sigma_0 = \min(\sigma_1, \ldots, \sigma_n)$.

This inequality shows that

$$\left\| \int_0^t a_{ij}^\alpha(\tau, z) \mathcal{D}^\alpha u_j(\tau, \cdot) d\tau \right\|_{m_i, r+\sigma|t|, q} \leq \frac{M}{\sigma_0} \max_{|\tau| \leq |t|} \|u_j(\tau, \cdot)\|_{m_j, r+\sigma|t|, q}$$

and hence,

$$\|(Bu)_i\|_{\delta; m_i, r, \sigma, q} \leq \frac{M}{\sigma_0} \sum_{j=1}^N \sup_{|t| \leq \delta} \|u_j(t, \cdot)\|_{m_j, r+\sigma|t|, q}.$$

Finally we see that $Bu(t, z) \in \mathcal{O}(\delta; \mathrm{Exp}_{m, r+\sigma|t|, q}(\mathbb{C}_z^n))$, moreover,

$$\|Bu\|_{\delta; m, r, \sigma, q} \leq \frac{M}{\sigma_0} \|u\|_{\delta; m, r, \sigma, q}. \tag{5.2.6}$$

It is important to note that constant $M > 0$ in (5.2.6) depends continuously on $r + \sigma|t|$ and therefore is bounded for bounded $r + \sigma|t|$.

Thus, choosing $\sigma_0 \gg 1$ and $\delta \ll 1$ (for instance, so that $\sigma_0 > M$ and $\sigma_j \delta < 1$, $j = 1, \ldots, n$) we find from (5.2.6) that the operator Bu is a contraction in $\mathcal{O}(\delta; \operatorname{Exp}_{m,r+\sigma|t|,q}(\mathbf{C}_z^n))$. Consequently, the operator on the right hand side of (5.2.3) is also a contraction in $\mathcal{O}(\delta; \operatorname{Exp}_{m,r+\sigma|t|,q}(\mathbf{C}_z^n))$. Thus there exists a unique fixed point of this operator and hence, there exists a unique solution of the Cauchy problem (5.2.1), (5.2.2) in the space $\mathcal{O}(\delta; \operatorname{Exp}_{m,r+\sigma|t|,q}(\mathbf{C}_z^n))$. The estimate

$$\|u\|_{\delta;m,r,\sigma,q} \leq M(r)\Big[\|h\|_{\delta;m+q,r,\sigma,q} + \|\phi\|_{m,r,q}\Big]$$

is clear. The sufficiency is proved.

Necessity. Suppose that the Cauchy problem (5.2.1), (5.2.2) is well-posed in the scale $\operatorname{Exp}_{m,r+\sigma|t|,q}(\mathbf{C}_z^n)$. We show that the coefficients $a_{ij}^\alpha(t,z)$ are polynomials with

$$\deg a_{ij}^\alpha(t,z) \leq m_i - m_j - |\alpha|(q-1) + q. \tag{5.2.7}$$

As before, the proof will be carried out in two steps. First we establish merely that the entire functions $a_{ij}^\alpha(t,z)$ are polynomials in z; then we determine their exact degrees. Both procedures use a special choice of the initial data.

Step 1. We fix an integer j $(1 \leq j \leq N)$ and set

$$u_j(0,z) = \phi_j(z) \equiv \exp(\zeta_1 z_1^q + \ldots + \zeta_n z_n^q),$$
$$u_i(0,z) = 0, \quad i \neq j \quad (i = 1,2,\ldots,N).$$

Clearly, $|\phi_j(z)| \leq (1+|z|)^{m_j} \exp r z_+^q$, where $r = (|\zeta_1|,\ldots,|\zeta_n|)$ (the complex numbers $\zeta_1,\ldots,\zeta_n$ will be chosen below). Then there exist $\delta > 0$ and a vector $\sigma = (\sigma_1,\ldots,\sigma_n)$, $\sigma_1 > 0,\ldots$, $\sigma_n > 0$, such that for the corresponding solution $u(t,z)$ (we do not use j in the notation) the inequality

$$|u_j(t,z)| \leq M(1+|z|)^{m_j} \exp(r + \sigma|t|)z_+^q, \quad |t| < \delta,$$

holds. Here $M = M(r) > 0$ is a constant which is bounded for bounded r.

It is obvious that

$$\mathcal{D}^\alpha \phi_j(z) = [q^{|\alpha|}\zeta^\alpha z^{(q-1)\alpha} + \ldots] \exp(\zeta_1 z_1^q + \ldots + \zeta_n z_n^q)$$

where the dots denote the group of the terms of type ζ^γ such that $|\gamma| < |\alpha|$.

We now use a special case of Lemma 5.2.1. Namely, if $u(t,z) \in \mathcal{O}(\delta; \operatorname{Exp}_{m,r+\sigma|t|,q}(\mathbf{C}_z^n))$, then

$$|u_i'(0,z)| \leq M(1+|z|)^{m_j+q} \exp r z_+^q$$

where the constant $M > 0$ depends on δ and σ.

By virtue of these facts and setting $t = 0$ in system (5.2.1), we obtain the inequality

$$\left| \sum_{|\alpha|=m_{ij}} a_{ij}^{\alpha}(0,z)[q^{|\alpha|}\zeta^{\alpha}z^{(q-1)\alpha} + \ldots] \right| \leq$$

$$\leq M(1+|z|)^{m_i+q} |\exp(rz_+^q - \zeta_1 z_1^q - \ldots - \zeta_n z_n^q)|. \tag{5.2.8}$$

We shall deduce from (5.2.8) that the entire functions $a_{ij}^{\alpha}(0,z)$ are polynomials in $z_1, \ldots, z_n$ separately. As we noted earlier, this will then imply that the $a_{ij}^{\alpha}(0,z)$ are polynomials in $z_1, \ldots, z_n$ jointly.

In fact, we fix $z_2, \ldots, z_n$, that is, we set $z_2 = z_{20}, \ldots, z_n = z_{n0}$, where $z_{20}, \ldots, z_{n0}$ are arbitrary complex numbers. Then for the variable z_1 we obtain the inequality

$$\left| \sum_{|\alpha|\leq m_{ij}} a_{ij}^{\alpha}(0;z_1,z_0')[q^{|\alpha|}\zeta^{\alpha}z_1^{(q-1)\alpha_1}(z_0')^{(q-1)\alpha'} + \ldots] \right| \leq$$

$$\leq M(z_0') |e^{r_1|z_1|^q - \zeta_1 z_1^q}|(1+|z_1|)^{m_i+q},$$

where $M(z_0') > 0$ is a constant; $z_0' = (z_{20}, \ldots, z_{n0})$, $\alpha' = (\alpha_2, \ldots, \alpha_n)$.

We now choose $\zeta_j = \tau_j$, where the $\tau_j \in [0,1]$ $(j = 2, \ldots, n)$ are arbitrary, and $\zeta_1 = \tau_1/z_{10}^q$, $\tau_1 \in [0,1]$, such that $|z_{10}| \geq 1$. Then, clearly, $r_1 = \tau_1/|z_{10}|^q$ and consequently (setting $z_1 = z_{10}$ and multiplying the last inequality by $z_{10}^{m_{ij}}$) we obtain

$$\left| \sum_{|\alpha|\leq m_{ij}} a_{ij}^{\alpha}(0;z_{10},z_0')[q^{|\alpha|}\tau^{\alpha}z_{10}^{m_{ij}-\alpha_1}(z_0')^{(q-1)\alpha'} + \ldots] \right| \leq$$

$$\leq M(z_0')\,(1+|z_{10}|)^{m_i+q+m_{ij}}.$$

(It is important to observe that since $|\zeta_j| \leq 1$, one can regard the constant $M(z_0')$ as being independent of ζ_j, that is, independent of τ_j, $j = 1, \ldots, n$.)

Further, since z_{10} $(|z_{10}| \geq 1)$ is arbitrary, the entire function

$$P(z_1;z_0',\tau) \equiv \sum_{|\alpha|\leq m_{ij}} a_{ij}^{\alpha}(0,z_1,z_0')[q^{|\alpha|}\tau^{\alpha}z_1^{m_{ij}-\alpha_1}(z_0')^{(q-1)\alpha'} + \ldots]$$

is a polynomial in z_1 for all τ_j, $0 \leq \tau_j \leq 1$, $j = 1, \ldots, n$. Moreover, its degree is clearly independent of z_0'.

Since for any α such that $|\alpha| = m_{ij}$

$$\mathcal{D}_\tau^\alpha P(z_1; z_0', \tau) \equiv a_{ij}^\alpha(0; z_1, z_0')q^{m_{ij}}\alpha! z_1^{m_{ij}-\alpha_1}(z_0')^{(q-1)\alpha'},$$

these functions are also polynomials for any fixed $z_0' = (z_{20}, \ldots, z_{n0})$. Hence it follows immediately that the functions $a_{ij}^\alpha(0, z_1, z_0')$ are polynomials as well.

In the same way one can prove that coefficients $a_{ij}^\alpha(0, z)$, $|\alpha| = m_{ij}$, are polynomials in the other variables $z_2, \ldots, z_n$. Hence $a_{ij}^\alpha(0, z)$, $|\alpha| = m_{ij}$, are polynomials in $z_1, \ldots, z_n$ jointly.

Finally, it remains to remark that after proving our assertion for $a_{ij}^\alpha(0, z)$, where $|\alpha| = m_{ij}$, we can prove that $a_{ij}^\alpha(0, z)$ is a polynomial for $|\alpha| = m_{ij} - 1$ by analogy with the previous case; we then prove it for $|\alpha| = m_{ij} - 2$, and so on. Eventually we will have proved that all the coefficients $a_{ij}^\alpha(0, z)$, $|\alpha| \leq m_{ij}$, are polynomials. The first step is finished.

Step 2. Here we determine the degrees of the polynomials $a_{ij}^\alpha(0, z)$. For this we turn again to the Cauchy problem (5.2.1), (5.2.2) and set (j is a fixed number)

$$u_j(0, z) = \phi_j(z) \equiv (z_1 + \ldots + z_n)^{m_j} \exp(\zeta_1 z_1^q + \ldots + \zeta_n z_n),$$
$$u_i(0, z) = \phi_k(z) \equiv 0, \quad i \neq j.$$

Then, by analogy with (5.2.8), we have

$$\left| \sum_{|\alpha| \leq m_{ij}} a_{ij}^\alpha(0, z)[q^{|\alpha|}\zeta^\alpha(z_1 + \ldots + z_n)^{m_j} z^{(q-1)\alpha} + \ldots] \right| \leq$$

$$\leq M(r)(1 + |z|)^{m_i + q}|\exp(rz_+^q - \zeta_1 z_1^q - \ldots - \zeta_n z_n^q)|,$$

where $r = (|\zeta_1|, \ldots, |\zeta_n|)$ and $\ldots$ denotes the terms involving ζ^β and z^γ such that $|\beta| < |\alpha|$ and $|\gamma| < m_j + (q - 1)|\alpha|$ respectively. Setting here $\zeta_k = \tau_k$ ($0 \leq \tau_k \leq 1$) and $z_k = x_k$, $x_k \geq 0$ ($k = 1, \ldots, n$), we obtain

$$\left| \sum_{|\alpha| \leq m_{ij}} a_{ij}^\alpha(0, x)[q^{|\alpha|}\tau^\alpha(x_1 + \cdots + x_n)^{m_j} x^{(q-1)\alpha} + \cdots] \right| \leq M(r)\,(1 + |x|)^{m_i + q}.$$

This inequality shows that the polynomial

$$P(\tau, x) \equiv \sum_{|\alpha| \leq m_{ij}} a_{ij}^\alpha(0, x)[q^{|\alpha|}\tau^\alpha(x_1 + \cdots + x_n)^{m_j} x^{(q-1)\alpha} + \cdots]$$

is of degree at most $m_i + q$ in x. Hence, calculating the derivatives of $P(\tau, x)$ with respect to τ, we obtain for $|\alpha| = m_{ij}$ the inequality

$$\deg a_{ij}^\alpha(0, x)(x_1 + \cdots + x_n)^{m_j} x^{(q-1)\alpha} \le m_i + q.$$

Consequently

$$\deg a_{ij}^\alpha(0, x) \le m_i - m_j - |\alpha|(q-1) + q.$$

Repeating the above argument, we then obtain the last inequality for $|\alpha| = m_{ij} - 1$, $|\alpha| = m_{ij} - 2$, and so on.

Finally, it remains to note that it is clearly possible to reduce the investigation of the necessity of conditions (5.2.7) at the point $t_0 \ne 0$ to the case $t_0 = 0$ by substituting $t \leftrightarrow t - t_0$.

Thus, the desired inequalities are completely proved. The theorem is also proved.

3. THE STRUCTURE OF SYSTEMS WITH $\deg a_{ij}^\alpha \le m_i - m_j - |\alpha|(q-1) + q$

Clearly, the structure of such systems

$$u_i' - \sum_{j=1}^N A_{ij}(t, z, D) u_j = h_i(t, z), \quad (i = 1, \ldots, N), \tag{5.2.9}$$

(with polynomial coefficients) depends on the values of the parameter q of the scale $\mathrm{Exp}_{m, r + \sigma|t|, q}(\mathbf{C}_z^n)$ that is on the order of growth in the argument $z \in \mathbf{C}^n$ of the entire solution.

Let us first consider the case $q = 1$ that is the case of the scale of exponential type functions.

I. The case $q = 1$. In this case for any α, the conditions given in the title assume the form

$$\deg a_{ij}^\alpha(t, z) \le m_i - m_j + 1. \tag{5.2.10}$$

The structure of the matrix satisfying such inequalities was investigated in §5.1. We use these results.

1) *Isotropic growth* ($m_1 = \cdots = m_N$). In this case the restrictions (5.2.10) on the degrees of the polynomials $a_{ij}^\alpha(t, z)$ (α is arbitrary) give the systems (5.2.9), where the $A_{ij}(t, z, D)$ are arbitrary differential operators with linear coefficients in z.

2) Let $m_1 < m_2 < \ldots < m_N$ with $m_{i+1} = m_i + 1$, $i = 1, 2, \ldots, N - 1$. In this case for each fixed α, the coefficient matrix $a_{ij}^\alpha(t, z)$ is

$$\begin{bmatrix} a_{11}^\alpha(t, z) & b_{12}(t) & 0 & \cdots & 0 \\ a_{21}^\alpha(t, z) & a_{22}^\alpha(t, z) & b_{23}(t) & \cdots & 0 \\ \vdots & \vdots & \vdots & \ddots & \vdots \\ \cdots & \cdots & \cdots & \cdots & b_{N-1,N}(t) \\ a_{N1}^\alpha(t, z) & a_{N2}^\alpha(t, z) & a_{N3}^\alpha(t, z) & \cdots & a_{NN}^\alpha(t, z) \end{bmatrix}$$

where the $a_{ij}^\alpha(t, z)$ are polynomials in z of degree at most $i - j + 1$ (for convenience we denote the coefficients, not depending on z, by $b_{i,i+1}(t)$). The corresponding system has the form

$$u_1' - A_{11}(t, z, \mathcal{D})u_1 - B_{12}(t, \mathcal{D})u_2 \qquad\qquad\qquad = h_1(t, z)$$

$$u_2' - A_{21}(t, z, \mathcal{D})u_1 - A_{22}(t, z, \mathcal{D})u_2 - B_{23}(t, \mathcal{D})u_3 \qquad = h_2(t, z)$$

$$u_N' - A_{N1}(t, z, \mathcal{D})u_1 - \qquad \cdots \qquad \cdots \qquad - A_{NN}(t, z, \mathcal{D})u_N = h_N(t, z),$$

where

$$A_{ij}(t, z, \mathcal{D}) \equiv \sum_\alpha a_{ij}^\alpha(t, z)\mathcal{D}^\alpha$$

and

$$B_{i,i+1}(t, \mathcal{D}) \equiv \sum_\alpha b_{i,i+1}^\alpha(t)\mathcal{D}^\alpha$$

are differential operators of arbitrary finite order.

3) Let us consider the Cauchy problem for the single differential equation of higher order N

$$u^{(N)} - \sum_{j=0}^{N-1} A_j(t, z, \mathcal{D})u^{(j)} = h(t, z), \tag{5.2.11}$$

$$u^{(j)}(0, z) = \phi_j, \quad z \in \mathbb{C}^n \quad (0 \leq j \leq N - 1), \tag{5.2.12}$$

where the

$$A_j(t, z, \mathcal{D}) \equiv \sum_\alpha a_j^\alpha(t, z)\mathcal{D}^\alpha$$

are operators with coefficients that are analytic in the domain G; $\phi_j(z) \in \text{Exp}_{m_1+j,r,1}(\mathbf{C}_z^n)$, that is,

$$|\phi_j(z)| \leq M_j(1+|z|)^{m_1+j} \exp(r_1|z_1| + \cdots + r_n|z_n|)$$

($M_j > 0$ are certain constants, $m_1 \geq 0$ is an integer).

Let $u(t,z) \in \mathcal{O}(\delta; \text{Exp}_{m_1,r+\sigma|t|,1}(\mathbf{C}_z^n))$ be a solution of the Cauchy problem (5.2.11), (5.2.12). Then for any $\delta_1 < \delta$ we have the inclusions $u^{(j)}(t,z) \in \mathcal{O}(\delta_1; \text{Exp}_{m_1+j,r+\sigma|t|,1}(\mathbf{C}_z^n))$. Hence the usual substitution

$$u_1 = u, \ u_2 = u', \ldots, \ u_N = u^{(N-1)}$$

gives the Cauchy problem for the system

$$u' - \begin{bmatrix} 0 & 1 & 0 & \ldots & 0 \\ 0 & 0 & 1 & \ldots & 0 \\ \vdots & \vdots & \vdots & & \vdots \\ 0 & 0 & 0 & \ldots & 1 \\ A_0 & A_1 & A_2 & \ldots & A_{N-1} \end{bmatrix} u = \begin{bmatrix} 0 \\ 0 \\ \vdots \\ 0 \\ h \end{bmatrix} \tag{5.2.13}$$

in the scale $\text{Exp}_{m,r+\sigma|t|,1}(\mathbf{C}_z^n)$, where $m = (m_1, m_1+1, \ldots, m_1+N-1)$. Conversely, it is clear that if the Cauchy problem for the system (5.2.13) is well-posed in the scale $\text{Exp}_{m,r+\sigma|t|,1}(\mathbf{C}_z^n)$, where $m = (m_1, m_1+1, \ldots, m_1+N-1)$, then the corresponding higher order equation (5.2.11) is well posed in the scale $\text{Exp}_{m_1,r+\sigma|t|,1}(\mathbf{C}_z^n)$. In accordance with the main Theorem 5.2.1, the well-posedness of the Cauchy problem for the system (5.2.13) implies that for any α

$$\deg a_j^\alpha(t,z) \leq N - j.$$

This means that the characteristic polynomial of the equation (5.2.11)

$$L(\lambda, z, \zeta) \equiv \lambda^N - \sum_{j=0}^{N-1} A_j(t, z, \zeta)\lambda^j$$

is a Kovalevskaya polynomial in the variables λ and z. Thus, the Cauchy problem for a single higher order equation is well posed in the scale of exponential functions if and only if this equation is a dual Kovalevskaya equation.

4) We assume that $m_1 \leq \ldots \leq m_k < m_{k+1} \leq \ldots \leq m_N$ with $m_{k+1} - m_k > 1$ (where k is some number). Then the matrix of the system (5.2.9) has the following structure

$$
\begin{bmatrix}
\begin{array}{c} a_{ij}^{\alpha}(t, z) \\[4pt] (1 \leq i, j \leq k) \end{array} & 0 \\[20pt]
\hline
\begin{array}{c} a_{ij}^{\alpha}(t, z) \\[4pt] (k+1 \leq i \leq N,\ 1 \leq j \leq k) \end{array} & \begin{array}{c} a_{ij}^{\alpha}(t, z) \\[4pt] (k+1 \leq i, j \leq N) \end{array}
\end{bmatrix}
$$

where the $a_{ij}^{\alpha}(t, z)$ are polynomials of degree at most $m_i - m_j + 1$. The system

$$
u_i' - \sum_{j=1}^{k} A_{ij}(t, z, \mathcal{D}) u_j = h_i(t, z) \quad (1 \leq i \leq k) \tag{5.2.14}
$$

$$
u_i' - \sum_{j=1}^{k} A_{ij}(t, z, \mathcal{D}) u_j - \sum_{j=k+1}^{N} A_{ij}(t, z, \mathcal{D}) u_j = h_i(t, z) \quad (k+1 \leq i \leq N), \tag{5.2.15}
$$

where the

$$
A_{ij}(t, z, \mathcal{D}) \equiv \sum_{\alpha} a_{ij}^{\alpha}(t, z) \mathcal{D}^{\alpha}
$$

are arbitrary differential operators of any finite order, corresponds to this matrix.

Clearly we can successively solve the given system: first we use system (5.2.14) to express the components $u_1, \ldots, u_k$ in terms of $u_{k+1}, \ldots, u_N$. We then use these expressions for $u_1, \ldots, u_k$ to solve the system (5.2.15) in the unknown functions $u_{k+1}, \ldots, u_N$.

5) Finally, we turn to the general case $m_1 \leq \ldots \leq m_{k+1} = m_{k+2} = \ldots = m_{k+l} < m_{k+l+1} < \ldots < m_N$. The last example shows that, without loss of generality, one can assume that $m_{i+1} - m_i \leq 1$. In this case the coefficient matrix $a_{ij}^{\alpha}(t, z)$ is defined by Table 2 and, consequently, the corresponding system (5.2.9) is

$$
\begin{bmatrix}
a_{11}^{\alpha} & b_{12}^{\alpha} & 0 & & & & & & & & & & \\
 & a_{22}^{\alpha} & b_{23}^{\alpha} & & & & & & \text{\Large 0} & & & & \\
 & & \ddots & \ddots & & & & & & & & & \\
 & & & a_{k-1,k-1}^{\alpha} & b_{k-1,k}^{\alpha} & & & & & & & & \\
 & & & & a_{kk}^{\alpha} & b_{k,k+1}^{\alpha} & 0 & 0 & \cdots & 0 & \cdots & 0 & 0 \\
 & & & & a_{k+1,k+1}^{\alpha} & b_{k+1,k+2}^{\alpha} & 0 & \cdots & 0 & \cdots & 0 & \\
 & & & & a_{k+2,k+2}^{\alpha} & a_{k+1,k+2}^{\alpha} & b_{k,k+2}^{\alpha} & \cdots & 0 & \cdots & 0 & \\
 & & & & \vdots & \vdots & \ddots & \ddots & \vdots & \cdots & \vdots & \\
 & & & & a_{k+1,k+l}^{\alpha} & a_{k+2,k+l}^{\alpha} & \cdots & b_{k,k+l}^{\alpha} & 0 & \cdots & 0 & \\
 & & & & a_{k+l+1,k+l+1}^{\alpha} & b_{k+1,k+l+1}^{\alpha} & b_{k+2,k+l+1}^{\alpha} & 0 & \cdots & 0 & \\
 & & & & \ddots & \vdots & \vdots & \vdots & \cdots & \vdots & \\
 & & & & & b_{N-1,N}^{\alpha} & 0 & 0 & \cdots & 0 & \cdots & a_{N,N}^{\alpha}
\end{bmatrix}
$$

$$a_{ij}^{\alpha} \ (i > j)$$

(Here $a_{ij}^{\alpha} \equiv a_{ij}^{\alpha}(t,z)$, $b_{ij}^{\alpha} \equiv b_{ij}^{\alpha}(t)$)

Table 2

$$u_i' - \sum_{j=1}^{i} A_{ij}(t,z,\mathcal{D})u_j - \sum_{\alpha} b_{i,i+1}^{\alpha}(t)\mathcal{D}^{\alpha}u_{i+1} = h_i(t,z) \quad (i = 1,\ldots, k-1)$$

$$u_k' - \sum_{j=1}^{k} A_{kj}(t,z,\mathcal{D})u_j - \sum_{j=k+1}^{k+l} \left(\sum_{\alpha} b_{kj}^{\alpha}(t)\mathcal{D}^{\alpha}u_j\right) = h_k(t,z)$$

$$u_i' - \sum_{j=1}^{i+l} A_{ij}(t,z,\mathcal{D})u_j - \sum_{\alpha} b_{i,k+l+1}(t)\mathcal{D}^{\alpha}u_{k+l+1} = h_i(t,z) \quad (i = k+1,\ldots, k+l)$$

$$u_i' - \sum_{j=1}^{i} A_{ij}(t,z,\mathcal{D})u_j - \sum_{\alpha} b_{i,i+1}(t)\mathcal{D}^{\alpha}u_{i+1} = h_i(t,z) \quad (i = k+l+1,\ldots, N-1)$$

$$u_N' - \sum_{j=1}^{N} A_{Nj}(t,z,\mathcal{D})u_j = h_N(t,z),$$

where

$$A_{ij}(t,z,\mathcal{D}) \equiv \sum_{\alpha} a_{ij}^{\alpha}(t,z)\mathcal{D}^{\alpha}$$

are arbitrary differential operators.

Remark. Let us compare (at first only formally) the conditions for the local well-posedness of the Cauchy problem in the scale $D_{m,r-\sigma|t|}$ with the conditions for the well-posedness of the Cauchy problem in the scale $\mathrm{Exp}_{m,r+\sigma|t|,1}(\mathbf{C}_z^n)$. Namely, let

$$u' - A(t,z,\mathcal{D})u = h(t,z) \qquad (5.2.16)$$

be a system of partial differential equations with polynomial coefficients $a_{ij}^{\alpha}(t,z)$ in z. Then the conditions

$$\mathrm{ord}A_{ij}(t,z,\mathcal{D}) \le m_i - m_j + 1 \qquad (5.2.17)$$

are dual (in the Fourier transform sense) to the conditions

$$\deg a_{ij}^{\alpha}(t,z) \le m_i - m_j + 1 \qquad (5.2.18)$$

($|\alpha| = 0, 1, \ldots$) and conversely. Moreover, if system (5.2.16) satisfies conditions (5.2.17) then their Fourier-image

$$\tilde{u}' - A(t,\partial,\zeta)\tilde{u} = \tilde{h}(t,\zeta)$$

satisfies conditions (5.2.18). (Here $\zeta = (\zeta_1, \ldots, \zeta_n)$ are the dual arguments and $\partial = (\partial/\partial\zeta_1, \ldots, \partial/\partial\zeta_n)$.)

This observation suggests that there is a deep non-formal connnection between the analytic Cauchy problem and the exponential Cauchy problem. However, we can describe this connection only within the framework of the theory of pseudodifferential operators with complex arguments. This will be done below.

II. The case $q = 2$ (scale of Gaussian growth). In this case the Cauchy problem is correct under conditions

$$\deg a_{ij}^{\alpha}(t, z) \leq m_i - m_j - |\alpha| + 2. \tag{5.2.19}$$

First we note that these conditions give a restriction on the orders of the operators $A_{ij}(t, z, \mathcal{D})$, namely,

$$\operatorname{ord} A_{ij}(t, z, \mathcal{D}) \leq \max_{i,j}(m_i - m_j + 2).$$

If we assume (without loss of generality) that

$$m_1 \leq m_2 \leq \ldots \leq m_N,$$

then

$$\operatorname{ord} A_{ij}(t, z, \mathcal{D}) \leq m_N - m_1 + 2.$$

We further note that $|\alpha| > 2$ implies $\deg a_{ij}^{\alpha} < 0$ for any $i \leq j$; hence the corresponding coefficient matrices have the form

$$\begin{bmatrix} 0 & & & & \\ & 0 & & & \\ & & \ddots & & 0 \\ & a_{ij}^{\alpha}(t, z) & & \ddots & \\ & (i > j) & & & \\ & & & & 0 \end{bmatrix} \tag{5.2.20}$$

Therefore, the case of greatest interest is when $|\alpha| = 2, 1, 0$. If $|\alpha| = 2$ or $|\alpha| = 1$, then the above conditions give the inequalities

$$\deg a_{ij}^{\alpha}(t, z) \leq m_i - m_j$$

or
$$\deg a_{ij}^{\alpha}(t, z) \leq m_i - m_j + 1.$$

The structure of the corresponding matrices has already been investigated for both cases. For the matrix $a_{ij}^0(t, z)$ we have

$$\deg a_{ij}^0(t, z) \leq m_i - m_j + 2$$

and we see that its structure depends on the values of $m_{k+1} - m_k$ with $k = 1, \ldots, N$. In particular, if there exists k such that $m_{k+1} - m_k > 2$, then

$$(a_{ij}^0)_{N \times N} = \left[\begin{array}{c|c} \begin{array}{c} a_{ij}^0(t, z) \\ (1 \leq i, j \leq k) \end{array} & 0 \\ \hline \begin{array}{c} a_{ij}^0(t, z) \\ (k+1 \leq i \leq N, \; 1 \leq j \leq k) \end{array} & \begin{array}{c} a_{ij}^0(t, z) \\ (k+1 \leq i, j \leq N) \end{array} \end{array} \right]$$

We consider two typical special cases.

1). *Isotropic case* $(m_1 = \cdots = m_N)$. In this case we obtain from (5.2.18) the inequalities

$$\deg a_{ij}^{\alpha}(t, z) \leq -|\alpha| + 2.$$

From this it follows immediately that $|\alpha| \leq 2$, therefore the system (5.2.16) can be written in the form

$$u' - A_2(t, \mathcal{D})u - A_1(t, z, \mathcal{D})u - A_0(t, z)u = h(t, z),$$

where

$A_2(t, \mathcal{D})$ is a matrix of second order differential operators with coefficients not depending on z;

$A_1(t, z, \mathcal{D})$ is a matrix of first order differential operators, the coefficients of which are linear functions in z;

$A_0(t, z, \mathcal{D})$ is a matrix of functions, the components of which are polynomials in z of degree not greater than two.

These systems are precisely those for which the Cauchy problem is well-posed in the isotropic Gauss scale $\text{Exp}_{m,r+\sigma|t|,q}(\mathbf{C}_z^n)$.

2). *One higher order equation.* Let $u(t,z) \in \mathcal{O}(\delta; \text{Exp}_{m_1,r+\sigma|t|,2}(\mathbf{C}_z^n))$ $(m_1 \geq 0$ is an integer$)$ be a solution of the Cauchy problem

$$u^{(N)} - \sum_{j=0}^{N-1} A_j(t,z,\mathcal{D})u^{(j)} = h(t,z) \qquad (5.2.21)$$

$$u^{(j)}(0,z) = \phi_j(z), \quad z \in \mathbf{C}^n \ (0 \leq j \leq N-1), \qquad (5.2.22)$$

where

$$A_j(t,z,\mathcal{D}) \equiv \sum_\alpha a_j^\alpha(t,z)\mathcal{D}^\alpha$$

are differential operators of finite order. Then the vector-valued function $\vec{u}(t,z) = (u(t,z), u'(t,z), \ldots, u^{(N-1)}(t,z))$ is a solution of the system (5.2.13) in the scale $\text{Exp}_{m,r+\sigma|t|,2}(\mathbf{C}_z^n)$, where $m = (m_1, m_1+2, \ldots, m_1+2(N-1))$. Consequently, the degrees of the polynomials $a_j^\alpha(t,z)$ satisfy the conditions

$$\deg a_j^\alpha(t,z) \leq m_N - m_{j+1} - |\alpha| + 2 = 2(N-1) - |\alpha|. \qquad (5.2.23)$$

From this it follows that the orders of the operators $A_j(t,z,\mathcal{D})$ satisfy the inequalities

$$\text{ord} A_j(t,z,\mathcal{D}) \leq 2(N-j).$$

It is clear that, conversely, if the orders of operators $A_j(t,z,\mathcal{D})$ satisfy the last inequalities, then for any $m_1 \geq 0$ the Cauchy problem for the system (5.2.13) is well-posed in the scale $\text{Exp}_{m,r+\sigma|t|,2}(\mathbf{C}_z^n)$, where $m = (m_1, m_1+2, \ldots, m_1+2(N-1))$, if and only if the functions $a_j^\alpha(t,z)$ are polynomials in z and conditions (5.2.23) hold. Therefore, the first component $u_1(t,z) \in \text{Exp}_{m_1,r+\sigma|t|,2}(\mathbf{C}_z^n)$ is the solution of the Cauchy problem (5.2.21), (5.2.22) for a single higher order equation.

III. The case $q \geq 3$. This case is similar to the case $q = 2$. In fact, it follows immediately from the inequalities

$$\deg a_{ij}^\alpha(t,z) \leq m_i - m_j - |\alpha|(q-1) + q$$

that for $|\alpha| > 1$ the coefficient matrices $a_{ij}^\alpha(t,z)$ have the form (5.2.20). Therefore, only the cases $|\alpha| = 1$ and $|\alpha| = 0$ are of real interest. In the first case we have

$$\deg a_{ij}^\alpha(t,z) \leq m_i - m_j + 1,$$

while in the second case,

$$\deg a_{ij}^0(t, z) \le m_i - m_j + q.$$

It is clear that the corresponding matrix becomes more "filled out" as $q \ge 3$ increases.

In conclusion, we give here just two particular examples:

1). *Isotropic scale* $(m_1 = \ldots = m_N)$. Then the inequalities

$$\deg a_{ij}^\alpha(t, z) \le -|\alpha|(q - 1) + q$$

determine a non-trivial matrix only for $|\alpha| = 1$ and $\alpha = 0$. Hence, the Cauchy problem is well-posed in the isotropic scale $\text{Exp}_{m,r+\sigma|t|,q}(\mathbf{C}_z^n)$, where $q \ge 3$, only for systems of type

$$u' - A_1(t, z, \mathcal{D})u - A_0(t, z)u = h(t, z),$$

where $A_1(t, z, \mathcal{D})$ is a matrix of first order differential operators, the coefficients of which are linear functions in z, and $A_0(t, z)$ is a matrix of polynomials of degree not greater than q.

2). One *higher order equation*. Considerations analogous to the case $q = 2$ show that the Cauchy problem (5.2.21), (5.2.22) is well posed in the scale $\text{Exp}_{m,r+\sigma|t|,q}(\mathbf{C}_z^n)$ if and only if

$$\deg a_j^\alpha(t, z) \le q(N - j) - |\alpha|. \tag{5.2.24}$$

It is clear that in this case

$$\text{ord}\, A_j(t, z, \mathcal{D}) \le q(N - j). \tag{5.2.25}$$

Conversely, if the inequalities (5.2.25) hold, then the Cauchy problem for the system (5.2.13) is well posed in the scale $\text{Exp}_{m,r+\sigma|t|,q}(\mathbf{C}_z^n)$, where $m = (m_1, m_1 + q, \ldots, m_1 + q(N - 1))$, if and only if the inequalities (5.2.24) hold. Hence in this case, the first component $u_1(t, z)$ is the solution of the Cauchy problem for the single higher order equation (5.2.21), (5.2.22) in the scale $\text{Exp}_{m_1,r+\sigma|t|,q}(\mathbf{C}_z^n)$.

4. THE CAUCHY PROBLEM IN THE DUAL SPACE $\mathrm{Exp}'_{m,r-\sigma|t|,q}(\mathbf{C}^n_z)$

In this subsection we study the Cauchy problem dual to the Cauchy problem

$$v' - A(t, z, \mathcal{D})v = h(t, z), \tag{5.2.26}$$

$$v(t_0, z) = \phi(z), \tag{5.2.27}$$

where the solution $v(t, z)$ is sought as a function $v(t, \cdot)$ with values in the scale $\mathrm{Exp}_{m,r+\sigma|t-t_0|,q}(\mathbf{C}^n_z)$.

Before giving the precise statements concerning the dual problem, we give a preliminary discussion. First we recall the conditions under which the problem (5.2.26), (5.2.27) is being investigated:

1) the coefficients $a^\alpha_{ij}(t, z)$ of the operators $A_{ij}(t, z, \mathcal{D})$ are polynomials in z;

2) the degrees of these polynomials satisfy the conditions

$$\deg a^\alpha_{ij}(t, z) \leq m_i - m_j - |\alpha|(q - 1) + q.$$

The estimates obtained in order to prove the well-posedness of (5.2.26), (5.2.27) in the scale $\mathrm{Exp}_{m,r+\sigma|t-t_0|,q}(\mathbf{C}^n_z)$ show that conditions 1) and 2) guarantee the continuity of the map

$$A(t, z, \mathcal{D}) : \mathrm{Exp}_{m,r,q}(\mathbf{C}^n_z) \to \mathrm{Exp}_{m+q,r,q}(\mathbf{C}^n_z)$$

(t is fixed, r is arbitrary). Moreover, these conditions are necessary and sufficient for the continuity of this map. Taking this into account, we shall assume that conditions 1) and 2) hold. We only consider (for simplicity) the case $r = (r, \ldots, r)$ and $\sigma = (\sigma, \ldots, \sigma)$ and work with them as with numbers.

Further, for the well-posedness of the dual problem it is useful to emphasize the nature of the dependence of the numbers $\sigma > 0$ and $\delta > 0$ on the initial point $t_0 \in G$ and on the coefficients of the original system of differential equations. Namely, the calculations given in the course of the proof of Theorem 5.2.1 of this section, show that the value of the constant M from the norm estimates of the operator Bu

$$\|Bu\|_{\delta;m,r+\sigma|t-t_0|,q} \leq \frac{M}{\sigma}\|u\|_{\delta;m,r+\sigma|t-t_0|,q}$$

depends on the modulus of the coefficients of $a^\alpha_{ij}(t, z)$ for $|t - t_0| < \delta$ and on the multipliers of type $(r + \sigma|t - t_0|)^{|\alpha|} \leq (r + \sigma\delta)^{|\alpha|}$, where the indices

$\alpha = (\alpha_1, \ldots, \alpha_n)$ run through a finite set of values (the dependence on the unimportant quantities q, N etc. is not indicated). Thus, the numbers $\sigma > 0$ and $\delta > 0$ may be chosen by various methods but the following two conditions must be valid:

$$r + \sigma\delta \leq q(r), \quad M/\sigma < 1,$$

where $q(r) > 0$ is a function that is bounded for bounded r. In particular, if $0 \leq r \leq R < \infty$ and $t_0 \in K \subset G$, where K is a compact set, then the quantities $\sigma > 0$ and $\delta > 0$ may be chosen the same for any r and t_0.

Moreover, it is clear that for any fixed $r > 0$ and $t_0 \in G$ one can put $\delta = r/\sigma$, where $\sigma > 0$ is sufficiently large. Then this $\delta > 0$ is suitable for the statement about the well-posedness of the Cauchy problem with initial data $\phi(z) \in \mathrm{Exp}_{m,r,q}(\mathbf{C}_z^n)$ not only for $t = t_0$, but also for any $t = t_1$, where $|t - t_1| \leq \frac{r}{\sigma}$. In particular, this means that the solution $u(t, z)$ of the Cauchy problem with initial data for $t = t_1$ is defined for all t such that $|t - t_1| \leq r/\sigma$, including $t = t_0$.

We also note that the inclusion $u(t, z) \in \mathcal{O}(r/\sigma; \mathrm{Exp}_{m,r+\sigma|t-t_0|,q}(\mathbf{C}_z^n))$ implies the inclusion

$$A(t, z, \mathcal{D})u(t, z) \in \mathcal{O}(r/\sigma; \mathrm{Exp}_{m+q,r+\sigma|t-t_0|,q}(\mathbf{C}_z^n));$$

consequently for the solution $u(t, z)$ of system (5.2.26) we immediately see that $u'(t, z) \in \mathcal{O}(r/\sigma; \mathrm{Exp}_{m+q,r+\sigma|t-t_0|,q}(\mathbf{C}_z^n))$.

These facts will now be used for posing the dual Cauchy problem in the scale of exponential functionals.

Thus, let conditions 1) and 2) hold. Then

$$A^*(t, z, \mathcal{D}) : \mathrm{Exp}'_{m+q,r+\sigma|t-t_0|,q}(\mathbf{C}_z^n) \to \mathrm{Exp}'_{m,r+\sigma|t-t_0|,q}(\mathbf{C}_z^n)$$

(t is fixed). Moreover, for any $u(t, z) \in \mathcal{O}(r/\sigma; \mathrm{Exp}'_{m+q,r+\sigma|t-t_0|,q}(\mathbf{C}_z^n))$ the value $A^*(t, z, \mathcal{D})u$ is defined (as usual) by the formula

$$\langle A^*(t, z, \mathcal{D})u, v(z) \rangle = \langle u(t, z), A(t, z, \mathcal{D})v(z) \rangle,$$

where $v(z) \in \mathrm{Exp}_{m,r+\sigma|t-t_0|,q}(\mathbf{C}_z^n)$ is arbitrary.

It turns out that it is natural to seek the solution of the dual problem in the space of analytic functionals (in t, $|t - t_0| < r/\sigma$) such that

$$u(t, \cdot) \in \mathrm{Exp}'_{m+q,r-\sigma|t-t_0|,q}(\mathbf{C}_z^n)$$

and
$$u'(t, \cdot) \in \mathrm{Exp}'_{m, r - \sigma|t - t_0|, q}(\mathbf{C}^n_z).$$

Moreover, the functional $u'(t, \cdot)$ is defined by the equality

$$\langle u'(t, z), v(t, z) \rangle \overset{\mathrm{def}}{=} \frac{d}{dt} \langle u(t, z), v(t, z) \rangle - \langle u(t, z), v'(t, z) \rangle,$$

or, what is the same, by the integral relation

$$\int_{t_0}^t \langle u'(\tau, z), v(\tau, z) \rangle d\tau = \langle u(t, z), v(t, z) \rangle - \langle u(t_0, z), v(t_0, z) \rangle$$
$$- \int_{t_0}^t \langle u(\tau, z), v'(\tau, z) \rangle d\tau,$$

where, as usual, the integration is performed along the line segment from t_0 to t (note that for any $v(t, \cdot) \in \mathrm{Exp}_{m, r - \sigma|t - t_0|, q}(\mathbf{C}^n_z)$ the derivative $v'(t, \cdot) \in \mathrm{Exp}_{m + q, r - \sigma|t - t_0|, q}(\mathbf{C}^n_z)$).

Finally, the integral identity

$$\int_{t_0}^t \langle u', v \rangle d\tau + \int_{t_0}^t \langle A^*(\tau, z, \mathcal{D}) u, v \rangle d\tau \overset{\mathrm{def}}{=}$$

$$= \langle u(t, z), v(t, z) \rangle - \langle u(t_0, z), v(t_0, z) \rangle - \int_{t_0}^t \langle u, v' \rangle d\tau +$$

$$+ \int_{t_0}^t \langle u, A(\tau, z, \mathcal{D}) v \rangle d\tau = \int_{t_0}^t \langle h, v \rangle d\tau, \quad |t - t_0| < r/\sigma,$$

is taken in its concrete sense. This identity is the basis of the definition of the well-posedness of the Cauchy problem in the scale of exponential functionals.

Let us turn to the precise formulations.

We consider the Cauchy problem

$$u' + A^*(t, z, \mathcal{D}) u = h(t, z), \qquad (5.2.28)$$

$$u(t_0, z) = \phi(z), \quad t_0 \in G, \ z \in \mathbf{C}^n, \qquad (5.2.29)$$

where $A^*(t, z, \mathcal{D})$ is the matrix operator adjoint to the operator $A(t, z, \mathcal{D}) \equiv (A_{ij}(t, z, \mathcal{D}))_{N \times N}$. Here, the

$$A_{ij}(t, z, \mathcal{D}) \equiv \sum_\alpha a^\alpha_{ij}(t, z) \mathcal{D}^\alpha$$

are differential operators with coefficients $a_{ij}^{\alpha}(t, z)$ satisfying conditions 1) and 2) (see p. 151).

The solution of this problem is defined in the Banach space $\mathcal{O}(r/\sigma;$ $\text{Exp}'_{m+q,r-\sigma|t-t_0|,q}(\mathbb{C}_z^n))$, where $r > 0$ and $\sigma > 0$. The norm in this space is

$$\|u\|'_{r/\sigma;m+q,r,\sigma,q} \overset{\text{def}}{=} \sup_{|t-t_0|<r/\sigma} \|u(t,\cdot)\|'_{m+q,r-\sigma|t-t_0|,q},$$

$\|u(t,\cdot)\|'_{m+q,r-\sigma|t-t_0|,q}$ being the norm in the dual space $\text{Exp}'_{m+q,r-\sigma|t-t_0|,q}(\mathbb{C}_z^n)$.

Let $\phi(z) \in \text{Exp}'_{m+q,r,q}(\mathbb{C}_z^n)$ and $h(t, z) \in \mathcal{O}(r/\sigma; \text{Exp}'_{m+q,r-\sigma|t-t_0|,q}(\mathbb{C}_z^n))$, where $r > 0$ and $\sigma > 0$.

Definition 5.2.2. The functional $u(t, z) \in \mathcal{O}(r/\sigma; \text{Exp}'_{m+q,r-\sigma|t-t_0|,q}(\mathbb{C}_z^n))$ is said to be the generalized solution of the Cauchy problem (5.2.28), (5.2.29) if the following integral identity holds:

$$\langle u(t, z), v(t, z)\rangle - \int_{t_0}^{t} \langle u(\tau, z), v'(\tau, z) - A(\tau, z, \mathcal{D})v(\tau, z)\rangle d\tau =$$

$$= \langle \phi(z), v(t_0, z)\rangle + \int_{t_0}^{t} \langle h(\tau, z), v(\tau, z)\rangle d\tau, \tag{5.2.30}$$

where t is an arbitrary complex number such that $|t-t_0| < r/\sigma$. Here, as usual, the integration is taken along the line segment joining t_0 and t, and $v(t, z)$ is an arbitrary test function in $\mathcal{O}(r/\sigma; \text{Exp}_{m+q,r-\sigma|t-t_0|,q}(\mathbb{C}_z^n))$ such that

$$v'(\tau, z) - A(\tau, z, \mathcal{D})v(\tau, z) \in \mathcal{O}(r/\sigma; \text{Exp}_{m+q,r-\sigma|\tau-t_0|,q}(\mathbb{C}_z^n)), \quad \tau \in [t, t_0].$$

The main result of this section is the following theorem.

Theorem 5.2.2. Let coefficients $a_{ij}^{\alpha}(t, z)$ be polynomials in z such that

$$\deg a_{ij}^{\alpha}(t, z) \leq m_i - m_j - |\alpha|(q - 1) + q.$$

Then for any $t_0 \in G$ and any parameter $r > 0$ there exists $\sigma > 0$ such that for any $\phi(z) \in \text{Exp}'_{m+q,r,q}(\mathbb{C}_z^n)$ and $h(t, z) \in \mathcal{O}(r/\sigma; \text{Exp}'_{m+q,r-\sigma|t-t_0|,q}(\mathbb{C}_z^n))$ we have a unique generalized solution $u(t, z) \in \mathcal{O}(r/\sigma; \text{Exp}'_{m+q,r-\sigma|t-t_0|,q}(\mathbb{C}_z^n))$ of the Cauchy problem (5.2.28), (5.2.29).

Proof. The proof of this theorem is parallel to that of Theorem 4.2.2.

First we shall obtain a representation of the generalized solution which automatically implies the uniqueness of the solution. Secondly, we shall show that our representation in fact gives the desired solution.

Thus, let $u(t,z) \in \mathcal{O}(r/\sigma; \mathrm{Exp}'_{m+q,r-\sigma|t-t_0|,q}(\mathbf{C}_z^n))$ satisfy the integral identity (5.2.30), that is,

$$\langle u(t,z), v(t,z)\rangle = \int_0^t \langle u(\tau,z), v'(\tau,z) - A(\tau,z,\mathcal{D})v(\tau,z)\rangle d\tau +$$

$$+\langle \phi(z), v(0,z)\rangle + \int_0^t \langle h(\tau,z), v(\tau,z)\rangle d\tau \qquad (5.2.31)$$

(without loss of generality we can set $t_0 = 0$).

Evidently, the left hand side of this equality depends on the value of the function $v(\tau, z)$ (as a function of $\tau \in [t, 0]$) only for $\tau = t$. Therefore the function $v(\tau, z)$ on the right hand side may be chosen as a suitable extension of $v(t, z)$ on the interval $[t, 0]$. Namely, we choose $v(\tau, z)$ as the solution $v(t; \tau, z)$ of the "direct" Cauchy problem

$$v'_\tau(t; \tau, z) - A(\tau, z, \mathcal{D})v(t; \tau, z) = 0 \qquad (5.2.32)$$

$$v(t; \tau, z)\big|_{\tau=t} = v(t, z), \quad z \in \mathbf{C}^n, \qquad (5.2.33)$$

where $v(t, \cdot) \in \mathrm{Exp}_{m+q,r-\sigma|t|,q}(\mathbf{C}_z^n)$ is a test function.

According to Theorem 5.2.1, there exists a unique function $v(t; \tau, z) \in \mathcal{O}(r/\sigma; \mathrm{Exp}_{m+q,r-\sigma|\tau|,q}(\mathbf{C}_z^n))$ satisfying (5.2.32) and (5.2.33). In particular, $v(t; 0, z) \in \mathrm{Exp}_{m+q,r,q}(\mathbf{C}_z^n)$.

Hence

$$\langle u(t,z), v(t,z)\rangle = \langle \phi(z), v(t; 0, z)\rangle + \int_0^t \langle h(\tau, z), v(t; \tau, z)\rangle d\tau, \qquad (5.2.34)$$

where (we repeat) $v(t, z)$ is an arbitrary test function and $v(t; 0, z)$ is the value of the Cauchy problem (5.2.32), (5.2.33) for $\tau = 0$.

Formula (5.2.34) is, in fact, the desired representation of the generalized solution. As we have said, it implies the uniqueness of the solution.

Let us now show that the functional $u(t, z)$ defined by formula (5.2.34) satisfies the identity (5.2.30), that is, $u(t, z)$ is a solution of the Cauchy problem (5.2.28), (5.2.29) in the sense of Definition 5.2.2.

For this we need

Lemma 5.2.2. *Let* $w(\tau; \eta, z) \in \mathcal{O}(r/\sigma; \mathrm{Exp}_{m+q,r-\sigma|\eta|,q}(\mathbf{C}_z^n))$ *be a solution of the Cauchy problem*

$$w_\eta'(\tau; \eta, z) - A(\eta, z, \mathcal{D})w(\tau; \eta, z) = 0,$$

$$w(\tau; \eta, z)\big|_{\eta=\tau} = v'(\tau, z) - A(\tau, z, \mathcal{D})v(\tau, z),$$

where $v(\tau, z)$ *is a test function. Then for any* $\eta \in [\tau, 0]$ *the identity* $w(\tau; \eta, z) \equiv v_\tau'(\tau; \eta, z)$ *holds.*

To prove this, it is enough to repeat the arguments in the proof of the analogous Lemma 4.2.2. The lemma is proved.

The proof of the theorem is completed by performing the same calculations as in Theorem 4.2.2 (p. 116).

In conclusion, we note that it is not difficult to obtain the estimate

$$\|u(t, z)\|_{r/\sigma; m+q, r, \sigma, q}' \leq M(\|h(t, z)\|_{r/\sigma; m+q, r, \sigma, q}' + \|\phi(z)\|_{m+q, r, q}'),$$

where $M > 0$ is a constant. For this one must use formula (5.2.34) and the main estimate for the solutions of the "direct" Cauchy problem. The theorem is proved.

Example. We consider the problem

$$\frac{\partial u}{\partial t} - \frac{\partial}{\partial z}(zu) - u = 0, \quad u(0, z) = \delta(z).$$

In view of (5.2.34), the solution $u(t, z) \in \mathcal{O}(r/\sigma; \mathrm{Exp}_{m+q,r-\sigma|t|,q}'(\mathbf{C}_z^n))$ of this problem is defined by the formula

$$\langle u(t, z), v(t, z) \rangle = \langle \delta(z), v(t; \tau, z) \rangle\big|_{\tau=0},$$

where $v(t; \tau, z)$ is the solution of the problem

$$v_\tau'(t; \tau, z) - z\frac{\partial}{\partial z}v(t; \tau, z) + v(t; \tau, z) = 0, \quad v(t; \tau, z)\big|_{\tau=t} = v(t, z)$$

$(v(t, \cdot) \in \mathrm{Exp}_{m+q,r-\sigma|t|,q}(\mathbf{C}_z^n)$ is an arbitrary test function). It is easy to see that

$$v(t, z) = \exp(t - \tau)v(t, z\exp(\tau - t))$$

and therefore,

$$\langle u(t, z), v(t, z) \rangle = \langle \delta(z), e^t v(t, z\exp(-t)) \rangle = e^t v(t, 0).$$

This means that $u(t, z) = e^t \delta(z)$.

Chapter 6. PD-Operators with Variable Analytic Symbols

6.1. Basic definitions

In this section we investigate the exponential Cauchy problem for PD-equations with symbols that depend on $z \in \mathbf{C}^n$. This case is precisely the Fourier dual of the Cauchy-Kovalevskaya theory.

1. DEFINITION OF THE PD-OPERATOR $A(z, \mathcal{D})$

Let $A(z, \zeta)$ be an analytic function in $z \in \mathbf{C}^n$ and $\zeta \in \Omega$, where $\Omega \subset \mathbf{C}_\zeta^n$ is a Runge domain. We have

$$A(z, \zeta) = \sum_{|\alpha|=0}^{\infty} z^\alpha A_\alpha(\zeta), \qquad (6.1.1)$$

where $A_\alpha(\zeta) \in \mathcal{O}(\Omega)$.

In accordance with this expansion we set

$$A(z, \mathcal{D})u(z) \overset{\text{def}}{=} \sum_{|\alpha|=0}^{\infty} z^\alpha A_\alpha(\mathcal{D})u(z),$$

where the $A_\alpha(\mathcal{D})$ are PD-operators with symbols $A_\alpha(\zeta)$.

Let us suppose that $A(z, \zeta)$ is an entire function in z of minimal type. More precisely, for any $\epsilon > 0$ and any compact $K \subset \Omega$ there exists a constant $M > 0$ (depending, in general, on ϵ and K) such that for every $z \in \mathbf{C}^n$ and $\zeta \in K$

$$|A(z, \zeta)| \le M \exp \epsilon |z|. \qquad (6.1.2)$$

Proposition 6.1.1. *Under condition (6.1.2) the map*

$$A(z, \mathcal{D}) : \mathrm{Exp}_\Omega(\mathbf{C}_z^n) \to \mathrm{Exp}_\Omega(\mathbf{C}_z^n)$$

is well defined.

Proof. Indeed, let $u(z) \in \mathrm{Exp}_{\Omega}(\mathbf{C}_z^n)$, that is,

$$u(z) = \sum_{\lambda} e^{\lambda z} \phi_{\lambda}(z),$$

where $\phi_{\lambda}(z) \in \mathrm{Exp}_{R(\lambda)}(\mathbf{C}_z^n)$ (see §1.1). Then, taking into account the uniform convergence of the series (6.1.1) on any compact set $K \subset \Omega$ and the integral representation of a PD-operator (see subsection 3 of §1.2), we have (for any $z \in \mathbf{C}^n$)

$$A(z, \mathcal{D})u(z) = \frac{1}{(2\pi i)^n} \sum_{\lambda} e^{\lambda z} \int_{\Gamma_{\epsilon, \lambda}} A(z, \lambda + \zeta)[B\phi_{\lambda}](\zeta) e^{-z\zeta} d\zeta.$$

From this, using (6.1.2) we immediately find that $A(z, \mathcal{D})u(z) \in \mathrm{Exp}_{\Omega}(\mathbf{C}_z^n)$, if $\epsilon > 0$ is sufficiently small.

Next, let

$$u_{\nu}(z) \equiv \sum_{\lambda} u_{\lambda\nu}(z) \equiv \sum_{\lambda} e^{\lambda z} \phi_{\lambda\nu}(z) \to 0$$

in $\mathrm{Exp}_{\Omega}(\mathbf{C}_z^n)$. This means that there exist numbers $r = r(\lambda) < R(\lambda)$ such that

$$\sup_{z \in \mathbf{C}^n} |\phi_{\lambda\nu}(z)| \exp(-r|z|) \to 0 \quad (\nu \to \infty).$$

Consequently, the Borel transform $[B\phi_{\lambda\nu}](\zeta)$ tends to zero uniformly under the conditions $|\zeta_j| \geq r_1 > r$ $(j = 1, \ldots, n)$; in particular, $[B\phi_{\lambda\nu}](\zeta) \to 0$ uniformly on the contours $\Gamma_{\epsilon, \lambda}$. Hence it follows that

$$A(z, \mathcal{D})u_{\lambda\nu}(z) \to 0$$

and, consequently,

$$A(z, \mathcal{D})u_{\nu}(z) \to 0$$

in the space $\mathrm{Exp}_{\Omega}(\mathbf{C}_z^n)$. The proposition is proved.

Example. Consider the symbol

$$A(z, \zeta) = \sum_{|\alpha|=0}^{m} P_{\alpha}(z) A_{\alpha}(\zeta),$$

where the $P_{\alpha}(z)$ are polynomials. Then the PD-operator

$$A(z, \mathcal{D}) = \sum_{|\alpha|=0}^{m} P_{\alpha}(z) A_{\alpha}(\mathcal{D})$$

is a PD-operator with polynomial coefficients.

Remark. Of course, we can define the action of the PD-operator $A(z, \mathcal{D})$ as

$$A(z, \mathcal{D})u(z) = \sum_{\lambda}(\sum_{|\alpha|=0}^{\infty} a_{\alpha}(\lambda, z)(\mathcal{D} - \lambda I)^{\alpha}[e^{\lambda z}\phi_{\lambda}(z)]),$$

where $a_{\alpha}(\lambda, z) = \partial^{\alpha}A(z,0)/\alpha!$. The final results will be the same.

2. DEFINITIONS OF THE REQUISITE SPACES

In the next subsection the Cauchy problem for the PD-equations with variable symbols will be considered. We now single out certain spaces which are necessary for the study of this problem.

Let $u(z) = (u_1(z), \ldots, u_N(z))$ be an entire vector-valued function satisfying the inequality

$$|u_j(z)| \leq M_j(1 + |z|)^{m_j} \exp r|z|, \quad z \in \mathbf{C}^n,$$

where the $m_j \geq 0$ are integers, $r > 0$ and $M_j > 0$ are constants $(j = 1, \ldots, N)$.

Definition 6.1.1. We set

$$\mathrm{Exp}_{m,r}(\mathbf{C}_z^n) = \Big\{u(z) : \|u_j(z)\|_{m_j,r} \equiv$$

$$\equiv \sup_{z \in \mathbf{C}^n} |u_j(z)|(1 + |z|)^{-m_j} \exp(-r|z|) < \infty, \quad j = 1, \ldots, N\Big\}$$

(here $m = (m_1, \ldots, m_N)$).

It is not difficult to see that $\mathrm{Exp}_{m,r}(\mathbf{C}_z^n)$ is a Banach space with norm

$$\|u(z)\|_{m,r} = \|u_1(z)\|_{m_1,r} + \cdots + \|u_N(z)\|_{m_N,r}.$$

Then for $\zeta_0 \in \Omega$, we say that $u(z) \in \mathrm{Exp}_{m,r}(\zeta_0; \mathbf{C}_z^n)$ if $u(z)\exp(-\zeta_0 z) \in \mathrm{Exp}_{m,r}(\mathbf{C}_z^n)$.

These spaces will be used as the spaces of initial data for the Cauchy problem.

We now introduce the spaces of solutions of the Cauchy problem.

We shall consider solutions of two types. Namely, let $\delta > 0$. Then we set $\mathcal{O}(\delta; \mathrm{Exp}_{m,r}(\zeta_0; \mathbf{C}_z^n))$ as the space of functions analytic in t, $|t - t_0| < \delta$, such that $u(t, \cdot) \in \text{Exp}_{m,r}(\zeta_0; \mathbf{C}^n_z)$. The space $\mathcal{O}(\delta; \text{Exp}_{m,r}(\zeta_0; \mathbf{C}^n_z))$ is a Banach space with norm

$$\|u(t, z)\|_{\delta; m, r} \overset{\text{def}}{=} \sup_{|t - t_0| < \delta} \|u(t, \cdot) \exp(-\zeta_0 z)\|_{m, r}.$$

This means that the solution $u(t, z) \in \mathcal{O}(\delta; \text{Exp}_{m,r}(\zeta_0; \mathbf{C}^n_z))$ defines the group of shifts $u(t_0, \cdot) \to u(t, \cdot)$ with constant range of values $\text{Exp}_{m,r}(\zeta_0; \mathbf{C}^n_z)$.

In the second case we shall find the solution of the Cauchy problem such that $u(t, \cdot) \in \text{Exp}_{m, r + \sigma|t - t_0|}(\zeta_0; \mathbf{C}^n_z)$, where $\sigma > 0$. More precisely, let $\sigma > 0$, $R > 0$ and $r < R$.

Definition 6.1.2. We set

$$\mathcal{O}\left(\frac{R - r}{\sigma}; \text{Exp}_{m, r + \sigma|t - t_0|}(\zeta_0; \mathbf{C}^n_\zeta)\right) = \left\{u(t, z) : \|u(t, z)\|_{\frac{R-r}{\sigma}; m, r, \sigma} = \right.$$

$$\left. = \sum_{j=1}^n \sup_{|t - t_0| < \frac{R-r}{\sigma}} \|u_j(t, \cdot) \exp(-\zeta_0 z)\|_{m_j, r + \sigma|t - t_0|} < \infty\right\}.$$

It is a Banach space corresponding to the group of shifts $u(t_0, \cdot) \to u(t, \cdot)$ with variable range of values $\text{Exp}_{m, r + \sigma|t - t_0|}(\zeta_0; \mathbf{C}^n_z)$.

6.2. The Cauchy problem in the scale $\text{Exp}_{m,r}(\mathbf{C}^n_z)$

In this section we solve the Cauchy problem for the systems of PD-equations, which is dual (in the sense of the Fourier transform) to the Cauchy-Kovalevskaya problem in the scale $D_{m,r}$ (see §4.1).

We consider the Cauchy problem

$$u' - A(z, t, D)u = h(t, z) \quad (u' \equiv \partial u/\partial t) \tag{6.2.1}$$

$$u(t_0, z) = \phi(z), \quad z \in \mathbf{C}^n, \tag{6.2.2}$$

where $u = (u_1, \ldots, u_N)$ and $h = (h_1, \ldots, h_N)$ are vector-valued functions of the variables $t \in \mathbf{C}^1$ and $z \in \mathbf{C}^n$; $\phi(z) = (\phi_1(z), \ldots, \phi_N(z))$. Here $A(z, t, D) = (A_{ij}(z, t, D))_{N \times N}$ is a matrix of PD-operators with symbols $A_{ij}(z, t, \zeta)$ that are analytic in $\mathbf{C}^n_z \times V$, where $V \subset \mathbf{C}^{1+n}_{t, z}$ is a domain of variables $t \in \mathbf{C}^1$ and $z \in \mathbf{C}^n$.

We denote by Ω_t the section of V by the plane $t = $ const, that is, $\Omega_t = V \cap \{t = \text{const}\}$. We further suppose that for any $t = $ const the symbols

$A_{ij}(z,t,\zeta)$ are entire functions of minimal type satisfying condition (6.1.2). Then the PD-operators $A_{ij}(z,t,\mathcal{D})$ act in the spaces $\mathrm{Exp}_{\Omega_t}(\mathbf{C}_z^n)$ and one can pose the Cauchy problem in the classes of exponential functions.

Definition 6.2.1. We say that the Cauchy problem (6.2.1), (6.2.2) is locally well posed in the scale $\mathrm{Exp}_{m,r}(\mathbf{C}_z^n)$ if for any point $(t_0,\zeta_0) \in V$ there exist numbers $R > 0$ and $\delta > 0$ such that for every $r < R$ and arbitrary

$$h(t,z) \in \mathcal{O}(\delta; \mathrm{Exp}_{m,r}(\zeta_0; \mathbf{C}_z^n)), \quad \phi(z) \in \mathrm{Exp}_{m,r}(\zeta_0; \mathbf{C}_z^n)$$

the problem (6.2.1), (6.2.2) has a unique solution $u(t,z) \in \mathcal{O}(\delta; \mathrm{Exp}_{m,r}(\zeta_0; \mathbf{C}_z^n))$.

The main result of this section is the following theorem.

Theorem 6.2.1. *The Cauchy problem* (6.2.1), (6.2.2) *is locally well posed in the scale* $\mathrm{Exp}_{m,r}(\mathbf{C}_z^n)$ *if and only if the symbols* $A_{ij}(z,t,\zeta)$ *are polynomials in* z, *where for any* t *and* ζ

$$m_{ij} \equiv \deg A_{ij}(z,t,\zeta) \le m_i - m_j \tag{6.2.3}$$

(as before, we set $\deg A_{ij} = -\infty$ *if and only if* $A_{ij} \equiv 0$). *In addition, we have the estimate*

$$\|u(t,z)\|_{\delta;m,r} \le M(\|\phi(z)\|_{m,r} + \|h(t,z)\|_{\delta;m,r}), \tag{6.2.4}$$

where $M > 0$ *is a constant not depending on* $r < R$.

Proof. Sufficiency. First, we indicate the numbers $R > 0$ and $\delta > 0$ that define the class of solutions. Namely, let $R > 0$ and $\delta > 0$ be arbitrary numbers such that the polycylinder

$$U_{R,\delta}(\zeta_0,t_0) = \{(\zeta,t) : |\zeta_j - \zeta_{0j}| < R \quad (1 \le j \le n), \ |t - t_0| < \delta\}$$

lies strictly inside the domain V.

It is clear that if $U_{R,\delta}(\zeta_0,t_0) \subset V$, then for any $r < R$ and $\delta_0 < \delta$ we also have $U_{r,\delta_0}(\zeta_0,t_0) \subset V$ (which is important for the estimates below).

Then without loss of generality, we may choose $\zeta_0 = 0, t_0 = 0$ (by substituting $u \leftrightarrow u\exp(-\zeta_0 z)$, $t \leftrightarrow t - t_0$). Then the problem (6.2.1), (6.2.2) is equivalent to the integro-differential equation

$$u(t,z) = \int_0^t A(z,\tau,\mathcal{D})u(\tau,z)d\tau + \phi(z) + \int_0^t h(\tau,z)d\tau.$$

Let us show that under conditions (6.2.3) the operator

$$Bu(t, z) \equiv \int_0^t A(z, \tau, \mathcal{D}) u(\tau, z) d\tau$$

is a contraction in the space $\mathcal{O}(\delta; \mathrm{Exp}_{m,r}(\mathbf{C}_z^n))$ if $\delta > 0$ is sufficiently small.

Lemma 6.2.1. *If $u(z) \in \mathrm{Exp}_{m,r}(\mathbf{C}_z^n)$, then for $|\alpha| > m_j$ $(j = 1, \ldots, N)$ the following estimate holds:*

$$\|\mathcal{D}^\alpha u_j(z)\|_{m_j,r} \leq M(r + |\alpha|)^{m_j}(\alpha_1 \ldots \alpha_n)^{1/2} r^{|\alpha| - m_j} \|u_j(z)\|_{m_j,r}, \qquad (6.2.5)$$

where $M > 0$ is a constant not depending on r.

Proof. From Cauchy's formula we have, as usual,

$$|\mathcal{D}^\alpha u_j(z)| \leq \alpha! \frac{(1 + |a| + |z|)^{m_j}}{a^\alpha} \cdot e^{r(|a| + |z|)} \|u_j(z)\|_{m_j,r},$$

where the vector $a = (a_1, \ldots, a_n)$, $a_1 > 0, \ldots, a_n > 0$ is arbitrary. Setting $a = \alpha/r$ (for $r = 0$ the estimate is trivial) and using Stirling's formula, we obtain the inequality

$$|\mathcal{D}^\alpha u_j(z)| \leq M(r + |\alpha| + r|z|)^{m_j}(\alpha_1 \ldots \alpha_n)^{1/2} e^{r|z|} r^{|\alpha| - m_j} \|u_j(z)\|_{m_j,r}$$

and, consequently,

$$|\mathcal{D}^\alpha u_j(z)| \leq M(r + |\alpha|)^{m_j}(\alpha_1 \ldots \alpha_n)^{1/2}(1 + |z|)^{m_j} e^{r|z|} r^{|\alpha| - m_j} \|u_j(z)\|_{m_j,r}.$$

This is the required estimate of the lemma. Q.E.D.

We now return to the proof of our theorem. Namely, let $u(\tau, \cdot) \in \mathrm{Exp}_{m,r}(\mathbf{C}_z^n)$. Then by Lemma 6.2.1 for $|\tau| < \delta$, $r \leq R$ and for all α such that $|\alpha| \geq m_j$ $(1 \leq j \leq N)$ we obtain the inequality

$$\|\mathcal{D}^\alpha u_j(\tau, z)\|_{m_j,r} \leq M(r + |\alpha|)^{m_j}(\alpha_1 \ldots \alpha_n)^{1/2} r^{|\alpha| - m_j} \|u_j(\tau, z)\|_{m_j,r} \leq$$

$$\leq M(R + |\alpha|)^{m_j}(\alpha_1 \ldots \alpha_n)^{1/2} R^{|\alpha| - m_j} \|u_j(\tau, z)\|_{m_j,r}.$$

It is also clear that for $|\alpha| < m_j$

$$\|\mathcal{D}^\alpha u_j(\tau, z)\|_{m_j,r} \leq M \|u_j(\tau, z)\|_{m_j,r},$$

where $M = M(R, m_j)$ is a constant.

Now by hypothesis, the functions $A_{ij}(z, t, \zeta)$ have the form

$$A_{ij}(z, t, \zeta) = \sum_{|\beta| \leq m_{ij}} z^\beta A_{ij}^\beta(t, \zeta), \qquad (6.2.6)$$

where the $A_{ij}^\beta(t, \zeta)$ are analytic in the domain V. In particular, the $A_{ij}^\beta(t, \zeta)$ are analytic in $\overline{U}_{R,\delta}(0, 0)$, that is, for $|t| \leq \delta$ and $|\zeta_j| \leq R$, $j = 1, \ldots, n$. Bearing in mind this fact, we obtain for all $z \in \mathbf{C}^n$:

$$|A_{ij}^\beta(\tau, D)u_j(\tau, z)| \leq M\|u_j(\tau, z)\|_{m_j, r}(1 + |z|)^{m_j}e^{r|z|},$$

where $M > 0$ is a constant not depending on $r \leq R$.

Now from (6.2.6) we see that for $|t| \leq \delta$ and all $z \in \mathbf{C}^n$

$$|A_{ij}(z, \tau, D)u_j(\tau, z)| \leq M\|u_j(\tau, z)\|_{m_j, r}(1 + |z|)^{m_{ij}+m_j}e^{r|z|},$$

where $M > 0$ is a constant. Hence

$$\left| \int_0^t A_{ij}(z, \tau, D)u_j(\tau, z)d\tau \right| \leq M \max_{|\tau| \leq |t|} \|u_j(\tau, z)\|_{m_j, r} \times$$

$$\times (1 + |z|)^{m_{ij}+m_j}|t| \exp r|z| \leq M\delta \max_{|\tau| \leq \delta} \|u_j(\tau, z)\|_{m_j, r}(1 + |z|)^{m_i} \exp r|z|,$$

since, in view of the main condition (6.2.3), $m_{ij} + m_j \leq m_i$.

From this we immediately obtain the inequality

$$\left\| \int_0^t A_{ij}(z, \tau, D)u_j(\tau, z)d\tau \right\|_{m_i, r} \leq M\delta \max_{|\tau| \leq |t|} \|u_j(\tau, z)\|_{m_j, r}.$$

Thus for $Bu(t, z) = ((Bu)_1, \ldots, (Bu)_N)$ we have the estimate

$$\|(Bu)_i\|_{m_i, r} \leq \sum_{j=1}^N \left\| \int_0^t A_{ij}(z, \tau, D)u_j(\tau, z)d\tau \right\|_{m_j, r} \leq$$

$$\leq M\delta \sum_{j=1}^N \max_{|\tau| \leq |t|} \|u_j(\tau, z)\|_{m_j, r} \quad (i = 1, \ldots, N)$$

and therefore the estimate (in the norm of the space $\mathcal{O}(\delta; \mathrm{Exp}_{m,r}(\mathbf{C}_z^n))$)

$$\|Bu(t, z)\|_{\delta; m, r} \leq M\delta\|u(t, z)\|_{\delta; m, r}.$$

Since the constant $M > 0$ is independent of $\delta > 0$, it follows that for small $\delta > 0$ the operator Bu is a contraction in $\mathcal{O}(\delta; \mathrm{Exp}_{m,r}(\mathbf{C}_z^n))$.

To finish the proof it suffices to note that for any $h(t,z) \in \mathcal{O}(\delta; \mathrm{Exp}_{m,r}(\mathbf{C}_z^n))$ we also have

$$\int_0^t h(\tau, z)d\tau \in \mathcal{O}(\delta; \mathrm{Exp}_{m,r}(\mathbf{C}_z^n)).$$

It is clear also that $\phi(z) \in \mathcal{O}(\delta; \mathrm{Exp}_{m,r}(\mathbf{C}_z^n))$. Finally we conclude that the operator

$$Bu + \int_0^t h(\tau, z)d\tau + \phi(z)$$

has a unique fixed point, that is, the original Cauchy problem (6.2.1), (6.2.2) has a unique solution $u(t, z) \in \mathcal{O}(\delta; \mathrm{Exp}_{m,r}(\mathbf{C}_z^n))$, as required.

Necessity. We set $h(t, z) \equiv 0$ and $\phi(z) = (0, \dots, 1, \dots, 0)$, that is, $\phi_j(z) \equiv 1$, $\phi_i(z) \equiv 0$ for $i \neq j$, where j is a fixed number. Then for the corresponding solution $u(t, z) \in \mathcal{O}(\delta; \mathrm{Exp}_{m,r}(\mathbf{C}_z^n))$ of the problem (6.2.1), (6.2.2) we have for $t = 0$

$$u_i'(0, z) - A_{ij}(z, 0, 0) \equiv 0, \quad z \in \mathbf{C}^n. \tag{6.2.7}$$

In accordance with the main Definition 6.2.1, for all t, $|t| < \delta$,

$$|u_i(t, z)| \leq M(1 + |z|)^{m_i},$$

where $M > 0$ is a constant. This means that all the functions $u_i(t, z)$ are polynomials in $z \in \mathbf{C}^n$ of degree at most m_i. It follows from (6.2.7) that the $A_{ij}(z, 0, 0)$ are also polynomials with

$$\deg A_{ij}(z, 0, 0) \leq m_i.$$

As has already been noted, the case of the Cauchy problem in a neighbourhood of the point (t_0, ζ_0) reduces to the case $(0, 0)$ via the substitution $u \leftrightarrow u \exp(-\zeta_0 z)$, $t \leftrightarrow t - t_0$. Therefore, for all $(t, \zeta) \in V$ the functions $A_{ij}(z, t, \zeta)$ are polynomials in z with analytic coefficients in t and ζ, and

$$\deg A_{ij}(z, t, \zeta) \leq m_i.$$

It follows that all the derivatives (with respect to ζ) $\partial^\beta A_{ij}(z, t, \zeta)$ are also polynomials and

$$\deg \partial^\beta A_{ij}(z, t, \zeta) \leq m_i. \tag{6.2.8}$$

We now set (j is again a fixed number) $\phi_j(z) = z_k$ ($k = 1, \ldots, n$), $\phi_i(z) \equiv 0$ for $i \neq j$. Then for the corresponding solution $u(t, z)$ (we do not indicate the dependence of $u(t, z)$ on j and k) we have for all $z \in \mathbf{C}^n$

$$u_i'(0, z) - A_{ij}(z, 0, 0)z_k - \frac{\partial}{\partial \zeta_k} A_{ij}(z, 0, 0) = 0.$$

Hence, taking into account (6.2.8), we find that the functions $A_{ij}(z, 0, 0)z_k$ are polynomials in z of degree at most m_i, therefore

$$\deg A_{ij}(z, 0, 0) \leq m_i - 1.$$

As was mentioned above, from this it follows that $\deg A_{ij}(z, t, \zeta) \leq m_i - 1$ for all $(t, \zeta) \in V$.

Repeating these arguments for the initial data z^γ for $|\gamma| = 2, \ldots, m_j$ we finally obtain

$$\deg A_{ij}(z, t, \zeta) \leq m_i - m_j, \tag{6.2.3}$$

as required. The necessity and Theorem 6.2.1 are completely proved.

Remark. One can see from the proof of the necessity that the conditions (6.2.3) are a corollary of the solvability of the Cauchy problem (6.2.1), (6.2.2) with only polynomial initial data, that is, $\phi(z) \in \mathrm{Exp}_{m,r}(\mathbf{C}_z^n)$ with $r = 0$. Thus the well-posedness of the Cauchy problem for polynomial initial data implies the conditions on $\deg A_{ij}(z, t, \zeta)$ given above and therefore the well-posedness of the original Cauchy problem in the scale $\mathrm{Exp}_{m,r}(\mathbf{C}_z^n)$ for all $0 \leq r \leq R$.

Example. Let us consider the Cauchy problem for one equation of higher order

$$u^{(s)} - A_{s-1}(z, t, D)u^{(s-1)} - \ldots - A_0(z, t, D)u = h(t, z) \tag{6.2.9}$$

$$u(0, z) = \phi_0(z), \ldots, \quad u^{(s-1)}(0, z) = \phi_{s-1}(z), \tag{6.2.10}$$

where the $A_i(z, t, \zeta)$ are analytic functions in $\mathbf{C}_z^n \times V$ of minimal type in z. The substitution $u_1 = u$, $u_2 = u', \ldots, u_s = u^{(s-1)}$ gives the system

$$\vec{u}' - \begin{bmatrix} 0 & 1 & 0 & \cdots & 0 & 0 \\ 0 & 0 & 1 & \cdots & 0 & 0 \\ \vdots & \vdots & \vdots & & \vdots & \vdots \\ 0 & 0 & 0 & \cdots & 0 & 1 \\ A_0 & A_1 & A_2 & \cdots & A_{s-2} & A_{s-1} \end{bmatrix} \vec{u} = \begin{bmatrix} 0 \\ 0 \\ \vdots \\ 0 \\ h \end{bmatrix}. \tag{6.2.11}$$

In this case the restrictions (6.2.3) give the inequalities $0 \leq m_i - m_{i+1}$ ($i = 1, \ldots, s - 1$) and $\deg A_i \leq m_s - m_{i+1}$, from which it follows immediately that $\deg A_i \leq 0$ for all i and consequently, $m_i \equiv m_j \overset{\text{def}}{=} m_0$.

Conversely, if $\deg A_i \leq 0$, then by setting $m_i \equiv m_0$, where m_0 is a non-negative integer, we obtain for the system (6.2.11) the conditions (6.2.3).

Thus we see that the Cauchy problem (6.2.9), (6.2.10) is well posed in the scale $\mathrm{Exp}_{m_0,r}(\mathbf{C}_z^n)$ if and only if $A_i(z,t,\zeta) \equiv A_i(t,\zeta)$, that is, if the PD-operators $A_i \equiv A_i(t,\mathcal{D})$ do not depend on z.

6.3. The Cauchy problem in the scale $\mathrm{Exp}_{m,r+\sigma|t|}(\mathbf{C}_z^n)$

In this section we investigate the Cauchy problem for the system of PD-equations which is dual to the Cauchy-Kovalevskaya theory in the scale $D_{m,r-\sigma|t|}$ (see §4.2).

We again consider the Cauchy problem

$$u' - A(z,t,\mathcal{D})u = h(t,z), \tag{6.3.1}$$

$$u(t_0, z) = \phi(z). \tag{6.3.2}$$

The conditions under which the problem (6.3.1), (6.3.2) is considered are similar to those in §6.2. This time, however, we shall find the solution of this problem in the spaces $\mathcal{O}(\frac{R-r}{\sigma}; \mathrm{Exp}_{m,r+\sigma|t-t_0|}(\zeta_0; \mathbf{C}_z^n))$ (see Definition 6.1.2) that is, in the scale of functions $u(t,z) = (u_1(t,z), \ldots, u_N(t,z))$, satisfying inequalities

$$|u_j(t,z)| \leq M_j(1 + |z|)^{m_j} \exp(r + \sigma|t - t_0|),$$

where $\sigma > 0$. It turns out that in this case the conditions on the symbols $A_{ij}(z,t,\zeta)$ of the PD-operators $A_{ij}(z,t,\mathcal{D})$ are dual (in the Fourier transform sense) to the Leray-Volevich conditions.

Let us give the precise formulations.

Let $R > 0$, $\sigma > 0$ and $0 \leq r < R$. Further, let $(t_0, \zeta_0) \subset V$ be a fixed point.

Definition 6.3.1. The Cauchy problem (6.3.1), (6.3.2) is said to be well posed in the scale $\mathrm{Exp}_{m,r+\sigma|t|}(\mathbf{C}_z^n)$ if for any point $(t_0, \zeta_0) \in V$ there exist numbers $R > 0$ and $\sigma > 0$ such that for any $r < R$ and for any $\phi(z) \in \mathrm{Exp}_{m,r}(\mathbf{C}_z^n)$ and $h(t,z) \in \mathcal{O}\left(\frac{R-r}{\sigma}; \mathrm{Exp}_{m+1,r+\sigma|t-t_0|}(\zeta_0; \mathbf{C}_z^n)\right)$ this problem has a unique solution

$$u(t,z) \in \mathcal{O}\left(\frac{R-r}{\sigma}; \mathrm{Exp}_{m,r+\sigma|t-t_0|}(\zeta_0; \mathbf{C}_z^n)\right).$$

(Here $m + 1 = (m_1 + 1, \ldots, m_N + 1)$.)

The central result of the present chapter is the following.

Theorem 6.3.1. *The Cauchy problem* (6.3.1), (6.3.2) *is well posed in the scale* $\mathrm{Exp}_{m,r+\sigma|t|}(\mathbf{C}_z^n)$ *if and only if the symbols* $A_{ij}(z,t,\zeta)$ *are polynomials in z such that for all t and ζ*

$$\deg A_{ij}(z,t,\zeta) \leq m_i - m_j + 1 \quad (i,j = 1,\ldots,N) \tag{6.3.3}$$

(as before, $A_{ij} \equiv 0 \leftrightarrow \deg A_{ij} = -\infty$). In addition, the estimate

$$\|u(t,z)\|_{\frac{R-r}{\sigma};m,r,\sigma} \leq M\left(\|h(t,z)\|_{\frac{R-r}{\sigma};m+1,r,\sigma} + \|\phi(z)\|_{m,r}\right) \tag{6.3.4}$$

holds, where $M > 0$ does not depend on $r < R$.

Proof. The proof of this theorem runs parallel to that of Theorem 6.2.1, therefore we merely underline its main features.

First let us indicate the numbers $R > 0$ and $\sigma > 0$ by means of which the classes of solutions are defined. Namely, R and σ can be chosen as arbitrary numbers, such that the closed polycylinder

$$U_{R,R/\sigma}(t_0,\zeta_0) = \{|\zeta_j - \zeta_{0j}| \leq R \ (1 \leq j \leq n), \ |t - t_0| \leq R/\sigma\}$$

lies strictly in the interior of the original domain V.

Further, it is useful (for the estimates below) to note that once the numbers R and σ have been chosen, we then have for any $\sigma_1 > \sigma$

$$U_{R,R/\sigma_1}(t_0,\zeta_0) \subset U_{R,R/\sigma}(t_0,\zeta_0).$$

Then by setting $t_0 = 0, \zeta_0 = 0$, as usual, we obtain the equivalent integro-differential equation

$$u(t,z) = \int_0^t A(z,\tau,\mathcal{D})u(\tau,z)d\tau + \phi(z) + \int_0^t h(\tau,z)d\tau.$$

For the proof that the operator

$$Bu \equiv \int_0^t A(z,\tau,\mathcal{D})u(\tau,z)d\tau$$

is a contraction in the space $\mathcal{O}\left(\frac{R-r}{\sigma}; \mathrm{Exp}_{m,r+\sigma|t|}(\mathbf{C}_z^n)\right)$ we use Lemma 6.2.1 on the estimates of $\mathcal{D}^\alpha u$ (see estimate (6.2.5)). Namely, let $u(\tau, \cdot) \in \mathrm{Exp}_{m,r+\sigma|t|}(\mathbf{C}_z^n)$, where $|\tau| < (R-r)/\sigma$. Then in view of Lemma 6.2.1, the following inequalities hold:

$$\|\mathcal{D}^\alpha u_j(\tau, z)\|_{m_j, r+\sigma|\tau|} \le M(r+\sigma|\tau|+|\alpha|)^{m_j} (\alpha_1 \ldots \alpha_n)^{1/2} \times$$

$$\times (r+\sigma|\tau|)^{|\alpha|-m_j} \|u_j(\tau, z)\|_{m_j, r+\sigma|\tau|} \le M(R+|\alpha|)^{m_j} (\alpha_1 \ldots \alpha_n)^{1/2} \times$$

$$R^{|\alpha|-m_j} \|u_j(\tau, z)\|_{m_j, r+\sigma|\tau|},$$

since $r+\sigma|\tau| \le R$.

It is also clear that for $|\alpha| < m_j$

$$\|\mathcal{D}^\alpha u_j(\tau, z)\|_{m_j, r+\sigma|\tau|} \le M\|u_j(\tau, z)\|_{m_j, r+\sigma|\tau|},$$

where $M = M(R, m)$ is a constant.

Further, by hypothesis,

$$A_{ij}(z, \tau, \mathcal{D}) \equiv \sum_{|\beta| \le m_{ij}} z^\beta A_{ij}^\beta(t, \mathcal{D}), \qquad (6.3.5)$$

where the PD-operators $A_{ij}^\beta(t, \mathcal{D})$ have analytic symbols $A_{ij}^\beta(t, \zeta)$ in V. In particular, the $A_{ij}^\beta(t, \zeta)$ are analytic in the closed polycylinder $U_{R, R/\sigma}(0, 0)$, that is, for $|\zeta_j| \le R$ $(j = 1, \ldots, n)$ and for $|t| \le R/\sigma$. Consequently, from the last estimates we find that for all $z \in \mathbb{C}^n$

$$|A_{ij}^\beta(\tau, \mathcal{D})u_j(\tau, z)| \le M\|u_j(\tau, z)\|_{m_j, r+\sigma|\tau|} \cdot (1+|z|)^{m_j} e^{(r+\sigma|\tau|)|z|},$$

where $M > 0$ is a constant not depending on $r < R$ and $\sigma > 0$.

Now using (6.3.5), we see that for all $|t| \le (R-r)/\sigma$ and $z \in \mathbb{C}^n$

$$|A_{ij}(z, \tau, \mathcal{D})u_j(\tau, z)| \le M\|u_j(\tau, z)\|_{m_j, r+\sigma|\tau|} \cdot (1+|z|)^{m_{ij}+m_j} e^{(r+\sigma|\tau|)|z|}$$

and therefore,

$$\left| \int_0^t A_{ij}(z, \tau, \mathcal{D})u_j(\tau, z)d\tau \right| \le M \max_{|\tau| \le |t|} \|u_j(\tau, z)\|_{m_j, r+\sigma|\tau|} \times$$

$$\times (1+|z|)^{m_{ij}+m_j} \int_0^{|t|} \exp \sigma(1+|z|)|\tau|d|\tau| \cdot \exp r|z| \le$$

$$\le \frac{M}{\sigma} \max_{|\tau| \le |t|} \|u_j(\tau, z)\|_{m_j, r+\sigma|\tau|} \cdot (1+|z|)^{m_{ij}+m_j-1} \exp(r+\sigma|t|)|z| \le$$

$$\le \frac{M}{\sigma} \max_{|\tau| \le |t|} \|u_j(\tau, z)\|_{m_j, r+\sigma|\tau|} (1+|z|)^{m_i} \exp(r+\sigma|t|)|z|,$$

since $m_{ij} \leq m_i - m_j + 1$.

From this we immediately obtain the inequality

$$\left\| \int_0^t A_{ij}(z,\tau,\mathcal{D})u_j(\tau,z)d\tau \right\|_{m_i,r+\sigma|t|} \leq \frac{M}{\sigma} \max_{|\tau|\leq|t|} \|u_j(\tau,z)\|_{m_j,r+\sigma|\tau|}.$$

Thus for $Bu(t,z) \equiv \left((Bu)_1,\ldots,(Bu)_N\right)$ we have the estimate

$$\|(Bu)_i\|_{m_i,r+\sigma|t|} \leq \sum_{j=1}^N \left\| \int_0^t A_{ij}(z,\tau,\mathcal{D})u_j(\tau,z)d\tau \right\|_{m_j,r+\sigma|t|} \leq$$

$$\leq \frac{M}{\sigma} \sum_{j=1}^N \max_{|\tau|\leq|t|} \|u_j(\tau,z)\|_{m_j,r+\sigma|\tau|}, \quad i=1,\ldots,N,$$

and, consequently, the estimate

$$\|Bu(t,z)\|_{\frac{R-r}{\sigma};m,r,\sigma} \leq \frac{M}{\sigma} \|u(t,z)\|_{\frac{R-r}{\sigma};m,r,\sigma}$$

(in the norm of the space $\mathcal{O}\left(\frac{R-r}{\sigma};\mathrm{Exp}_{m,r+\sigma|t|}(\mathbf{C}_z^n)\right)$).

Since the constant $M > 0$ does not depend on σ, it follows that for sufficiently large $\sigma > 0$ the operator Bu is a contraction in $\mathcal{O}\left(\frac{R-r}{\sigma};\mathrm{Exp}_{m,r+\sigma|t|}(\mathbf{C}_z^n)\right)$.

To finish the proof, it suffices to note that for any $h(t,z) \in \mathcal{O}\left(\frac{R-r}{\sigma};\mathrm{Exp}_{m+1,r+\sigma|t|}(\mathbf{C}_z^n)\right)$ its primitive belongs to $\mathcal{O}\left(\frac{R-r}{\sigma};\mathrm{Exp}_{m,r+\sigma|t|}(\mathbf{C}_z^n)\right)$. It is also clear that $\phi(z) \in \mathcal{O}\left(\frac{R-r}{\sigma};\mathrm{Exp}_{m,r+\sigma|t|}(\mathbf{C}_z^n)\right)$.

Thus, finally, we have shown that the operator

$$Bu + \phi(z) + \int_0^t h(\tau,z)d\tau$$

is a contraction in $\mathcal{O}\left(\frac{R-r}{\sigma};\mathrm{Exp}_{m,r+\sigma|t|}(\mathbf{C}_z^n)\right)$ and therefore, has a unique fixed point $u(t,z) \in \mathcal{O}\left(\frac{R-r}{\sigma};\mathrm{Exp}_{m,r+\sigma|t|}(\mathbf{C}_z^n)\right)$. This clearly implies that the original Cauchy problem has a unique solution also. The sufficiency of the conditions of Theorem 6.3.1 is proved.

Necessity. Suppose that the Cauchy problem (6.3.1), (6.3.2) is well posed in the scale $\mathrm{Exp}_{m,r+\sigma|t|}(\mathbf{C}_z^n)$ in the sense of Definition 6.3.1. We must prove (6.3.3), that is, all the symbols $A_{ij}(z,t,\zeta)$ are polynomials in z such that for any $(t,\zeta) \in V$

$$\deg A_{ij}(z,t,\zeta) \leq m_i - m_j + 1. \tag{6.3.6}$$

Indeed, setting $h(t,z) \equiv 0$, $\phi(z) = (0,\ldots,1,\ldots,0)$, that is, $\phi_j(z) \equiv 1$, $\phi_i(z) \equiv 0$ $(i \neq j)$, where j is a fixed number, for the corresponding solution $u(t,z) \in \mathcal{O}\left(\frac{R}{\sigma}; \mathrm{Exp}_{m,\sigma|t|}(\mathbf{C}_z^n)\right)$ we have

$$u_i'(0,z) - A_{ij}(z,0,0) \equiv 0, \quad z \in \mathbf{C}_z^n. \tag{6.3.7}$$

Since for $|t| \leq R/\sigma$ and $z \in \mathbf{C}^n$

$$|u_i(t,z)| \leq M(1+|z|)^{m_i} \exp \sigma|tz|,$$

it follows from Cauchy's formula that

$$|u_i'(0,z)| \leq M(1+|z|)^{m_i+1},$$

where $M = M(R,\sigma)$ is a constant. Consequently (6.3.7) implies that

$$|A_{ij}(z,0,0)| \leq M(1+|z|)^{m_i+1},$$

that is, the entire function $A_{ij}(z,0,0)$ is a polynomial in z with

$$\deg A_{ij}(z,0,0) \leq m_i + 1.$$

Repeating the final argument in the necessity part of the proof of Theorem 6.2.1, we then obtain (in the same way) the desired inequalities (6.3.3). The theorem is completely proved.

Example. Let us consider the Cauchy problem for one equation of higher order $s \geq 1$

$$u^{(s)} - A_{s-1}(z,t,D)u^{(s-1)} - \ldots - A_0(z,t,D)u = h(t,z) \tag{6.3.8}$$

$$u(0,z) = \phi_0(z),\ldots, \quad u^{(s-1)}(0,z) = \phi_{s-1}(z), \tag{6.3.9}$$

where the $A_i(z,t,\zeta)$ are entire functions in z of minimal type and analytic in $(t,\zeta) \in V$.

After the standard substitution $u_1 = u$, $u_2 = u',\ldots,$ $u_s = u^{(s-1)}$ we clearly obtain the Cauchy problem for the system

$$\vec{u}' - \begin{bmatrix} 0 & 1 & 0 & \ldots & 0 & 0 \\ 0 & 0 & 1 & \ldots & 0 & 0 \\ \vdots & \vdots & \vdots & & \vdots & \vdots \\ 0 & 0 & 0 & \ldots & 0 & 1 \\ A_0 & A_1 & A_2 & \ldots & A_{s-2} & A_{s-1} \end{bmatrix} \vec{u} = \begin{bmatrix} 0 \\ 0 \\ \vdots \\ 0 \\ h \end{bmatrix}. \tag{6.3.10}$$

In this case the restriction (6.3.6) shows that $0 \leq m_i - m_{i+1} + 1$ $(i = 1, \ldots, s - 1)$ and $\deg A_i \leq m_s - m_{i+1} + 1$. These inequalities immediately imply that $\deg A_i \leq s - i$, which are the Kovalevskaya conditions on the z variables.

Conversely, if $\deg A_i \leq s - i$, then, reducing (6.3.8) to the system (6.3.10), it is enough to put $m_i = k + i - 1$, where $k \in \mathbf{N}$ is arbitrary. Thus the Cauchy problem (6.3.8), (6.3.9) is well posed in the scale $\mathrm{Exp}_{m, r + \sigma |t|}(\mathbf{C}_z^n)$ if and only if the characteristic symbol

$$\lambda^s - A_{s-1}(z, t, \zeta)\lambda^{s-1} - \ldots - A_0(z, t, \zeta)$$

is a Kovalevskaya polynomial in $z \in \mathbf{C}^n$.

Conclusion: The connection between the Cauchy exponential theory and the classical Cauchy-Kovalevskaya theory

As was shown (§2.2, Part I) there exists the following connection, via the Fourier transformation, between analytic functions and exponential functionals, on the one hand, and analytic functionals and exponential functions on the other. This connection finds its expression in the following diagram

$$
\begin{array}{ccc}
\mathcal{O}(G) & \underset{\mathcal{F}^{-1}}{\overset{\mathcal{F}}{\rightleftarrows}} & \mathrm{Exp}'_G(\mathbf{C}_\xi^n) \\
\downarrow * & & \downarrow * \\
\mathcal{O}'(G) & \underset{\mathcal{F}^{-1}}{\overset{\mathcal{F}}{\rightleftarrows}} & \mathrm{Exp}_G(\mathbf{C}_\xi^n)
\end{array}
$$

where $\mathcal{F}$ is the Fourier transformation and $*$ is the operation of taking the adjoint.

As follows from this diagram, there exists a connection between the corresponding Cauchy problems, which we indicate in the form of a table. Namely, let $A(t, z, \mathcal{D})$ be a matrix of differential operators with analytic coefficients (for simplicity we confine ourselves to a system of first order in $t \in \mathbf{C}^1$). Then we have the following table of connections:

$$\boxed{\begin{array}{c} u' - A(t,z,\mathcal{D})u = h(t,z) \\ u(t_0(z) = u_0(z) \\ \text{solutions are} \\ \text{analytic functions} \end{array}} \quad \overset{\mathcal{F}}{\underset{\mathcal{F}^{-1}}{\longrightarrow}} \quad \boxed{\begin{array}{c} \tilde{u}' - A(t,\partial,\zeta)\tilde{u} = \tilde{h}(t,\zeta) \\ \tilde{u}'(t_0,\zeta) = \tilde{u}_0(z) \\ \text{solutions are} \\ \text{exponential functionals} \end{array}}$$

$$\downarrow * \qquad\qquad\qquad\qquad\qquad\qquad\qquad \downarrow *$$

$$\boxed{\begin{array}{c} v' - A^*(t,z,\mathcal{D})v = h(t,z) \\ v(t_0,z) = v_0(z) \\ \text{solutions are} \\ \text{analytic functionals} \end{array}} \quad \overset{\mathcal{F}}{\underset{\mathcal{F}^{-1}}{\longleftarrow}} \quad \boxed{\begin{array}{c} \tilde{v}' - A^*(t,-\partial,\zeta)\tilde{v} = \tilde{f}(t,\zeta) \\ \tilde{v}'(t_0,\zeta) = \tilde{v}_0(\zeta) \\ \text{solutions are} \\ \text{exponential functions} \end{array}}$$

We recall that $\zeta = (\zeta_1, \dots, \zeta_n)$ are dual arguments for $z = (z_1, \dots, z_n)$; $\partial = (\partial/\partial\zeta_1, \dots, \partial/\partial\zeta_n)$ is the symbol of differentiation with respect to ζ.

As was noted, all these Cauchy problems are equivalent from the point of view of well-posedness. Namely, if one of them is well posed then every other Cauchy problem is well posed. For instance, after investigating the Cauchy problem in the classes of analytic functions we obtain the solvability of the Cauchy problem in the classes of exponential functionals and then the solvability of the Cauchy problem in the corresponding exponential functions and, finally, in the classes of analytic functionals.

The table opposite may be useful for a concrete realization of this scheme (see Taylor [1]).

Here, as usual, $M(r; u) = \max\limits_{|z|=r} |u(z)|$, moreover, $r \to +\infty$. Here, $\mathcal{E}$ denotes some space of entire functions, $\mathcal{F}^{-1}(\mathcal{E}')$ is the inverse image of the Fourier transformation of the space $\mathcal{E}'$ of linear functionals on $\mathcal{E}$. In accordance with the above diagram, Taylor's table is to be read according to the scheme

$$\mathcal{F}^{-1}(\mathcal{E}') \quad \longleftarrow \quad \mathcal{E}'$$
$$\uparrow$$
$$\mathcal{E}$$

In conclusion, let us note that in accordance with this scheme, if $\mathcal{E}$ is the space of entire functions of order $p > 1$, then $\mathcal{F}^{-1}(\mathcal{E}')$ is the space of

$$\mathcal{F}^{-1}(\mathcal{E}') \longleftarrow \mathcal{E}$$

$\mathcal{F}^{-1}(\mathcal{E}')$	$\mathcal{E}$
1. All entire functions	All entire functions of exponential type
2. All analytic functions for $\|z\| < R$	All entire functions of exponential type $< R$ (i.e. $\log M(r; u) \leq \tau r + O(1)$, $\tau < R$)
3. All entire functions of order ≤ 1 (i.e. $\log M(r; f) = O(r^{1+\epsilon})$ for any $\epsilon > 0$)	All entire functions of finite order (i.e. $\log M(r; u) = O(r^p)$ for some $p > 0$)
4. All entire functions of order $\leq q$, where $q^{-1} + p^{-1} = 1$ (i.e. $\log M(r; f) = O(r^{q-\epsilon})$ for any $\epsilon > 0$)	All entire functions of order $\leq p$ ($p \geq 1$) (i.e. $\log M(r; u) = O(r^{p+\epsilon})$ for any $\epsilon > 0$)
5. All entire functions of order $\leq q$, $q^{-1} + p^{-1} = 1$ and minimal type (i.e. $\log M(r; f) = o(r^q)$)	All entire functions of finite type (i.e. $\log M(r; u) = O(r^p), p \geq 1$)
6. All entire functions of order $\leq q$ and of type $\leq \sigma$, where $q^{-1} + p^{-1} = 1$, $\sigma = (pR)^{1-q}/q$ (i.e. $\log M(r; f) \leq (\sigma + \epsilon)^q + O(1)$ for every $\epsilon > 0$)	All entire functions of order $\leq p$ ($p > 1$) and of type $< R$ (i.e. $\log M(r; u) \leq (R - \epsilon)r^p + O(1)$ for every $\epsilon > 0$)
7. All entire functions such that $\log M(r; f) = o(r \log r)$	All entire functions such that $\log \log M(r; u) = O(r)$
8. All entire functions such that $\log M(r; f) = O(r \log r)$	All entire functions such that $\log \log M(r; u) = o(r)$

— * —

entire functions of order q, where $q^{-1} + p^{-1} = 1$. This means that the Fourier transformation of the Cauchy problem in the scale of type $\mathrm{Exp}_{m,r+\sigma|t|,q}(\mathbf{C}_z^n)$ gives the analogous Cauchy problem in the scale of entire functions of order q (see the above diagram). Thus, the Cauchy problem in the scale of entire functions of order greater than unity is invariant with respect to the Fourier transformation in the above sense.

Part III. PD-Operators with Real Arguments

Introduction

As is known, the theory of real PD-operators in recent years has become an effective tool for the analysis of partial differential equations and, in particular, equations of mathematical physics. The foundations of this theory were laid by Kohn, Nirenberg, Hörmander, Maslov and were then developed by many other mathematicians. One can now regard this theory as classical.

In the last Part III we also give a version of the theory of real PD-operators, more precisely, the theory of real PD-operators whose symbols are arbitrary analytic functions in $G \subset \mathbf{R}^n$.

By contrast with the classical theory, we do not impose any restrictions on the asymptotic behaviour of the symbols. Our approach to the construction of real analytic PD-operators is, roughly speaking, the "real trace" of the complex theory which has been developed in the two earlier parts of this book.

We now describe briefly some additional details concerning the contents of the last part.

As before, our main technical tool is the algebra of differential operators of infinite order. In its turn, this algebra is based on the theory of test and generalized functions $H^\infty(G)$ and $H^{-\infty}(G)$.

By definition, the test-function space $H^\infty(G)$ contains precisely those functions $u(x) \in L_2(\mathbf{R}^n)$ whose Fourier transforms have compact support in G. In other words, the space $H^\infty(G)$ consists of those functions $u(x) : \mathbf{R}^n \to \mathbf{C}^1$ that have an analytic continuation to the complex space $\mathbf{C}^n$ as entire functions of at most first order growth and of finite type. Locally (with respect to the dual variables $\xi \in \mathbf{R}^n$) $H^\infty(G)$ can be regarded as the inductive limit of the spaces of such functions, when the type of the entire functions tends to some limit. In conformity with the Paley-Wiener theorem this corresponds to the process of extending the supports of the Fourier transforms.

Thus, our approach to the construction of the space of generalized functions is based on the process of extension over compact sets in the Fourier dual space $\mathbf{R}^n_\xi$ rather than in the space $\mathbf{R}^n_x$, as is usually done. Thus, we extend over the spectral sets of the elementary operators $D_j \equiv i^{-1}\partial/\partial x_j$, $j = 1, \ldots, n$.

We also note in this context that this idea of construction of the space of generalized functions is in full agreement with the basic concept of the computational projection methods of Ritz, Galerkin and others. It is useful to note that in conformity with these methods the problem is projected onto the subspace of its eigenfunctions and adjoint functions, because the approximate problem is simplified greatly under such a projection.

By virtue of the duality $(H^\infty(G), H^{-\infty}(G))$, the construction of the PD-operator $A(\mathcal{D})$ with $A(\xi) \in \mathcal{O}(G)$ is given by analogy with the complex case.

Remark. Of course, the duality $(H^\infty(G), H^{-\infty}(G))$ is not the only way of constructing the theory of real analytic PD-operators. Recently Tran Duc Van [1], [2], [3], Trin Ngok Min and Tran Duc Van [1] gave another duality $(W^\infty, W^{-\infty})$, based on the Fourier transform in the Schwartz spaces (S, S') and developed a subsequent theory of PD-operators and its applications.

The original applications were given in Tran Duk Van and Dinh Nho Hào [1]; Tran Duk Van, Dinh Ngo Hào, Trinh Ngoc Minh and R. Gorenflo [1] and others.

It is clear that the PD-operators introduced here are identical with the usual PD-operators

$$A(\mathcal{D})u(x) = \frac{1}{(2\pi)^n} \int_G A(\xi)\tilde{u}(\xi)e^{ix\xi}\,d\xi$$

and one should be able to obtain all the results below using this integral representation. But in our opinion, the technique of differential operators of infinite order has a number of significant advantages, the principal one being that it does not require the introduction of dual variables. This is the reason why this technique gives a new insight into many problems that either have a long history (for example, the Cauchy problem for partial differential equations, see §§8.3, 8.4) or are relatively recent (for example, the Schrödinger equation for a relativistic free particle, see §8.5).

In conclusion we note the interesting papers of A. Hrennikov [1], [2], [3]. In these papers the author gives the development of the theory of analytic PD-operators in the cases of infinite dimensional space and Vladimirov-Volovich superanalysis.

Chapter 7. Spaces of Test Functions and Distributions

7.1. The test-function space $H^\infty(S_R)$

Here we introduce the test-function space $H^\infty(S_R)$, which is the basic local element of the duality $\{H^\infty(G), H^{-\infty}(G)\}$.

1. DEFINITION AND EXAMPLES OF TEST FUNCTIONS

Let $x \in \mathbf{R}^n$, $n \geq 1$ and $\xi \in \mathbf{R}^n$ be real variables. Further, let $R = (R_1, \ldots, R_n)$ be a real vector, where $R_1 > 0, \ldots, R_n > 0$ and let

$$a(\xi) \equiv \sum_{|\alpha|=0}^{\infty} a_\alpha \xi^{2\alpha}, \quad a_\alpha \geq 0, \tag{7.1.1}$$

be any function analytic in $S_R = \{\xi : |\xi_j| < R_j, \ 1 \leq j \leq n\}$. (As usual, $\alpha = (\alpha_1, \ldots, \alpha_n)$ is a multi-index, $\alpha_1 \geq 0, \ldots, \alpha_n \geq 0$; $|\alpha| = \alpha_1 + \ldots + \alpha_n$; $\xi^{2\alpha} = \xi_1^{2\alpha_1} \ldots \xi_n^{2\alpha_n}$.) We note that if $R_j = +\infty$, $1 \leq j \leq n$, then $a(\xi)$ is an entire function defined on the whole Euclidean space $\mathbf{R}^n$ with values in $\mathbf{C}^1$.

Definition 7.1.1. The test-function space $H^\infty(S_R)$ is the space of functions $\phi(x) \in L_2(\mathbf{R}^n)$ for which the following two conditions are satisfied:

1) the Fourier transform $\tilde{\phi}(\xi)$ of $\phi(x)$ equals zero (a.e.) for $\xi \notin S_R$, that is, $\tilde{\phi}(\xi) = 0$ (a.e.) outside S_R;

2) for every function $a(\xi)$ of type (7.1.1)

$$\|\phi\|_a^2 \equiv \int_{S_R} a(\xi)|\tilde{\phi}(\xi)|^2 d\xi < \infty.$$

Using integration over the whole space $\mathbf{R}^n$, we can define our space $H^\infty(S_R)$ in the following way. Namely, let

$$a(\xi) = \begin{cases} \sum_{|\alpha|=0}^{\infty} a_\alpha \xi^{2\alpha}, & \xi \in S_R; \\ +\infty, & \xi \notin S_R \end{cases} \tag{7.1.2}$$

be an arbitrary function (obviously $a(\xi)$ is analytic in S_R). Then

$$H^\infty(S_R) = \left\{ \phi(x) \in L_2(\mathbf{R}^n) : \|\phi\|_a^2 \equiv \int_{\mathbf{R}^n} a(\xi)|\tilde{\phi}(\xi)|^2 d\xi < \infty \right\}.$$

This means that the finiteness of the norm $\|\phi\|_a$ implies the equality $\tilde{\phi}(\xi) \equiv 0$ for (almost all) $\xi \notin S_R$ and consequently, the integration is, in fact, only over S_R.

We note the special case $R_j = +\infty$ ($1 \leq j \leq n$) which is of frequent occurrence in mathematical physics and in the theory of partial differential equations. In this case there is no condition 1) in the definition of the space $H^\infty(S_R) \equiv H^\infty(\mathbf{R}^n)$. More precisely,

$$H^\infty\{\mathbf{R}^n\} = \left\{ u(x) \in L_2(\mathbf{R}^n) : \|\phi\|_a^2 \equiv \int_{\mathbf{R}^n} a(\xi)|\tilde{\phi}(\xi)|^2 d\xi < \infty \right\}$$

for any entire function $a(\xi)$ of type (7.1.2). In other words, the inclusion $\phi(x) \in H^\infty(\mathbf{R}^n)$ is equivalent to the inclusions $\tilde{\phi}(\xi) \in L_{2,a}(\mathbf{R}^n)$ where $L_{2,a}(\mathbf{R}^n)$ is Lebesgue space with arbitrary weight $a(\xi) \geq 0$.

Let us now give some simple examples of test functions $\phi(x) \in H^\infty(S_R)$:

$$\phi(x) = \frac{\sin rx}{x}, \quad \phi(x) = \frac{e^{irx} - 1}{x}, \quad \phi(x) = \frac{\sin^2 \frac{r}{2} x}{x},$$

where $r < R$. After some simple calculations we see that the Fourier transforms of these functions are

$$F\left[\frac{\sin rx}{x}\right] = \begin{cases} \frac{1}{2}, & x \in (-r, r); \\ 0, & x \notin (-r, r); \end{cases}$$

$$F\left[\frac{e^{irx}}{x}\right] = \begin{cases} i, & x \in (0, r); \\ 0, & x \notin (0, r); \end{cases}$$

$$F\left[\frac{\sin^2 \frac{r}{2} x}{x}\right] = \begin{cases} -i, & x \in (0, r); \\ i, & x \in (-r, 0); \\ 0, & x \notin (-r, r); \end{cases}$$

that is, they all have compact support. Moreover, it is evident that

$$\|\phi\|_a^2 = \int_{S_R} a(\xi)|\tilde{\phi}(\xi)|^2 d\xi < \infty,$$

that is, $\phi(x) \in H^\infty(S_R)$.

In general, if $\operatorname{supp} \tilde{\phi}(\xi) \subset [-r, r]$ and $\phi(x) \in L_2(\mathbf{R}^n)$, then $\phi(x) = F^{-1}[\tilde{\phi}]$ is a function in any $H^\infty(S_R)$, where $R > r$. We shall prove below that the converse also holds.

2. DESCRIPTION OF $H^\infty(S_R)$ IN THE x-VARIABLES

To describe the space $H^\infty(S_R)$ in the above sense, we prove the following proposition about $\phi(x) \in L_2(\mathbf{R}^n)$.

Proposition 7.1.1. *The inclusion* $\phi(x) \in H^\infty(S_R)$ *holds if and only if there are numbers* $r_j < R_j$ $(1 \le j \le n)$ *such that* supp $\tilde\phi(\xi) \in S_r$ *(which is strictly smaller than* S_R*).*

Proof. The sufficiency is trivial. Consequently, we need to prove only the necessity.

Assume the contrary. Then there exists an increasing sequence $S_{R_n} \to S_R$ as $n \to \infty$ such that for all $n = 0, 1, \ldots$

$$\int_{S_{R_{n+1}} \setminus S_{R_n}} |\tilde\phi(\xi)|^2 d\xi = \alpha_n \to +\infty.$$

We choose a sequence of positive numbers β_n such that $\alpha_0\beta_0 + \alpha_1\beta_1 + \ldots < \infty$. Then we can construct a function $a(\xi)$ such that

$$\|\phi\|_a^2 \equiv \int_{S_R} a(\xi)|\tilde\phi(\xi)|^2 d\xi \ge \sum_{n=0}^{\infty} \alpha_n\beta_n = +\infty,$$

with $a(\xi)$ analytic in S_R.

In order to do this, we set

$$a_0(\xi) = \sum_{|\alpha|=0}^{m_0} \frac{1}{R_0^{2\alpha}} \xi^{2\alpha}$$

(here, as usual, $R_0^{2\alpha} = R_{01}^{2\alpha_1} \ldots R_{0n}^{2\alpha_n}$ and $\xi^{2\alpha} = \xi_1^{2\alpha_1} \ldots \xi_n^{2\alpha_n}$), where m_0 is chosen so that $a_0(R_0) \ge \beta_0$. Similarly, we set

$$a_1(\xi) = \sum_{|\alpha|=m_0+1}^{m_1} \frac{1}{R_1^{2\alpha}} \xi^{2\alpha},$$

where m_1 is chosen so that $a_1(R_1) \ge \beta_1$ and so on. Finally we set

$$a(\xi) = a_0(\xi) + a_1(\xi) + \ldots.$$

It is easy to see that $a(\xi)$ is analytic in S_R and, at the same time,

$$\|\phi\|_a^2 \equiv \int_{S_R} a(\xi)|\tilde\phi(\xi)|^2 d\xi = \sum_{n=0}^{\infty} \int_{S_{R_{n+1}} \setminus S_{R_n}} a(\xi)|\tilde\phi(\xi)|^2 d\xi \ge$$

$$\ge \sum_{n=0}^{\infty} \alpha_n\beta_n = +\infty.$$

This fact implies that $\phi(x) \notin H^\infty(S_R)$, which contradicts our assumption. The necessity is proved, which completes the proof of Proposition 7.1.1.

Strictly speaking, we should also consider the case supp $\tilde{\phi}(\xi) \notin S_R$. But in this case, it is evident that $\phi(x) \notin H^\infty(S_R)$.

Using the Paley-Wiener theorem, we obtain from Proposition 7.1.1 the following corollary.

Corollary. *A function $\phi(x) \in L_2(\mathbf{R}^n)$ is a test function in $H^\infty(S_R)$ if and only if there exists an entire function $\phi(z)$, $z = x + iy \in \mathbf{C}^1$, such that $\phi(x) = \phi(z)\big|_{z=x}$ and*

$$|\phi(z)| \le C \exp(r_1|y_1| + \ldots + r_n|y_n|),$$

where $r_j < R_j$ $(1 \le j \le n)$ and $C > 0$ is a constant.

This is the desired description of $H^\infty(S_R)$ in the x-representation.

Remark (the density of $H^\infty(S_R)$ in $L_2(\mathbf{R}^n)$). As is well known, the functions with compact support are dense in $L_2(\mathbf{R}^n)$. Consequently, if $R_j = +\infty$ $(1 \le j \le n)$, then the Fourier transform $FH^\infty(\mathbf{R}^n)$ of $H^\infty(\mathbf{R}^n)$ is dense in $L_2(\mathbf{R}^n)$. Hence by Parseval's equality $H^\infty(\mathbf{R}^n)$ is also dense in $L_2(\mathbf{R}^n)$.

For $S_R \ne \mathbf{R}^n$ the space $H^\infty(S_R)$ is dense only in the subspace of $L_2(\mathbf{R}^n)$ consisting of those functions in $L_2(\mathbf{R}^n)$ that have an analytic continuation $\phi(z)$ to $\mathbf{C}^n$ and for which

$$|\phi(z)| \le C \exp(r_1|y_1| + \ldots + r_n|y_n|),$$

where $C > 0$ is a constant, $r_j < R_j$ $(1 \le j \le n)$.

3. CONVERGENCE IN $H^\infty(S_R)$

Definition 7.1.2. We say that a sequence $\phi_m(x)$, $m = 1, 2, \ldots$, converges to $\phi(x)$ in $H^\infty(S_R)$ if $\|\phi(x) - \phi_m(x)\|_a \to 0$ for any weight function $a(\xi)$ of type (7.1.1). One can characterize the convergence without the use of the weight functions $a(\xi)$ as follows:

Proposition 7.1.2. *A sequence $\phi_m(x)$ converges to $\phi(x)$ in $H^\infty(S_R)$ if and only if there are numbers $r_j < R_j$ $(1 \le j \le n)$ for which the two conditions*

1) $\text{supp } \tilde{\phi}_m(\xi) \subset S_r = \{\xi : |\xi_j| < r_j \ (1 \le j \le n)\}$;

2) $\tilde{\phi}_m(\xi) \to \tilde{\phi}(\xi)$ in $L_2(\mathbf{R}^n)$

hold.

Proof. Clearly we need to prove only the necessity.

Necessity. First we note that without loss of generality we may assume that supp $\tilde{\phi}(\xi) \subset S_r$. We can also put $\phi(x) \equiv 0$, that is, we can assume that $\phi_m(x) \to 0$ in $H^\infty(S_R)$.

Suppose now that the supp $\tilde{\phi}_m(\xi)$, $m = 1, 2, \ldots$, are not contained in any parallelepiped S_r, where $r < R$. Then there exists a sequence of vectors $R_n \uparrow R$ (monotonically) such that for any $m = 1, 2, \ldots$ we can find a function $\phi_{sm}(x)$ such that

$$\int_{S_{R_{m+1}} \setminus S_{R_m}} |\tilde{\phi}_{sm}(\xi)|^2 d\xi = \alpha_m > 0.$$

Choosing $\beta_n = 1/\alpha_n$, we use the weight function $a(\xi)$ from the proof of Proposition 7.1.1. Then from the construction of $a(\xi)$ we obtain

$$\|\phi_{sm}(x)\|_a^2 \geq \alpha_n \beta_n = 1,$$

which contradicts the fact that $\phi_m(x) \to 0$ in $H^\infty(S_R)$. The necessity of condition 1) is proved. The necessity of 2) is evident. Proposition 7.1.2 is proved.

4. INVARIANCE OF $H^\infty(S_R)$ UNDER DIFFERENTIAL OPERATORS OF INFINITE ORDER

Let

$$A(\mathcal{D}) \equiv \sum_{|\alpha|=0}^{\infty} a_\alpha \mathcal{D}^\alpha, \quad a_\alpha \in \mathbb{C}^1, \qquad (7.1.3)$$

be a differential operator with constant coefficients, $\mathcal{D} = \left(\frac{1}{i} \frac{\partial}{\partial x_1}, \ldots, \frac{1}{i} \frac{\partial}{\partial x_n} \right)$. Let us suppose that the symbol of this operator

$$A(\xi) \equiv \sum_{|\alpha|=0}^{\infty} a_\alpha \xi^\alpha, \quad \xi \in \mathbb{R}^n,$$

is an analytic function in S_R.

From the definition of $H^\infty(S_R)$ and by Propositions 7.1.1 and 7.1.2, we have the following.

Main property of $H^\infty(S_R)$. *For any $\phi(x) \in H^\infty(S_R)$ the action of $A(D)$ is well-defined for any differential operator of infinite order (7.1.3) and $A(D)\phi(x) \in H^\infty(S_R)$. The map*

$$A(D) : H^\infty(S_R) \to H^\infty(S_R)$$

is continuous.

5. AN EXAMPLE

In this subsection we consider an example which is useful in the investigation of the Cauchy problem. Namely, we consider the function

$$u(t,x) = \exp\left(at\frac{d^2}{dx^2}\right)\phi(x),$$

where $t \in \mathbf{R}^1$, $a \in \mathbf{C}^1$, $|a| = 1$, and $\phi(x) \in H^\infty(\mathbf{R}^1)$ is arbitrary. According to the main property of $H^\infty(\mathbf{R}^1)$, we have the inclusion $u(t,x) \in H^\infty(\mathbf{R}^1)$ for all $t \in \mathbf{R}^1$. We prove that $u(t,x)$ can be written in the form of the Poisson-type integral

$$u(t,x) = \begin{cases} \dfrac{1}{2(\pi at)^{1/2}} \displaystyle\int_{-i\sqrt{a}\infty}^{i\sqrt{a}\infty} e^{-\frac{s^2}{4at}}\phi(x-s)ds, & t < 0; \\[3mm] \phi(x), & t = 0; \\[3mm] \dfrac{1}{2(\pi at)^{1/2}} \displaystyle\int_{-\sqrt{a}\infty}^{\sqrt{a}\infty} e^{-\frac{s^2}{4at}}\phi(x-s), & t > 0, \end{cases}$$

where the contour $(-i\sqrt{a}\infty, i\sqrt{a}\infty)$ is, for example, the straight line in the plane $z = x + iy$ joining the two points $-i\sqrt{a}$ and $i\sqrt{a}$ (similarly for the contour $(-\sqrt{a}\infty, \sqrt{a}\infty)$).

First we suppose that $t > 0$. Since $\phi(x)$ is an entire function,

$$\frac{1}{2(\pi at)^{1/2}} \int_{-\sqrt{a}\infty}^{\sqrt{a}\infty} e^{-\frac{s^2}{4at}}\phi(x-s)ds = \frac{1}{2(\pi at)^{1/2}} \int_{-\sqrt{a}\infty}^{\sqrt{a}\infty} e^{-\frac{s^2}{4at}} \times$$

$$\times \sum_{n=0}^{\infty} \frac{\phi^{(n)}(x)}{n!}(-s)^n ds = \sum_{n=0}^{\infty} \frac{\phi^{(n)}(x)}{n!} \cdot \frac{1}{2(\pi at)^{1/2}} \int_{-\sqrt{a}\infty}^{\sqrt{a}\infty} e^{-\frac{s^2}{4at}}(-s)^n ds, \quad (7.1.4)$$

where $\phi^{(n)}(x) = d^n\phi(x)/dx^n$. Then, on substituting $s = 2(at)^{1/2}\eta$, $\eta \in \mathbf{R}^1$, we obtain

$$\frac{1}{2(\pi at)^{1/2}} \int_{-\sqrt{a}\infty}^{\sqrt{a}\infty} e^{-\frac{s^2}{4at}}(-s)^n ds = \frac{1}{\sqrt{\pi}}(-2\sqrt{at})^n \int_{-\infty}^{\infty} e^{-\eta^2}\eta^n d\eta =$$

$$= \begin{cases} 0, & \text{if } n = 2m+1; \\[2mm] \frac{1}{\sqrt{\pi}}(4at)^m\Gamma(m+\tfrac{1}{2}), & \text{if } n = 2m \ (m = 0, 1, \ldots) \end{cases}$$

where $\Gamma(m + \frac{1}{2})$ is the gamma function. Taking into account the fact that

$$\Gamma(m + \frac{1}{2}) = (m - \frac{1}{2})(m - \frac{3}{2}) \cdots \frac{3}{2} \cdot \frac{1}{2}\Gamma(\frac{1}{2}) = \frac{(2m-1)!\sqrt{\pi}}{2^{2m-1}(m-1)!},$$

we immediately deduce from the last formula and from (7.1.4) that for $t > 0$

$$\frac{1}{2(\pi a t)^{1/2}} \int_{-\sqrt{a}\infty}^{\sqrt{a}\infty} e^{-\frac{s^2}{4at}} \phi(x - s)ds = \sum_{m=0}^{\infty} \frac{\phi^{(2m)}(x)}{m!}(at)^m = \exp\left(at\frac{d^2}{dx^2}\right)\phi(x),$$

as required.

The computations for $t < 0$ are completely analogous. The fact that this integral takes the value $\phi(x)$ for $t = 0$ can be proved by the classical arguments usually applied in the study of the Poisson integral. It should be noted also that all the operations used in our calculations are justified and that the integrals exist. We shall not dwell on these points.

Finally we note that $u(t, x)$ can be written in the form

$$u(t, x) = \frac{1}{2(\pi a t)^{1/2}} \int_{-\sqrt{a}\,\text{sign}\,t\infty}^{\sqrt{a}\,\text{sign}\,t\infty} e^{-\frac{s^2}{4at}} \phi(x - s)ds,$$

or, what is the same, after substitution $s = \sqrt{a}\,\text{sign}\,t\xi$

$$u(t, x) = \frac{1}{2(\pi|t|)^{1/2}} \int_{-\infty}^{\infty} e^{-\frac{\xi^2}{4|t|}} \phi(x - \sqrt{a}\,\text{sign}\,t\xi)d\xi.$$

7.2. The generalized function space $H^{-\infty}(S_R)$

1. DEFINITION OF $H^{-\infty}(S_R)$. MAIN PROPERTY

We denote by $H^{-\infty}(S_R)$ the space of generalized functions on $H^{\infty}(S_R)$, that is, the space of all continuous linear functionals defined on the test-function space $H^{\infty}(S_R)$. This space has all the standard properties of spaces of distributions: linearity, continuity with respect to differentiation, completeness and so on. We single out the following original property of $H^{-\infty}(S_R)$ which is most important (for us).

Proposition 7.2.1. *The space $H^{-\infty}(S_R)$ is invariant under differential operators of infinite order with analytic symbols in S_R.*

Indeed, let $h(x) \in H^{-\infty}(S_R)$ be arbitrary and let $A(\mathcal{D})$ be a differential operator with analytic symbol $A(\xi)$. Then

$$\langle A(\mathcal{D})h(x), \phi(x) \rangle \stackrel{\text{def}}{=} \langle h(x), A(-\mathcal{D})\phi(x) \rangle, \quad \phi(x) \in H^{\infty}(S_R),$$

and this expression is well defined because $A(-\mathcal{D})\phi(x) \in H^{\infty}(S_R)$ if $\phi(x) \in H^{\infty}(S_R)$.

We now introduce the Fourier transform of distributions $h(x) \in H^{-\infty}(S_R)$.

Definition 7.2.1. We define the Fourier transform $\tilde{h}(\xi)$ by the formula

$$\langle \tilde{h}(\xi), \tilde{\phi}(-\xi) \rangle = (2\pi)^n \langle h(x), \phi(x) \rangle, \tag{7.2.1}$$

where $\phi(x) \in H^{\infty}(S_R)$ is arbitrary and $\tilde{\phi}(\xi)$ is the classical Fourier transform of $\phi(x)$.

Remark. If $\phi(x)$ and $h(x)$ are ordinary functions in $L_2(\mathbf{R}^n)$, then formula (7.2.1) expresses one of the forms of the well-known unitary property of the Fourier transformation operator, namely,

$$\int \tilde{h}(\xi)\phi(\xi)d\xi = (2\pi)^n \int h(x)\tilde{\phi}(x)dx.$$

In view of the compactness of supp $\tilde{\phi}(\xi)$, it follows from Proposition 7.1.2 that $\tilde{h}(\xi)$, defined by formula (7.2.1) is a bounded linear functional on the space

$$L_{2,\text{comp}}(S_R) = \{\tilde{\phi}(\xi) \in L_2(\mathbf{R}^n) : \text{supp } \tilde{\phi}(\xi) \subset S_R\}.$$

Thus we can identify this functional with some function $\tilde{h}(\xi) \in L_{2,\text{loc}}(\mathbf{R}^n)$. Since $\tilde{\phi}(\xi) \equiv 0$ for $\xi \notin S_R$, this function $\tilde{h}(\xi)$ is uniquely determined by (7.2.1) only for $\xi \in S_R$, that is, $\tilde{h}(\xi) \in L_{2,\text{loc}}(S_R)$. Moreover, the values of $\tilde{h}(\xi)$ outside S_R are not important and may be chosen arbitrarily. This means that

$$\tilde{h}(\xi) = \begin{cases} \tilde{h}(\xi) \in L_{2,\text{loc}}(S_R) & \text{if } \xi \in S_R, \\ \text{irrelevant}, & \text{if } \xi \notin S_R \end{cases}$$

(if $R = +\infty$, the second part is absent). For definiteness, let us agree to fix the Fourier transform $\tilde{h}(\xi)$ as $\tilde{h}(\xi) \in L_{2,\text{loc}}(S_R)$ and such that $\tilde{h}(\xi) \equiv 0$ outside S_R.

The Fourier transform introduced here clearly has all the classical properties of the usual Fourier transform. In particular, we note that if

$$h(x) = A(\mathcal{D})f(x), \qquad (7.2.2)$$

where $f(x) \in L_2(\mathbf{R}^n)$ and $A(\xi) \in \mathcal{O}(S_R)$, then

$$\tilde{h}(\xi) = A(\xi)\tilde{f}(\xi) \qquad (7.2.3)$$

and conversely.

We end this subsection by describing the elements of the space $H^{-\infty}(S_R)$ in terms of distributions of infinite order.

Proposition 7.2.2. *Every functional $h(x) \in H^{-\infty}(S_R)$ can be represented in the form*
$$h(x) = A(-\mathcal{D})f(x),$$
where $A(\xi)$ is an analytic function in S_R and $f(x) \in L_2(\mathbf{R}^n)$ is a function (defined by $h(x)$) such that $\tilde{f}(\xi) \equiv 0$ outside S_R.

Proof. In accordance with Definition 7.2.1 we have $\tilde{h}(\xi) \in L_{2,\mathrm{loc}}(S_R)$. Then in view of (7.1.2) and (7.1.3) it suffices to prove that every function $\tilde{h}(\xi) \in L_{2,\mathrm{loc}}(S_R)$ can be represented in the form

$$\tilde{h}(\xi) = A(\xi)\tilde{f}(\xi),$$

where $\tilde{f}(\xi) \in L_2(S_R)$ and $\tilde{f}(\xi) = 0$ outside S_R.

To this end, we consider a monotone increasing sequence of parallelepipeds (see the proof of Proposition 7.2.1) and let

$$A(\xi) = \sum_{n=0}^{\infty} \sum_{|\alpha|=m_n+1}^{m_{n+1}} \frac{1}{R_n^{\alpha}} \xi^{2\alpha} \equiv \sum_{n=0}^{\infty} a_n(\xi).$$

It is not difficult to see that $a(\xi)$ is analytic in S_R. It only remains to choose the numbers m_n so that $\tilde{h}(\xi)/A(\xi) \equiv \tilde{f}(\xi)$ belongs to $L_2(S_R)$. For this it is sufficient to choose them so that

$$\frac{1}{a_n^2(R_n)} \int_{S_{R_{n+1}} \setminus S_{R_n}} |\tilde{h}(\xi)|^2 d\xi \le \alpha_n,$$

where $\alpha_0 + \alpha_1 + \ldots < \infty$. The proposition is proved.

2. EXAMPLES OF FUNCTIONALS IN $H^{-\infty}(S_R)$

Example 1. Clearly each $h(x) \in L_2(\mathbf{R}^n)$ determines a regular functional on $H^\infty(S_R)$ by the formula

$$\langle h(x), \phi(x) \rangle \overset{\text{def}}{=} \int h(x)\phi(x)dx.$$

This formula gives an imbedding $L_2(\mathbf{R}^n) \subset H^{-\infty}(S_R)$ and, in particular, an imbedding $H^\infty(S_R) \subset H^{-\infty}(S_R)$.

Example 2. The delta function $\delta(x)$ gives an example of a singular functional on $H^\infty(S_R)$ in accordance with the formula

$$\langle \delta(x), \phi(x) \rangle = \phi(0).$$

This formula clearly determines a well defined functional on $H^\infty(S_R)$, since $\phi(x) \in H^\infty(S_R)$ is continuous.

Example 3 (Complex shift of $L_2(\mathbf{R}^n)$). Let $h(x) \in L_2(\mathbf{R}^n)$ and let $a = (a_1, \ldots, a_n)$ be an arbitrary complex vector. The shift of $h(x) \in L_2(\mathbf{R}^n)$ by the vector a is defined as the functional

$$h(x+a) \overset{\text{def}}{=} \sum_{|\alpha|=0}^{\infty} \frac{a^\alpha}{\alpha!} \mathcal{D}^\alpha h(x),$$

or, what is the same,

$$\langle h(x+a), \phi(x) \rangle \overset{\text{def}}{=} \Big\langle \sum_{|\alpha|=0}^{\infty} \frac{a^\alpha}{\alpha!} \mathcal{D}^\alpha h(x), \phi(x) \Big\rangle =$$

$$= \Big\langle h(x), \sum_{|\alpha|=0}^{\infty} \frac{(-1)^{|\alpha|}a^\alpha}{\alpha!} \mathcal{D}^\alpha \phi(x) \Big\rangle = \langle h(x), \phi(x-a) \rangle.$$

This definition is correct, because each $\phi(x) \in H^\infty(S_R)$ has an analytic continuation $\phi(z)$ to $\mathbf{C}^n$ such that for any $y \in \mathbf{R}^n$ $\phi(x+iy) \in L_2(\mathbf{R}^n_x)$ (see, for example, Ronkin [1], p.272).

Example 4. Consider the functional

$$E(t,x) = \exp\Big(at\frac{d^2}{dx^2}\Big)\delta(x) = \sum_{n=0}^{\infty} \frac{(at)^n}{n!}\delta^{(2n)}(x),$$

where $a \in \mathbf{C}^1$ and $t \in \mathbf{R}^1$. The action of this functional on $\phi(x) \in H^\infty(S_R)$ is given by formula

$$\langle E(t,x), \phi(x) \rangle = \langle \delta(x), \exp\left(at\frac{d^2}{dx^2}\right)\phi(x) \rangle.$$

Taking into account the example in subsection 5 of §7.1, we obtain

$$\langle E(t,x), \phi(x) \rangle = \frac{1}{2\sqrt{\pi a t}} \int_{-\sqrt{a\,\text{sign}\,t}\infty}^{\sqrt{a\,\text{sign}\,t}\infty} \exp\left(-\frac{s^2}{4at}\right)\phi(-s)ds.$$

Consequently, $E(t,x)$ is the analytic functional determined by the function

$$\frac{1}{2\sqrt{\pi a t}} \exp\left(-\frac{s^2}{4at}\right)$$

with contour of integration $(-\sqrt{a\,\text{sign}\,t}\infty, \sqrt{a\,\text{sign}\,t}\infty)$

Example 5. Consider the functional

$$E(t,x) = \frac{\sinh at\frac{d}{dx}}{a\frac{d}{dx}}\delta(x) \equiv \sum_{n=0}^{\infty} \frac{t(at)^{2n}}{(2n+1)!}\delta^{(2n)}(x),$$

where $a \in \mathbf{C}^1$, $t \in \mathbf{R}^1$. The action of this functional on $\phi(x) \in H^\infty(S_R)$ is given by the formula

$$\langle E(t,x), \phi(x) \rangle = \left\langle \delta(x), \frac{\exp\left(at\frac{d}{dx}\right) - \exp\left(-at\frac{d}{dx}\right)}{2a\frac{d}{dx}}\phi(x) \right\rangle$$

$$= \left\langle \delta(x), \frac{1}{2a}\int_{x-at}^{x+at}\phi(\tau)d\tau \right\rangle = \frac{1}{2a}\int_{-at}^{at}\phi(\tau)d\tau. \tag{7.2.4}$$

Note that when $\phi(x)$ decays rapidly enough as $x \to +\infty$ (for more details, see later), this functional can be represented as

$$E(t,x) = \frac{1}{2a}[\theta(x+at) - \theta(x-at)],$$

where $\theta(x \pm at)$ is the (in general, complex) shift of the Heaviside function

$$\langle \theta(x+h), \phi(x) \rangle \stackrel{\text{def}}{=} \langle \theta(x), \phi(x-a) \rangle = \int_0^\infty \phi(x-h)dx.$$

Indeed, in accordance with this definition

$$\frac{1}{2a}\langle \theta(x+at) - \theta(x-at), \phi(x) \rangle =$$

$$= \frac{1}{2a}\int_0^\infty \phi(x-at)dx - \frac{1}{2a}\int_0^\infty \phi(x+at)dx. \tag{7.2.5}$$

We take the following contour of integration:

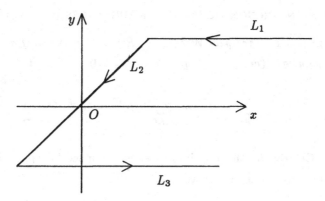

and suppose that $\phi(x)$ decays as $x \to +\infty$, so that Cauchy's theorem holds for the whole contour $L = L_1 \cup L_2 \cup L_3$, that is,

$$\int_L \phi(z)dz = 0.$$

This formula is clearly equivalent to

$$\int_{+\infty}^0 \phi(x - at)dx - \int_0^{+\infty} \phi(x + at)dx = \int_{-at}^{at} \phi(\tau)d\tau.$$

Comparing this equality with (7.2.4) and (7.2.5), we immediately obtain

$$E(t, x) = \frac{1}{2a}[\theta(x + at) - \theta(x - at)].$$

This completes the proof.

7.3. Sobolev spaces of infinite order $W^\infty\{a_\alpha, p\}(\mathbf{R}^n)$

In this section we introduce the Banach spaces $W^\infty\{a_\alpha, p\}(\mathbf{R}^n)$, with norm defined by the series

$$\|u\|_\infty^p \overset{\text{def}}{=} \sum_{|\alpha|=0}^\infty a_\alpha\|\mathcal{D}^\alpha u\|_p^p,$$

where $a_\alpha \geq 0$, $p \geq 1$ are arbitrary numbers and $\|\cdot\|_p$ is the norm in the Lebesgue space $L_p(\mathbf{R}^n)$. These spaces are called Sobolev spaces of infinite order.

By contrast with Sobolev spaces of finite order, the first question that arises in the study of the spaces $W^\infty\{a_\alpha, p\}(\mathbf{R}^n)$ is the question of their non-triviality (or non-emptiness), that is, whether there exists a function $u(x)$ such that $\|u\|_\infty < \infty$.

There is the following result in this direction.

1. CRITERION FOR NON-TRIVIALITY OF THE SPACES $W^\infty\{a_\alpha, p\}(\mathbf{R}^n)$

Theorem 7.3.1. *The space $W^\infty\{a_\alpha, p\}(\mathbf{R}^n)$ is non-trivial if and only if there exists a point $q = (q_1, \ldots, q_n)$, $q_1 > 0, \ldots, q_n > 0$, such that*

$$\sum_{|\alpha|=0}^\infty a_\alpha q^{\alpha p} \equiv \sum_{|\alpha|=0}^\infty a_\alpha q_1^{\alpha_1 p} \ldots q_n^{\alpha_n p} < \infty \qquad (*)$$

Remark. Clearly the theorem implies that the space $W^\infty\{a_\alpha, p\}\{\mathbf{R}^n\}$ is non-trivial precisely when the function

$$\phi(z) = \sum_{N=0}^\infty b_N z^N \quad (b_N = \sum_{|\alpha|=N} a_\alpha)$$

is an analytic function at the point $z = 0$, or, what is the same

$$\varlimsup_{N \to \infty} \left(\sum_{|\alpha|=N} a_\alpha \right)^{1/N} < \infty.$$

Proof of theorem. Sufficiency. Suppose that condition $(*)$ holds. We shall construct a function $u(x) \in W^\infty\{a_\alpha, p\}(\mathbf{R}^n)$ that is not identically zero. In fact, we take a function $\tilde{u}(\xi) \in C_0^\infty(G)$, where

$$G = \{\xi \in \mathbf{R}^n : |\xi| \le \xi_0 < q_0\}, \quad q_0 = \min(q_1, \ldots, q_n).$$

We then show that

$$u(x) = \int_G \tilde{u}(\xi) e^{ix\xi} d\xi$$

is a suitable function. In fact, after simple calculations we obtain the inequality

$$|D^\alpha u(x)| \le K \xi_0^{|\alpha|} |\alpha|^m \prod_{k=1}^m \min\left(1, |x_k|^{-m/n}\right),$$

where the multi-indices α and $m \in N$ are arbitrary; $K = K(m)$ is a constant depending on m, q_0 and $\max_{|\beta| \le m} |D^\beta \tilde{u}(\xi)|$, but not depending on α. Choosing $m = n + 1$ we find that $D^\alpha u(x) \in L_p(\mathbf{R}^n)$ and for any α

$$\|D^\alpha u(x)\|_p \le K \xi_0^{|\alpha|} |\alpha|^m,$$

where $K > 0$ is a constant.

From the latter inequality and using condition $(*)$ we obtain

$$\|u\|_\infty^p \equiv \sum_{|\alpha|=0}^\infty a_\alpha \|\mathcal{D}^\alpha u\|_p^p \leq K^p \sum_{|\alpha|=0}^\infty \xi_0^{|\alpha|p} |\alpha|^{mp} < \infty,$$

for $\xi_0 < q_0$ (recall that $q_0 = \min(q_1, \ldots, q_n)$, where $q_1 > 0, \ldots, q_n > 0$ are from condition $(*)$).

This means that $u(x) \in W^\infty\{a_\alpha, p\}(\mathbf{R}^n)$. The sufficiency of condition $(*)$ is proved.

Necessity of condition $(*)$. We show that if for any $q = (q_1, \ldots, q_n)$, $q_1 > 0, \ldots,$ $q_n > 0$, the series $(*)$ diverges, then every function $u(x) \in W^\infty\{a_\alpha, p\}(\mathbf{R}^n)$ is identically zero.

Our proof is based on the following lemma, which (in our opinion) is of independent interest.

Lemma 7.3.1. *Suppose that the series* $(*)$ *diverges for fixed* $q = (q_1, \ldots, q_n)$, $q_1 > 0, \ldots,$ $q_n > 0$, *that is,*

$$\sum_{|\alpha|=0}^\infty a_\alpha q^{\alpha p} = +\infty. \tag{**}$$

Then for every function $u(x) \in W^\infty\{a_\alpha, p\}(\mathbf{R}^n)$ *the Fourier transform* $\tilde{u}(\xi)$ *is an ordinary generalized function. Moreover,* $\tilde{u}(\xi) = 0$ *if* $\xi \in G_q$, *where*

$$G_q = \{\xi \in \mathbf{R}^n : |\xi_j| > q_j, \ j = 1, \ldots, n\}.$$

Proof of lemma. First we consider the case $1 \leq p \leq 2$. As is well known, if $u(x) \in L_p(\mathbf{R}^n)$, where $1 \leq p \leq 2$, then its Fourier transform $\tilde{u}(\xi)$ is an ordinary function; moreover, $\tilde{u}(\xi) \in L_{p'}(\mathbf{R}^n)$, where $p' = p/(p-1)$, and

$$\|\tilde{u}(\xi)\|_{p'} \leq \|u(\xi)\|_p \tag{7.3.1}$$

(see, for example, Zygmund [1], Stein and Weiss [1], and others). Consequently, for all α the Fourier transform $(D^\alpha u(x))^\sim = (i\xi)^\alpha \tilde{u}(\xi)$ is an ordinary function; moreover, $(D^\alpha u(x))^\sim \in L_{p'}(\mathbf{R}^n)$. Hence it follows that $\tilde{u}(\xi)$ for $\xi_j \neq 0$,

$j = 1, \ldots, n$, is also an ordinary function (we do not exclude the case $a_0 = 0$, therefore we cannot assert the regularity of $\tilde{u}(\xi)$ for $\xi_j = 0$, since $\tilde{u}(\xi)$ is *a priori* only a generalized function).

Taking into account the inequality (7.3.1) we obtain

$$\sum_{|\alpha|=0}^{\infty} a_\alpha \|\xi^\alpha \tilde{u}(\xi)\|_{p'}^p \leq \sum_{|\alpha|=0}^{\infty} a_\alpha \|\mathcal{D}^\alpha u\|_p^p < \infty, \qquad (7.3.2)$$

since $u(x) \in W^\infty\{a_\alpha, p\}(\mathbf{R}^n)$.

On the other hand, clearly,

$$\sum_{|\alpha|=0}^{\infty} a_\alpha \|\xi^\alpha \tilde{u}(\xi)\|_{p'}^p \geq \sum_{|\alpha|=0}^{\infty} a_\alpha q^{\alpha p} \beta_q^p, \qquad (7.3.3)$$

where

$$\beta_q = \left(\int_{G_q} |\tilde{u}(\xi)|^{p'} d\xi \right)^{1/p'}.$$

Since the series (**) diverges, inequalities (7.3.2) and (7.3.3) are consistent if and only if $\beta_q = 0$. Obviously, the latter is possible only if $\tilde{u}(\xi) = 0$ in G_q. The lemma is proved for the case $p \leq 2$.

We now consider the case $p > 2$. In this case the Fourier transform $\tilde{u}(\xi)$ is the generalized function defined by

$$\langle \tilde{u}(\xi), \tilde{\phi}(\xi) \rangle = (2\pi)^n \langle u(x), \phi(x) \rangle, \qquad (7.3.4)$$

where $\phi(x) \in C_0^\infty(\mathbf{R}^n)$ and $\tilde{\phi}(\xi)$ is the classical Fourier transform of the function $\phi(x)$ (see, for example, Gel'fand and Shilov [1]). Since $u(x) \in L_p(\mathbf{R}^n)$ (without loss of generality one can assume that $a_0 > 0$) and the functions with compact support are dense in $L_{p'}(\mathbf{R}^n)$, equality (7.3.4) holds for any function $\phi(x) \in L_{p'}(\mathbf{R}^n)$ (we emphasize that $p' = p/(p-1) < 2$). Consequently, equality (7.3.4) determines $\tilde{u}(\xi)$ as a functional on the space V_p, which is the image of $L_{p'}(\mathbf{R}^n)$ under the Fourier transform $\mathcal{F} : \phi(x) \to \tilde{\phi}(\xi)$. The space V_p does not coincide with $L_p(\mathbf{R}^n)$ but (as is obvious) contains the smooth functions of compact support. Thus, the Fourier transform $\tilde{u}(\xi)$ of the function $u(x) \in L_p(\mathbf{R}^n)$ is the usual generalized function, that is, the functional on $\mathcal{D}$, where $\mathcal{D}$ is the classical test function space.

We now show (and this is important) that $\tilde{u}(\xi) = 0$ in the domain G_q. Indeed, let G be an arbitrary domain with a smooth boundary Γ that is contained in the parallelepiped $K_s^+ = \{\xi \in \mathbf{R}^n : \xi_j \geq s_j > q_j, \ j = 1, \ldots, n\}$. Choosing in (7.3.4)

$$\phi(\xi) = \tilde{v}(\xi)\tilde{w}(\xi),$$

where $\tilde{v}(\xi) \in C_0^\infty(G)$ is a fixed function and $\tilde{w}(\xi) \in C_0^\infty(G)$ is an arbitrary function, we obtain the equality

$$\langle \tilde{v}(\xi)\tilde{u}(\xi), \tilde{w}(\xi) \rangle = (2\pi)^n \langle u(x), \phi(x) \rangle, \tag{7.3.5}$$

where $\phi(x) = F^{-1}[\tilde{v}(\xi)\tilde{w}(\xi)] = v(x) * w(x) \in L_{p'}(\mathbf{R}^n)$. The functional $\tilde{v}(\xi)\tilde{w}(\xi)$ clearly has compact support. Thus, this functional as a generalized function has finite order of singularity, that is, there exist a number $m \in \mathbf{N}$ and ordinary functions $h_\alpha(\xi)$, $|\alpha| \leq m$, such that the function $\tilde{v}(\xi)\tilde{w}(\xi)$ can be represented in the form

$$\tilde{v}(\xi)\tilde{w}(\xi) = \sum_{|\alpha| \leq m} \mathcal{D}^\alpha h_\alpha(x). \tag{7.3.6}$$

Without loss of generality we may suppose that the functions $h_\alpha(x) \in C^1(G)$. Hence, it follows that there exists a function $\tilde{z}(\xi) \in C^m(G)$, $\mathcal{D}^\alpha \tilde{z}|_\Gamma = 0$, $|\alpha| \leq m - 1$, such that

$$\tilde{v}(\xi)\tilde{w}(\xi) = L_{2m}\tilde{z}(\xi) \equiv \sum_{|\alpha| \leq m} (-1)^{|\alpha|} \mathcal{D}^{2\alpha}\tilde{z}(\xi)$$

(it is clear that the function $\tilde{z}(\xi)$ is the solution of the boundary value problem

$$L_{2m}\tilde{z}(\xi) = \sum_{|\alpha| \leq m} \mathcal{D}^\alpha h_\alpha(\xi), \quad \xi \in G,$$

$$\mathcal{D}^\alpha \tilde{z}|_\Gamma = 0, \quad |\alpha| \leq m - 1;$$

the advantage of the latter representation as compared to (7.3.6) is that it is unique).

From (7.3.5) we now have the equality

$$\langle \tilde{v}(\xi)\tilde{u}(\xi), \tilde{w}(\xi) \rangle = \langle L_{2m}\tilde{z}(\xi), \tilde{w}(\xi) \rangle = (2\pi)^n \langle u(x), \phi(x) \rangle,$$

or, equivalently,

$$\langle \tilde{z}(\xi), L_{2m}\tilde{w}(\xi) \rangle = (2\pi)^n \langle u(x), \phi(x) \rangle, \tag{7.3.7}$$

where $\tilde{w}(\xi) \in C_0^\infty(G)$ is an arbitrary function. It is clear that the left side of (7.3.7) admits closure up to an arbitrary function $\tilde{w}(\xi) \in C^{2m}(G) \cap C_0^m(G)$. Therefore, the right side of (7.3.7) admits closure up to an arbitrary function of the same type also. It should be noted in this connection that

$$\phi(x) = v(x) * w(x) \in L_{p'}(\mathbf{R}^n).$$

The same arguments for the function $\mathcal{D}^\alpha u(x)$ give us the equality

$$\langle \tilde{z}(\xi), L_{2m}(\xi^\alpha \tilde{w}(\xi)) \rangle = (2\pi)^n (-1)^{|\alpha|} \langle \mathcal{D}^\alpha u(x), \phi(x) \rangle, \qquad (7.3.8)$$

where $\tilde{w}(\xi)$ belongs to the space $C^{2m}(G) \cap C_0^m(G)$ also. We take the function $\tilde{w}(\xi) \equiv \tilde{w}_\alpha(\xi)$ such that

$$L_{2m}(\xi^\alpha \tilde{w}_\alpha(\xi)) = q^\alpha \tilde{z}(\xi), \quad \xi \in G.$$

Then from (7.3.8) we shall have the inequality

$$q^\alpha \langle \tilde{z}(\xi), \tilde{z}(\xi) \rangle \le (2\pi)^n \|\mathcal{D}^\alpha u\|_p \cdot \|\phi_\alpha\|_{p'}, \qquad (7.3.9)$$

where $\phi_\alpha(x) = v(x) * w_\alpha(x)$.

Since supp $\tilde{w}_\alpha(\xi) \subset G \subset K_s^+ \subset G_q$ and $s_j > q_j$, $j = 1, \dots, n$, it follows that for large α

$$(2\pi)^n \|\phi_\alpha(x)\|_{p'} \le (2\pi)^n \|v(x)\|_1 \cdot \|w_\alpha(x)\|_{p'} \le 1;$$

therefore, without loss of generality we can suppose that $(2\pi)^n \|\phi_\alpha(x)\|_{p'} \le 1$ for all indices α. In that case from (7.3.9) we have the inequality

$$\sum_{|\alpha|=0}^\infty a_\alpha q^{\alpha p} \langle \tilde{z}(\xi), \tilde{z}(\xi) \rangle^p \le \sum_{|\alpha|=0}^\infty a_\alpha \|\mathcal{D}^\alpha u\|_p^p < \infty,$$

which by conditions $(**)$ reduces to the identity $\tilde{z}(\xi) = 0$ in the domain G. Therefore, $\tilde{u}(\xi) \equiv 0$ in G also.

Since the domain $G \subset K_s^+$ and parallelepiped $K_s^+ = \{\xi \in \mathbf{R}^n : \xi_j > s_j > q_j, 1 \le j \le n\}$ are arbitrary, then $\tilde{u}(\xi) \equiv 0$ in the domain $G_q^+ = \{\xi \in \mathbf{R}^n : \xi_j > q_j, 1 \le j \le n\}$.

In the same way it can be proved that $\tilde{u}(\xi) \equiv 0$ in the other open octants of G_q. The lemma is completely proved.

We now finish the proof of our theorem. Namely, let $u(x)$ be a function in $W^\infty\{a_\alpha, p\}(\mathbf{R}^n)$ and suppose that the series $(**)$ is divergent for some $q = (q_1, \ldots, q_n)$, where $q_1 > 0, \ldots, q_n > 0$. Then in view of Lemma 7.3.1 the Fourier transform $\tilde{u}(\xi)$ is concentrated on the hyperplanes $\xi_j = 0$, $j = 1, \ldots, n$. We shall prove that actually $\tilde{u}(\xi) = 0$ everywhere. Indeed, let $\tilde{\nu}(\xi) \in C_0^\infty(\mathbf{R}^n)$ be a function whose support has a non-empty intersection with the hyperplane $\xi_j = 0$ (j is fixed) but does not intersect any other coordinate hyperplanes of codimension one or more. The function $\tilde{v}(\xi) \equiv \tilde{\nu}(\xi)\tilde{u}(\xi)$ can be represented as a tensor product

$$\tilde{v}(\xi) = \sum_{k=0}^{m} \tilde{v}_k(\xi_1, \ldots, \xi_{j-1}, \xi_{j+1}, \ldots, \xi_n) \otimes \delta^{(k)}(\xi_j), \qquad (7.3.10)$$

where the $\tilde{v}_k$ are generalized functions and $m \in \mathbf{N}$.

Further, since $u(x) \in L_p(\mathbf{R}^n)$ we also have $v(x) = \nu(x) * u(x) \in L_p(\mathbf{R}^n)$ (the notation is clear).

On the other hand, using (7.3.10) we obtain

$$v(x) = \sum_{k=0}^{m} v_k(x_1, \ldots, x_{k-1}, x_{k+1}, \ldots, x_n) \otimes (-ix_j)^k$$

so that the inclusion $v(x) \in L_p(\mathbf{R}^n)$ is possible if and only if $v_k = 0$, $k = 0, 1, \ldots, m$. It follows that $\tilde{v}(\xi) = 0$ and we immediately have that $\tilde{u}(\xi) \equiv 0$ on the hyperplanes $\xi_j = 0$, $1 \le j \le n$, except on their intersection. Thus the function $\tilde{u}(\xi)$ is concentrated on coordinate hyperplanes of codimension two or more.

Repeating the above arguments, we see that the generalized function $\tilde{u}(\xi)$ is concentrated at the point $\xi = 0$. It is well known that in this case the function $\tilde{u}(\xi)$ is a finite linear combination of the δ-function and its derivatives. Therefore $u(x)$ is a polynomial and, since $u(x) \in L_p(\mathbf{R}^n)$, we have $u(x) \equiv 0$. This proves the necessity of condition $(*)$ and hence Theorem 7.3.1 as well.

2. THE DISTRIBUTION SPACE $W^{-\infty}\{a_\alpha, p'\}(\mathbf{R}^n)$

We set

$$W^{-\infty}\{a_\alpha, p'\}(\mathbf{R}^n) = \left\{ h(x) : h(x) = \sum_{|\alpha|=0}^{\infty} (-1)^{|\alpha|} a_\alpha \mathcal{D}^\alpha h_\alpha(x) \right\},$$

where $h_\alpha(x) \in L_{p'}(\mathbf{R}^n)$, $p' = p/(p-1)$ and

$$\sum_{|\alpha|=0}^{\infty} a_\alpha \|h_\alpha(x)\|_{p'}^{p'} < \infty.$$

Using Young's inequality it is easy to see that for any function $v(x) \in W^\infty\{a_\alpha, p\}(\mathbf{R}^n)$ the equality

$$\langle h(x), v(x) \rangle \overset{\text{def}}{=} \sum_{|\alpha|=0}^{\infty} a_\alpha \int_{\mathbf{R}^n} h_\alpha(x) \mathcal{D}^\alpha v(x) dx$$

is well defined. This means that $h(x)$ is a generalized function on the test-function space $W^\infty\{a_\alpha, p\}(\mathbf{R}^n)$ (in general, of infinite order of singularity), that is,

$$W^{-\infty}\{a_\alpha, p'\}(\mathbf{R}^n) \subset \left(W^\infty\{a_\alpha, p\}(\mathbf{R}^n)\right)^*,$$

where $\left(W^\infty\{a_\alpha, p\}(\mathbf{R}^n)\right)^*$ denotes the dual space.

We show that the reverse inclusion

$$\left(W^\infty\{a_\alpha, p\}(\mathbf{R}^n)\right)^* \subset W^{-\infty}\{a_\alpha, p'\}(\mathbf{R}^n)$$

also holds.

Indeed, let us consider the equation

$$L(\mathcal{D})u(x) \equiv \sum_{|\alpha|=0}^{\infty} (-1)^{|\alpha|} a_\alpha \mathcal{D}^\alpha u(x) = h,$$

where $h \in \left(W^\infty\{a_\alpha, p\}(\mathbf{R}^n)\right)^*$ and $u(x) \in W^\infty\{a_\alpha, p\}(\mathbf{R}^n)$ is an unknown function. It is easy to verify that the operator

$$L(\mathcal{D}) : W^\infty\{a_\alpha, p\}(\mathbf{R}^n) \to \left(W^\infty\{a_\alpha, p\}(\mathbf{R}^n)\right)^*$$

is a monotone continuous operator. Then in accordance with the well-known theory of monotone operators (see, for example, Minty [1], Browder [1], Leray and Lions [1] et al) the operator $L(\mathcal{D})$ defines a homeomorphism

$$L(\mathcal{D}) : W^\infty\{a_\alpha, p\}(\mathbf{R}^n) \to \left(W^\infty\{a_\alpha, p\}(\mathbf{R}^n)\right)^*.$$

This means, in particular, that every element $h \in \left(W^\infty\{a_\alpha, p\}(\mathbf{R}^n)\right)^*$ may be represented in the form

$$h = \sum_{|\alpha|=0}^{\infty} (-1)^{|\alpha|} a_\alpha \mathcal{D}^\alpha h_\alpha(x),$$

where $h_\alpha(x) \equiv \mathcal{D}^\alpha u(x) \in L_{p'}(\mathbf{R}^n)$ and

$$\sum_{|\alpha|=0}^{\infty} a_\alpha \|h_\alpha(x)\|_{p'}^{p'} < \infty.$$

Thus, we have obtained

Theorem 7.3.2. *The equality*

$$\left(W^\infty \{a_\alpha, p\}(\mathbf{R}^n)\right)^* = W^{-\infty}\{a_\alpha, p'\}(\mathbf{R}^n)$$

holds. In other words, every functional on $W^\infty\{a_\alpha, p\}(\mathbf{R}^n)$ is a distribution of infinite order of singularity.

Chapter 8. Analytic PD-operators with Real Arguments. Applications

8.1. Algebra of PD-operators with analytic symbols

In this section we construct the spaces of test functions and generalized functions $\{H^\infty(G), H^{-\infty}(G)\}$, which form the basis for the correct construction of the algebra of pseudodifferential (PD) operators with analytic symbols. As will be shown in what follows, the local elements of the theory of the duality $\{H^\infty(G), H^{-\infty}(G)\}$ are given by the $\{H^\infty(S_R), H^{-\infty}(S_R)\}$-theory, constructed in §§7.1, 7.2.

1. THE SPACE $H^\infty(G)$

Let $G \in \mathbf{R}_\xi^n$ be a domain and let $\lambda \in G$ be any fixed point. We denote by

$$S_R(\lambda) = \{\xi : |\xi_j - \lambda_j| < R_j, \ j = 1, \ldots, n\}$$

the parallelepiped centred at λ of polyradius $R = (R_1, \ldots, R_n)$ and completely contained in G. In accordance with the definition of $H^\infty(S_R)$ (see §7.1) we set

$$H^\infty(S_R(\lambda)) = \left\{ u(x) \in L_2(\mathbf{R}^n) : \int_{S_R(\lambda)} a(\xi) |\tilde{u}(\xi)|^2 d\xi < \infty \right\},$$

where

$$a(\xi) = \sum_{|\alpha|=0}^{\infty} a_\alpha(\lambda)(\xi - \lambda)^\alpha$$

is any function analytic in $S_R(\lambda)$.

Clearly, the function $u(x) \in H^\infty(S_R(\lambda))$ if and only if $\exp(-i\lambda x)u(x) \in H^\infty(S_R)$. Thus we can write (symbolically)

$$H^\infty(S_R(\lambda)) = \exp(i\lambda x)H^\infty(S_R),$$

196

where $H^\infty(S_R)$ is the test-function space investigated in §7.1. Then, by definition, the sequence $u_k(x) \to 0$ in the space $H^\infty(S_R(\lambda))$ if the sequence $\exp(-i\lambda x)u_k(x) \to 0$ in $H^\infty(S_R)$.

We now turn to a general domain G. Namely,

Definition 8.1.1. We define

$$H^\infty(G) = \left\{ u(x) \in L_2(\mathbf{R}^n) : u(x) = \sum_{\lambda \in G} u_\lambda(x),\ u_\lambda(x) \in H^\infty(S_R(\lambda)) \right\},$$

where the symbol $\sum_{\lambda \in G}$ denotes the summation over all finite sets of $\lambda \in G$.

Convergence in $H^\infty(G)$: we say that $u_k(x) \to 0$ $(k \to \infty)$ in $H^\infty(G)$ if there exists a set of $\lambda \in G$, not depending on $k = 1, 2, \ldots$, such that

$$u_k(x) = \sum_{\lambda \in G} u_{k\lambda}(x)$$

and $u_{k\lambda}(x) \to 0$ in $H^\infty(S_R(\lambda))$.

It is not difficult to prove that the space $H^\infty(G)$ is complete with respect to this convergence.

The space $H^\infty(G)$ is the main test-function space for the construction of the subsequent theory of PD-operators with analytic symbols.

Remark. If $G = \mathbf{R}^n_\xi$, then the space $H^\infty(\mathbf{R}^n)$ is dense in $L_2(\mathbf{R}^n_x)$. This follows from the density in $L_2(\mathbf{R}^n_\xi)$ of the set of functions with compact support and Parseval's equality. If $G \neq \mathbf{R}^n_\xi$ and $\mathrm{mes}(\mathbf{R}^n_\xi \backslash G) = 0$ then $H^\infty(G)$ is also dense in $L_2(\mathbf{R}^n_x)$. Otherwise there is no dense imbedding $H^\infty(G) \subset L_2(\mathbf{R}^n_x)$.

We conclude this subsection with a description of the space $H^\infty(G)$ in terms of the Fourier transform, that is, in the ξ-representation.

Proposition 8.1.1 *The function* $u(x) \in L_2(\mathbf{R}^n)$ *belongs to the space* $H^\infty(G)$ *if and only if its Fourier transform* $\tilde{u}(\xi)$ *has compact support in* G.

Proof. The implication $\Rightarrow$ (necessity) is obvious. In fact, each term $u_\lambda(x)$ belongs to $H^\infty(S_R(\lambda))$, therefore $\mathrm{supp}\,\tilde{u}(\xi) \subset \cup_\lambda S_R(\lambda) \subset G$.

Conversely, suppose that $\mathrm{supp}\,\tilde{u}(\xi) \subset K \subset G$, where K is compact. Then there exists a finite set of points $\lambda_i \in G$ $(i = 1, \ldots, N)$ and measurable sets $K_i \ni \lambda_i$ such that:

1) $K = \bigcup_{1 \le i \le N} K_i$; 2) $K_i \subset S_R(\lambda_i)$, $i = 1, \ldots, N$,

where $S_R(\lambda_i)$ is the parallelipiped of convergence of the Taylor expansion of $A(\xi)$ with centre λ_j.

Let $\chi_i(\xi)$ be the characteristic function of K_i. Then

$$\tilde{u}(\xi) = \sum_{i=1}^{N} \chi_i(\xi)\tilde{u}(\xi),$$

and hence

$$u(x) = \sum_{i=1}^{N} u_{\lambda_i}(x),$$

where $u_{\lambda_i}(x) \in H^\infty(S_R(\lambda_i))$ is the image of the inverse Fourier transform of $\chi_i(\xi)\tilde{u}(\xi)$. That proves our proposition.

2. THE ACTION OF PD-OPERATORS

First let

$$A(\xi) = \sum_{|\alpha|=0}^{\infty} a_\alpha(\lambda)(\xi - \lambda)^\alpha \qquad (8.1.1)$$

be analytic for $\xi \in S_R(\lambda)$. Then, by definition, for every $u_\lambda(x) \in H^\infty(S_R(\lambda))$

$$A(\mathcal{D})u_\lambda(x) = \sum_{|\alpha|=0}^{\infty} a_\alpha(\lambda)(\mathcal{D} - \lambda I)^\alpha u_\lambda(x). \qquad (8.1.2)$$

By the shift formula $(\mathcal{D} - \lambda I)^\alpha[e^{i\lambda x}u_\lambda(x)] = e^{\lambda x}\mathcal{D}^\alpha u_\lambda(x)$, this is well-defined; moreover

$$A(\mathcal{D})u_\lambda(x) = e^{i\lambda x} \sum_{|\alpha|=0}^{\infty} a_\alpha(\lambda)\mathcal{D}^\alpha[e^{-i\lambda x}u_\lambda(x)] \in H^\infty(S_R(\lambda)),$$

since $e^{i\lambda x}u_\lambda(x) \in H^\infty(S_R)$.

Consequently, we can assert that the space $H^\infty(S_R(\lambda))$ is invariant under the action of differential operators (8.1.2) whose symbols are analytic in $S_R(\lambda)$ in the sense of Weierstrass.

Now let G be an arbitrary domain in $\mathbf{R}^n_\xi$ and let $A(\xi) : G \to \mathbf{C}^1$ be any given analytic function. We associate with this function a PD-operator $A(\mathcal{D})$ by means of the formal substitution

$$\xi \leftrightarrow \mathcal{D} = \left(\frac{1}{i}\frac{\partial}{\partial x_1}, \ldots, \frac{1}{i}\frac{\partial}{\partial x_n}\right).$$

We now describe the domain of definition of $A(\mathcal{D})$ and also its action.

In order to do this, we first suppose that $u(x)$ belongs to $H^\infty(S_R(\lambda))$, where $\lambda \in G$ and R are such that

$$A(\xi) = \sum_{|\alpha|=0}^\infty a_\alpha(\lambda)(\xi - \lambda)^\alpha, \quad \xi \in S_R(\lambda).$$

(This assumption is essential since (by contrast with the complex case) we do not set $R = \operatorname{dist}(\lambda, \partial G)$.)

Then, as before,

$$A(\mathcal{D})u(x) = \sum_{|\alpha|=0}^\infty a_\alpha(\lambda)(\mathcal{D} - \lambda I)^\alpha u(x). \qquad (8.1.3)$$

It is not difficult to prove that the values of $A(\mathcal{D})u(x)$ are independent of the choice of λ. Namely, let $u(x) \in H^\infty(S_R(\lambda))$ also belong to $H^\infty(S_R(\lambda_0))$ where $\lambda_0 \neq \lambda$. Then, by definition,

$$A(\mathcal{D})u(x) \equiv \sum_{|\alpha|=0}^\infty a_\alpha(\lambda_0)(\mathcal{D} - \lambda_0 I)^\alpha u(x). \qquad (8.1.4)$$

Taking the Fourier transform of (8.1.3) and (8.1.4) we obtain the equality

$$\left(A(\mathcal{D})u(x)\right)^\sim = \sum_{|\alpha|=0}^\infty a_\alpha(\lambda)(\xi - \lambda)^\alpha \tilde{u}(\xi) =$$

$$= \sum_{|\alpha|=0}^\infty a_\alpha(\lambda_0)(\xi - \lambda_0)^\alpha \tilde{u}(\xi) \equiv A(\xi)\tilde{u}(\xi).$$

This means that the Fourier transform of $A(\mathcal{D})u(x)$ is the same in both cases. Hence the functions $A(\mathcal{D})u(x)$ are also the same in both cases. Consequently, our definition is correct.

Now suppose that $u(x)$ can be represented as

$$u(x) = \sum_{\lambda \in G} u_\lambda(x), \qquad (8.1.5)$$

where $u_\lambda(x) \in H^\infty(S_R(\lambda))$. Then we set

$$A(\mathcal{D})u(x) = \sum_{\lambda \in G} A(\mathcal{D})u_\lambda(x). \qquad (8.1.6)$$

This definition is also correct, since the value of $A(\mathcal{D})u(x)$ is independent of the representation (8.1.5). This easily follows from the theory of the Fourier transform.

For these reasons it seems natural to give the following definition.

Definition 8.1.2. In accordance with (8.1.6), if $u(x) \in H^\infty(G)$, then by definition,

$$A(\mathcal{D})u(x) = \sum_{\lambda \in G} \left(\sum_{|\alpha|=0}^{\infty} a_\alpha(\lambda)(\mathcal{D} - \lambda I)^\alpha u_\lambda(x) \right).$$

As a corollary of the preceding arguments we can assert that for any $u(x) \in H^\infty(G)$ the operator $A(\mathcal{D})u(x)$ is uniquely determined and $A(\mathcal{D})u(x) \in H^\infty(G)$. Furthermore, the map

$$A(\mathcal{D}) : H^\infty(G) \to H^\infty(G)$$

is continuous.

As a result of these arguments we obtain the following corollary.

Corollary. *The set of operators $A(\mathcal{D})$ with symbols analytic in G and with domain of definition $H^\infty(G)$ forms a non-formal operator algebra isomorphic to the algebra of analytic functions in G. This isomorphism is defined by $A(\mathcal{D}) \leftrightarrow A(\xi)$. Moreover*

$$A(\mathcal{D}) \pm B(\mathcal{D}) \leftrightarrow A(\xi) \pm B(\xi),$$
$$\lambda A(\mathcal{D}) \leftrightarrow \lambda A(\xi),$$
$$A(\mathcal{D}) \circ B(\mathcal{D}) \leftrightarrow A(\xi)B(\xi).$$

In particular, if $A^{-1}(\xi)$ is also analytic in G, then $B(\mathcal{D}) = I/A(\mathcal{D})$ is the inverse operator to $A(\mathcal{D})$.

Remark. According to the above, the action of $A(\mathcal{D})u(x)$, where $u(x) \in H^\infty(G)$, may be given in terms of the Fourier transform by the formula

$$A(\mathcal{D})u(x) = \frac{1}{(2\pi)^n} \int_G A(\xi)\tilde{u}(\xi)e^{ix\xi}d\xi.$$

Clearly the operator $A(\mathcal{D})u(x)$ is the usual pseudodifferential operator with symbol $A(\xi)$ defined in the domain G (see, for example, the collection of papers "PD-operators", Mir, Moscow, 1967).

3. EXAMPLES

We begin with the following observation. The space $H^\infty(G)$ constructed in such a way can serve as the domain of definition for any operator $A(\mathcal{D})$ whose symbol is analytic in G. When we have some specific operator $A(\mathcal{D})$, then its domain of definition can be extended to the wider space

$$H_A^\infty(G) = \Big\{ u(x) \in L_2(\mathbf{R}^n) : u(x) = \sum_{\lambda \in G} u_\lambda(x),$$

$$\sum_{|\alpha|=0}^{\infty} a_\alpha(\lambda)(\mathcal{D} - \lambda I)^\alpha u_\lambda(x) \in L_2(\mathbf{R}^n) \Big\}.$$

The elements of this space are sums of terms $u_\lambda(x)$ whose Fourier transforms may be concentrated in $\overline{S_R(\lambda)}$. In particular, for $G = \mathbf{R}^n_\xi$ the support of the Fourier transform of $u(x) \in H_A^\infty(G)$ may be arbitrary.

In all the following examples we consider the maximum domain of definition of the operators $A(\mathcal{D})$.

Example 1. Let $A(\mathcal{D})$ be the differential operator of infinite order

$$A(\mathcal{D}) \equiv \sum_{|\alpha|=0}^{\infty} a_\alpha \mathcal{D}^\alpha,$$

whose symbol is an entire function. In this case $G = \mathbf{R}^n_\xi$. Therefore we can take as the domain of definition of $A(\mathcal{D})$ the space

$$H_A^\infty(\mathbf{R}^n) = \Big\{ u(x) \in L_2(\mathbf{R}^n) : \sum_{|\alpha|=0}^{\infty} a_\alpha \mathcal{D}^\alpha u(x) \in L_2(\mathbf{R}^n) \Big\}$$

or the Sobolev space of infinite order

$$W_A^\infty = \Big\{ u(x) \in L_2(\mathbf{R}^n) : \sum_{|\alpha|=0}^{\infty} |a_\alpha| \, \|\mathcal{D}^\alpha u\|_{L_2(\mathbf{R}^n)} < \infty \Big\}.$$

(For the fact that these spaces are non-trivial see §8.2.)

In both cases the maps

$$A(\mathcal{D}) : H_A^\infty(\mathbf{R}^n) \to L_2(\mathbf{R}^n)$$

and

$$A(\mathcal{D}) : W_A^\infty \to L_2(\mathbf{R}^n)$$

are continuous.

Example 2. The operator $A(\mathcal{D}) = (I + \mathcal{D}^2)^{1/2} \equiv (I - \Delta)^{1/2}$ is defined by the symbol $A(\xi) = (1 + \xi^2)^{1/2}$. Clearly, $A(\xi)$ is analytic in the whole space $\mathbf{R}^n$. Hence, $H^\infty(\mathbf{R}^n)$ consists of all functions $u(x) \in L_2(\mathbf{R}^n)$ the Fourier transform of which has compact support, that is, it is the space $H^\infty(S_R)$, where $R = +\infty$.

Using the Fourier transform, it is easy to extend the domain of definition of $(I - \Delta)^{1/2}$ to the space

$$H_a^\infty(\mathbf{R}^n) = \left\{ u(x) \in L_2(\mathbf{R}^n) : \int_{\mathbf{R}^n} (1 + \xi^2) |\tilde{u}(\xi)|^2 d\xi < \infty \right\},$$

that is, the space

$$W_2^1(\mathbf{R}^n) = \left\{ u(x) : \|u\|_{L_2(\mathbf{R}^n)}^2 + \sum_{i=1}^n \left\| \frac{\partial u}{\partial x_i} \right\|_{L_2(\mathbf{R}^n)}^2 < \infty \right\}.$$

Example 3. Let $A(\xi) = |\xi|$, $\xi \in \mathbf{R}^1$. This function is analytic in $\mathbf{R}^1 \backslash \{0\}$, so that $G = (-\infty, 0) \cup (0, +\infty)$. It follows that for $u(x) \in H^1(\mathbf{R}^1)$

$$|\mathcal{D}|u(x) = \mathcal{D}u(x), \quad \text{if supp } \tilde{u}(\xi) \in (0, +\infty),$$
$$|\mathcal{D}|u(x) = -\mathcal{D}u(x), \quad \text{if supp } \tilde{u}(\xi) \in (-\infty, 0).$$

We write $\tilde{u}(\xi) \in L_2(\mathbf{R}^1)$ in the form

$$\tilde{u}(\xi) = \tilde{u}_+(\xi) + \tilde{u}_-(\xi),$$

where

$$\tilde{u}_+(\xi) = \begin{cases} \tilde{u}(\xi), & \xi > 0 \\ 0, & \xi < 0 \end{cases}, \quad \tilde{u}_-(\xi) = \begin{cases} 0, & \xi > 0, \\ \tilde{u}(\xi), & \xi < 0. \end{cases}$$

This sum corresponds to the representation of $u(x) \in L_2(\mathbf{R}^n)$ as the sum of two orthogonal terms

$$u(x) = u_+(x) + u_-(x),$$

where $u_\pm(x) = F^{-1}\tilde{u}_\pm(\xi)$; F^{-1} is the inverse Fourier transform.

It is clear that the action of $|\mathcal{D}|u(x)$ on any function $u(x) \in H^1(\mathbf{R}^1)$ is defined by

$$|\mathcal{D}|u(x) = \mathcal{D}u_+(x) - \mathcal{D}u_-(x).$$

Example 4. The function $A(\xi) = 1/\xi^2$ is the symbol of the operator $-I/\Delta$. In this case $G = \mathbf{R}^n \backslash \{0\}$, hence

$$H^\infty(G) = \{u(x) \in L_2(\mathbf{R}^n) : \text{supp } \tilde{u}(\xi) \in \mathbf{R}^n \backslash \{0\}\}.$$

The action of $-I/\Delta$ is locally the action of a differential operator of infinite order for any $u(x) \in H^\infty(G)$.

4. THE DUAL THEORY

First we consider the space of generalized functions $\{H^\infty(S_R(\lambda))\}^*$, that is, the space of linear bounded functionals on $H^\infty(S_R(\lambda))$.

Now we claim that this space is invariant under the operators $A(-\mathcal{D})$, where $A(\xi)$ is analytic on $S_R(\lambda)$. In fact, if $u(x) \in \{H^\infty(S_R(\lambda))\}^*$, then for any $\phi(x) \in H^\infty(S_R(\lambda))$

$$\langle A(-\mathcal{D})u(x), \phi(x) \rangle \overset{\text{def}}{=} \langle u(x), A(\mathcal{D})\phi(x) \rangle.$$

In view of the continuity of

$$A(\mathcal{D}) : H^\infty(S_R(\lambda)) \rightarrow H^\infty(S_R(\lambda))$$

this means that $A(-\mathcal{D})u(x)$ is a continuous functional on $H^\infty(S_R(\lambda))$.

We now describe the structure of the distribution $u(x) \in \{H^\infty(S_R(\lambda))\}^*$. Clearly, if $A(\xi)$ is analytic in $S_R(\lambda)$ then for any $u_0(x) \in L_2(\mathbf{R}^n)$ we have $A(-\mathcal{D})u_0(x) \in \{H^\infty(S_R(\lambda))\}^*$ and

$$\langle A(-\mathcal{D})u_0(x), \phi(x) \rangle = \int_{\mathbf{R}^n} u_0(x)A(\mathcal{D})\phi(x)dx, \qquad (8.1.7)$$

where $\phi(x) \in H^\infty(S_R(\lambda))$ is any test function. Using the unitarity of the Fourier transform, we obtain

$$\langle A(-\mathcal{D})u_0(x), \phi(x) \rangle = \int_{S_R(\lambda)} \tilde{u}_0(-\xi)A(\xi)\tilde{\phi}(\xi)d\xi,$$

because supp $\tilde{\phi}(\xi) \subset S_R(\lambda)$.

It follows from the latter equality that the values of $\tilde{u}_0(\xi)$ outside $S_R(\lambda)$ are not important and may be taken equal to zero.

Conversely, by analogy with the arguments in Proposition 7.2.1, one can verify that any functional $u(x) \in \{H^\infty(S_R(\lambda))\}^*$ may be represented in the form

$$u(x) = A(-\mathcal{D})u_0(x) \equiv \sum_{|\alpha|=0}^{\infty} a_\alpha(\lambda)(-1)^{|\alpha|}(\mathcal{D} + \lambda I)^\alpha u_0(x),$$

where $u_0(x) \in L_2(\mathbf{R}^n)$ is such that $\tilde{u}_0(\xi) \equiv 0$ outside $S_R(-\lambda)$ and $A(\xi)$ is analytic on $S_R(\lambda)$.

Thus we obtained the following

Proposition 8.1.2. *The space* $\{H^\infty(S_R(\lambda))\}^*$ *is given by*

$$\{H^\infty(S_R(\lambda))\}^* = \left\{ u(x) : u(x) = \sum_{|\alpha|=0}^{\infty} (-1)^{|\alpha|} a_\alpha(\lambda)(\mathcal{D} + \lambda I)^\alpha u_0(x) \right\},$$

where $u_0(x) \in L_2(\mathbf{R}^n)$ *and* $\tilde{u}_0(\xi) \equiv 0$ *for* $|\xi_j + \lambda_j| > R_j$, $j = 1, \ldots, n$.

General case. We now consider an arbitrary domain $G \subset \mathbf{R}_\xi^n$. Let $H^\infty(G)$ be the corresponding test-function space. By definition of the PD-operator $A(\mathcal{D})$ with analytic symbol in G and in accordance with the local description of $\{H^\infty(G)\}^*$ obtained in subsection 2, we find that

$$\{H^\infty(G)\}^* = \{u(x) : u(x) = A(-\mathcal{D})u_0(x), \ u_0(x) \in L_2(\mathbf{R}^n)\},$$

where $A(-\mathcal{D})$ is a PD-operator with symbol $A(\xi) \in \mathcal{O}(G)$, that is, it is analytic in G. Moreover, we may choose the function $u_0(x)$ so that $\mathrm{supp}\,\tilde{u}_0(\xi) \subset -G$, where $-G = \{\xi : -\xi \in G\}$.

It is clear that the space $\{H^\infty(G)\}^*$ is invariant under the action of any such operator $A(-\mathcal{D})$.

Finally we note that it is convenient to use the following notation:

$$H^{+\infty}(G) \equiv H^\infty(G), \quad H^{-\infty} \equiv \{H^\infty(-G)\}^*.$$

In this notation the maps

$$A(\mathcal{D}) : H^{\pm\infty}(G) \to H^{\pm\infty}(G)$$

are continuous and give a non-formal algebra of pseudodifferential operators isomorphic to the algebra of analytic functions in G. Here "+" corresponds to "+" and "−" corresponds to "−".

5. A POSSIBLE GENERALIZATION

The algebra of differential operators of infinite order in the spaces $H^{\pm\infty}(G)$ constructed above, uses the space $L_{2,\text{comp}}(\mathbf{R}^n_\xi)$ in the Fourier ξ-representation, that is the duality $\langle L_{2,\text{comp}}, L_{2,\text{loc}}\rangle$. However, it is *a posteriori* clear that similar algebras may be also obtained for certain other dualities, for example, $\langle \mathcal{D}, \mathcal{D}'\rangle$ or $\langle L_{p,\text{comp}}, L_{p,\text{loc}}\rangle$, $p' = p/(p-1)$, $p > 1$. Moreover, no difficulties in principle arise in these considerations.

One can give another approach to the construction of the test-function spaces (and the dual spaces) that are invariant under the action of pseudodifferential operators with analytic symbols. One of these approaches is based on estimates of the derivatives of the function.

Namely, let $r = (r_1, \ldots, r_n)$ be a vector, with $r_1 > 0, \ldots, r_n > 0$. We set

$$E_r(\mathbf{R}^n_x) = \{\phi(x) \in C^\infty(\mathbf{R}^n) : |\mathcal{D}^\alpha \phi(0)| \le M r^\alpha, \ |\alpha| = 0, 1, \ldots\},$$

where $M > 0$ is a constant depending in general on $\phi(x)$, but not depending on α.

Further, let $R = (R_1, \ldots, R_n)$ be a vector, $R_1 > 0, \ldots, R_n > 0$. The function $\phi(x)$ is said to belong to $\text{Exp}_R(\mathbf{R}^n_x)$ if there exists a vector $r < R$ such that $\phi(x) \in E_r(\mathbf{R}^n_x)$.

Proposition 8.1.3. *The space* $\text{Exp}_R(\mathbf{C}^n_z)$ *is the complex continuation of the space* $\text{Exp}_R(\mathbf{R}^n_x)$ *and, conversely,*

$$\text{Exp}_R(\mathbf{R}^n_x) = \text{Exp}_R(\mathbf{C}^n_z)\big|_{z=x}.$$

Proof. Indeed, if $\phi(x) \in \text{Exp}_R(\mathbf{R}^n_x)$ then for some $r < R$ the inequalities $|\mathcal{D}^\alpha \phi(0)| \le M r^\alpha$, $|\alpha| = 0, 1, \ldots$ hold. This means that the Taylor series of this function converges for all $x \in \mathbf{R}^n$ and therefore defines an entire function

$$\phi(z) = \sum_{|\alpha|=0}^\infty \frac{\mathcal{D}^\alpha \phi(0)}{\alpha!} z^\alpha.$$

Moreover,

$$|\phi(z)| \le M \sum_{|\alpha|=0}^\infty \frac{r^\alpha |z^\alpha|}{\alpha!} \le M \exp rz_+,$$

where $rz_+ = r_1|z_1| + \ldots + r_n|z_n|$.

Conversely, if $\phi(z) \in \operatorname{Exp}_R(\mathbf{C}_z^n)$ then, according to Proposition 1.1.2 there exists a vector $r < R$ such that the inequalities (1.1.2) hold. Here, the constant $M > 0$ depends on $\phi(z)$, but not on α.

In particular, putting $z \equiv x$ and then $x = 0$ we obtain

$$\left. |\mathcal{D}^\alpha \phi(x)| \right|_{x=0} \leq M r^\alpha, \quad |\alpha| = 0, 1, \ldots,$$

that is,

$$\phi(x) \in E_r(\mathbf{R}_x^n) \subset \operatorname{Exp}_R(\mathbf{R}_x^n).$$

The theorem is proved.

8.2. PD-equations in the whole Euclidean space

We consider the equation

$$A(\mathcal{D})u(x) = h(x), \quad x \in \mathbf{R}^n, \tag{8.2.1}$$

where $A(\mathcal{D})$ is a pseudodifferential operator with analytic symbol $A(\xi)$.

We obtain from the results of §8.1 the following theorem.

Theorem 8.2.1. *If the functions $A(\xi)$ and $A^{-1}(\xi)$ are analytic in some domain $G \subset \mathbf{R}_\xi^n$, then for any $h(x) \in H^{+\infty}(G)$ equation (8.2.1) has the unique solution*

$$u(x) = \frac{I}{A(\mathcal{D})} h(x). \tag{8.2.2}$$

Example 1. We consider the Helmholtz equation ($\omega \in \mathbf{C}^1$ is a parameter)

$$\Delta u(x) + \omega^2 u(x) = h(x), \quad x \in \mathbf{R}^n.$$

Obviously, the symbols of the Helmholtz operator and its inverse operator $I/(\Delta + \omega^2 I)$ are analytic functions for $\xi^2 \neq \omega^2$, $\xi \in \mathbf{R}^n$. Hence, if ω is not real, then the whole space $\mathbf{R}^n$ is the common domain G of analyticity for the symbols of both operators. In this case (that is, $\omega \notin \mathbf{R}^1$) the space $H^{+\infty}(G)$ consists of all functions $h(x) \in L_2(\mathbf{R}^n)$ that have analytic continuations as entire functions of first order of growth and of any finite type. Thus, if $h(x)$ is such a function, then there exists a unique solution of (8.2.2); moreover

$$u(x) = \frac{I}{\Delta + \omega^2 I} h(x). \tag{8.2.3}$$

Now let ω be real. Then we have $G = \mathbf{R}_\xi^n \backslash S$, where S is the sphere $\xi^2 = \omega^2$. Hence formula (8.2.3) gives the solution only for those functions $h(x) \in L_2(\mathbf{R}^n)$ whose Fourier transform is equal to zero for $\xi^2 = \omega^2$.

We now consider the dual theory.

Let $H^{-\infty}(G)$ be the space of generalized functions over $H^\infty(G)$, that is, the conjugate space to $H^\infty(G)$.

Theorem 8.2.2. *Let $A(\xi)$ and $A^{-1}(\xi)$ be analytic in a domain G. Then for any function $h(x) \in H^{-\infty}(G)$, equation (8.2.1) has a unique solution defined by (8.2.2). This solution also belongs to $H^{-\infty}(G)$.*

The proof is clear.

Example 2. In this example we discuss the problem of the existence of a fundamental solution for the operator $A(\mathcal{D})$. Recall that, by definition, $E(x)$ is called a fundamental solution of the operator $A(\mathcal{D})$ if

$$A(\mathcal{D})E(x) = \delta(x), \quad x \in \mathbf{R}^n,$$

where $\delta(x)$ is the delta function.

Assume that both functions $A(\xi)$ and $A^{-1}(\xi)$ have a common domain of analyticity G which is non-empty, that is $G \neq \varnothing$. It is clear that $\delta(x) \in H^{-\infty}(G)$. Hence by Theorem 8.2.2 the fundamental solution $E(x)$ exists as a unique functional on $H^\infty(G)$ and is given by the formula

$$E(x) = \frac{I}{A(\mathcal{D})} \delta(x).$$

Example 3. We consider equation (8.2.1) with $h(x) = P_m(x)\exp(i\lambda x)$, where $\lambda \in G$ is arbitrary and $P_m(x)$ is a polynomial of degree $m \geq 0$. Then $\operatorname{supp} \tilde{h}(\xi) = \lambda$, so that $h(x) \in H^{+\infty}(G)$ and hence

$$u(x) = \frac{I}{A(\mathcal{D})} h(x) = \sum_{|\alpha|=0}^\infty b_\alpha(\lambda)(\mathcal{D} - \lambda I)^\alpha \left[P_m(x)\exp(i\lambda x)\right] =$$

$$= \exp(i\lambda x) \sum_{|\alpha|=0}^\infty b_\alpha(\lambda)\mathcal{D}^\alpha P_m(x) \equiv \exp(i\lambda x)Q_m(x),$$

where $Q_m(x)$ is a polynomial of degree m.

Thus, every equation of the form (8.2.1), having a quasipolynomial right hand side, has a unique solution in the form of a quasipolynomial as well. This result may be treated (by analogy with ordinary differential equations) as an inspection method (non-resonance case).

Example 4. We consider the equation

$$u(x+1) + u(x-1) = h(x), \quad x \in \mathbf{R}^1. \tag{8.2.4}$$

In accordance with Taylor's formula, we can write (8.2.4) in the form of a differential equation of infinite order. We obtain

$$2\cosh\left(\frac{d}{dx}\right)u(x) = h(x), \quad x \in \mathbf{R}^1. \tag{8.2.5}$$

We set $G = \mathbf{R}^1 \backslash \{\frac{\pi}{2} + k\pi, \ k = 0, \pm 1, \ldots\}$. Then $\cosh(i\xi) \neq 0$ for any $\xi \in G$ and consequently, for any $h(x) \in H^\infty(G)$ there exists the unique solution of equation (8.2.5)

$$u(x) = \tfrac{1}{2}\mathrm{sech}\left(\frac{d}{dx}\right)h(x).$$

If $\mathrm{supp}\,\tilde{h}(\xi) \in \,]-\pi/2, \pi/2\,[$, then

$$u(x) = \frac{1}{2}\sum_{n=0}^{\infty} \frac{(-1)^n E_{2n}}{(2n)!} \frac{d^{2n}}{dx^{2n}} h(x),$$

where the E_{2n} are the Euler numbers: $E_0 = 1$, $E_2 = -1$, $E_4 = 5$, etc. (see, for example, the handbook of Abramowitz and Stegun [1], p.810).

In particular, if $h(x) \equiv P_m(x)$, where $P_m(x)$ is polynomial, then the solution

$$u(x) = \tfrac{1}{2}\mathrm{sech}\left(\frac{d}{dx}\right)P_m(x)$$

is also a polynomial of the same degree.

8.3. The Cauchy problem

1. THE CAUCHY PROBLEM IN THE SPACE $H^\infty(G)$

We consider the Cauchy problem

$$\frac{\partial^m u}{\partial t^m} + \sum_{k=0}^{m-1} A_k(t, \mathcal{D}) \frac{\partial^k u}{\partial t^k} = h(t, x), \tag{8.3.1}$$

$$\frac{\partial^k u}{\partial t^k}(0, x) = \phi_k(x), \quad k = 0, 1, \ldots, m-1, \tag{8.3.2}$$

($t \in \mathbf{R}^1$, $x \in \mathbf{R}^n$). Here $A_k(t, \mathcal{D})$ are PD-operators whose symbols $A_k(t, \xi)$ are analytic for $t \in \mathbf{R}^1$ and $\xi \in G \subset \mathbf{R}^n$; G is a domain.

Our aim is to prove the well-posedness of the problem (8.3.1), (8.3.2).

We denote by $C(\mathbf{R}^1; H^\infty(G))$ the space whose elements are continuous functions in $t \in \mathbf{R}^1$ and such that for each $t \in \mathbf{R}^1$ we have the inclusion

$$u(t, \cdot) \in H^\infty(G).$$

For $k \geq 1$ we set

$$C^k(t; H^\infty(G)) = \Big\{ u(t, x) : u(t, x) \in C(\mathbf{R}^1; H^\infty(G)), \ldots$$

$$\ldots, u^{(k)}(t, x) \in C(\mathbf{R}^1; H^\infty(G)) \Big\}.$$

The main result of this section is

Theorem 8.3.1. *For any $\phi_k(x) \in H^\infty(G)$ and $h(t, x) \in C(\mathbf{R}^1; H^\infty(G))$ there exists a unique solution $u(t, x) \in C^m(\mathbf{R}^1; H^\infty(G))$ of the problem* (8.3.1), (8.3.2).

Proof. First we remark that by Duhamel's principle one can set $h(t, x) \equiv 0$. Indeed, if $U(t, s, x)$ is a solution of the problem

$$L\Big(\frac{\partial}{\partial t}, \mathcal{D}\Big) U \equiv \frac{\partial^m U}{\partial t^m} + \sum_{k=0}^{m-1} A_k(t, \mathcal{D}) \frac{\partial^k U}{\partial t^k} = 0,$$

$$\frac{\partial^k U}{\partial t^k}(0, s, x) = 0, \quad 0 \leq k \leq m - 2,$$

$$\frac{\partial^{m-1} U}{\partial t^{m-1}}(0, s, x) = h(s, x),$$

where $s \in {]}0, t{[}$ is arbitrary, then (as is easy to verify) the solution $u(t, x)$ of the problem

$$L\Big(\frac{\partial}{\partial t}\Big) u = h(t, x), \quad \frac{\partial^k u}{\partial t^k}(0, x) = 0, \quad k = 0, 1, \ldots, m - 1,$$

is given by the formula

$$u(t, x) = \int_0^t U(t - s, s, x) ds.$$

Taking into account this remark, we can consider the Cauchy problem for the homogeneous equation

$$L\left(\frac{\partial}{\partial t}, \mathcal{D}\right) u(t, x) = 0 \qquad (8.3.3)$$

with the non-homogeneous initial data

$$\frac{\partial^k u}{\partial t^k}(0, x) = \phi_k(x), \quad k = 0, 1, \ldots, m-1, \qquad (8.3.4)$$

where $\phi_k(x) \in H^\infty(G)$. In order to solve it, we formally set $\mathcal{D} \leftrightarrow \xi$, $\xi = (\xi_1, \ldots, \xi_n)$, and consider the Cauchy problem for the ordinary differential equation

$$L\left(\frac{\partial}{\partial t}, \xi\right) u_j(t, \xi) \equiv u_j^{(m)} + \sum_{k=0}^{m-1} A_k(t, \xi) u_j^{(k)} = 0,$$

$$u_j^{(k)}(0, \xi) = \delta_{kj} \quad (0 \leq k, j \leq m-1),$$

where δ_{kj} is the Kronecker symbol and ξ is a real parameter.

Let us note that, by hypothesis, the symbols $A_k(t, \xi)$ of all the PD-operators $A_k(t, \mathcal{D})$ are analytic in ξ in the domain G. Consequently, using the classical theorem on the analytic dependence on a parameter, we can assert that the solutions $u_j(t, \xi)$, $j = 0, 1, \ldots, m-1$, are analytic functions in $\xi \in G$ (see, for example, Kamke [1]). Therefore we can associate with each function $u_j(t, \xi)$ the PD-operator $u_j(t, \mathcal{D})$ whose symbol is $u_j(t, \xi)$ itself.

After this it is clear that the formula

$$u(t, x) = \sum_{j=0}^{m-1} u_j(t, \mathcal{D}) \phi_j(x) \qquad (8.3.5)$$

gives the desired solution of the problem (8.3.3), (8.3.4).

It remains to prove the uniqueness of this solution. For this we note that the Fourier transform $\tilde{u}(t, \xi)$ of an arbitrary solution $u(t, x) \in C^m(\mathbf{R}^1; H^\infty(G))$ satisfies the problem for the ordinary differential equation

$$\tilde{u}^{(m)} + \sum_{k=0}^{m-1} A_k(t, \xi) \tilde{u}^{(k)} = \tilde{h}(t, \xi),$$

$$\tilde{u}^{(k)}(0, \xi) = \tilde{\phi}_k(\xi), \quad k = 0, 1, \ldots, m-1.$$

Consequently, $\tilde{u}(t, \xi)$ is determined uniquely. Then the main problem (8.3.1), (8.3.2) has only one possible solution. The theorem is proved.

2. THE CAUCHY PROBLEM IN THE DUAL SPACE $H^{-\infty}(G)$

In this subsection we also consider the problem

$$L\left(\frac{\partial}{\partial t}, \mathcal{D}\right)u \equiv \frac{\partial^m u}{\partial t^m} + \sum_{k=0}^{m-1} A_k(t, \mathcal{D})\frac{\partial^k u}{\partial t^k} = h(t, x) \qquad (8.3.6)$$

$$\frac{\partial^k u}{\partial t^k}(0, x) = \phi_k(x), \quad k = 0, 1, \ldots, m-1, \qquad (8.3.7)$$

but this time, $\phi_k(x)$ and $h(t, x)$ are generalized functions.

Thus, let $\phi_k(x) \in H^{-\infty}(G)$, $h(t, x) \in C(\mathbf{R}^1; H^{-\infty}(G))$.

Theorem 8.3.2. *Under these conditions there exists a unique solution $u(t, x)$ in $C^m(\mathbf{R}^1; H^{-\infty}(G))$, which is given by the same formula (8.3.5).*

Proof. By Duhamel's principle it suffices to consider the case $h(t, x) \equiv 0$. Namely, we prove that for any $\phi_k(x) \in H^{-\infty}(G)$ formula (8.3.5), that is,

$$u(t, x) = \sum_{j=0}^{m-1} u_j(t, \mathcal{D})\phi_k(x) \qquad (8.3.8)$$

gives the generalized solution of our problem (recall that $u_j(t, \mathcal{D})$ are the PD-operators whose symbols $u_j(t, \xi)$ are the solutions of the ordinary differential equation

$$L\left(\frac{d}{dt}, \xi\right)u_j(t, \xi) = 0, \quad u_j^{(k)}(0, \xi) = \delta_{kj} \quad (0 \leq k, j \leq m-1)$$

where δ_{kj} is the Kronecker symbol).

Indeed, it is not difficult to verify that if the function

$$w(t, x) = U(t, \mathcal{D})\phi(x), \quad \phi(x) \in H^\infty(G),$$

$(U(t, \mathcal{D})$ is a PD-operator) is a solution of the homogeneous equation

$$\frac{\partial^m w}{\partial t^m} + \sum_{k=0}^{m-1} A_k(t, \mathcal{D})\frac{\partial^k w}{\partial t^k} = 0,$$

then the function

$$w^*(t, x) = U(t, -\mathcal{D})\phi(x), \quad \phi(x) \in H^\infty(-G),$$

satisfies the equation

$$L^*\left(\frac{\partial}{\partial t}, \mathcal{D}\right)w^* \equiv \frac{\partial^m w^*}{\partial t^m} + \sum_{k=0}^{m-1} A_k(t, -\mathcal{D})\frac{\partial^k w^*}{\partial t^k} = 0.$$

Consequently, for any $\phi_j(x) \in H^{-\infty}(G)$ and for any $v(x) \in H^\infty(-G)$ we have

$$\left\langle L\left(\frac{\partial}{\partial t}, \mathcal{D}\right)[u_j(t, \mathcal{D})\phi_j(x)], v\right\rangle = \left\langle u_j^{(m)}(t, \mathcal{D})\phi_j(x), v(x)\right\rangle +$$

$$+ \sum_{k=0}^{m-1} \left\langle A_k(t, \mathcal{D})u_j^{(k)}(t, \mathcal{D})\phi_j(x), v(x)\right\rangle =$$

$$= \left\langle \phi_j(x), u_j^{(m)}(t, -\mathcal{D})v(x) + \sum_{k=0}^{m-1} A_k(t, -\mathcal{D})u_j^{(k)}(t, -\mathcal{D})v(x)\right\rangle =$$

$$= \langle \phi_j(x), L^*(t, \mathcal{D})u_j(t, -\mathcal{D})v(x)\rangle = 0.$$

This means that the function (8.3.8) is a generalized solution of the original equation (8.3.6).

Let us verify that this solution satisfies the initial conditions (8.3.7) over $H^\infty(-G)$. Indeed, in accordance with the construction of the operators $u_j(t, \mathcal{D})$ we have

$$\frac{d^k}{dt^k}u_j(t, \mathcal{D})\Big|_{t=0} = \delta_{kj}I,$$

where I is the identity operator in $H^\infty(G)$. Then

$$\frac{d^k}{dt^k}u_j(t, -\mathcal{D})\Big|_{t=0} = \delta_{kj}I,$$

where I is the identity operator in $H^\infty(-G)$; consequently, for any $v(x) \in H^\infty(-G)$

$$\left\langle \frac{d^k}{dt^k}u_j(t, \mathcal{D})\phi_j(x), v(x)\right\rangle\Big|_{t=0} =$$

$$= \left\langle \phi_j(x), \frac{d^k}{dt^k}u_j(t, -\mathcal{D})v(x)\right\rangle\Big|_{t=0} = \langle \phi_j(x), \delta_{kj}v(x)\rangle,$$

which means that

$$\frac{d^k}{dt^k}u(t,x)\Big|_{t=0} = \phi_k(x) \text{ in } H^{-\infty}(G).$$

Thus the existence of the solution of the original problem (8.3.6), (8.3.7) is established.

Uniqueness. In order to prove the uniqueness of the generalized solution we note that, in accordance with the results of subsection 3 of §8.1, every functional $h(x) \in H^{-\infty}(G)$ (we recall that $H^{-\infty}(G) = (H^\infty(-G))^*$) can be represented in the form

$$h(x) = A_0(\mathcal{D})u_0(x),$$

where $A_0(\mathcal{D})$ is a PD-operator whose symbol $A_0(\xi)$ is analytic in G and $u_0(x) \in L_2(\mathbf{R}^n)$ is such that $\tilde{u}_0(\xi) \equiv 0$ outside G. Therefore, the Fourier transform $\tilde{h}(\xi) = A_0(\xi)\tilde{u}_0(\xi)$, $\xi \in G$, is an ordinary function.

Further, if $u(t,x) \in C^m(\mathbf{R}^1; H^{-\infty}(G))$ is a generalized solution of the original Cauchy problem (8.3.6), (8.3.7), then its Fourier x-transform $\tilde{u}(t,\xi)$ satisfies (as an ordinary function) the ordinary problem

$$\tilde{u}^{(m)} + \sum_{k=0}^{m-1} A_k(t,\xi)\tilde{u}^{(k)} = \tilde{h}(t,\xi),$$

$$\tilde{u}^{(k)}(0,\xi) = \tilde{\phi}_k(\xi), \quad 0 \le k \le m-1,$$

and is therefore unique. The theorem is completely proved.

3. ON THE EXISTENCE OF THE FUNDAMENTAL SOLUTION OF THE CAUCHY PROBLEM

We begin with the well-known definition.

Definition 8.3.1. The generalized function $E(t,x) \in C^m(\mathbf{R}^1, H^{-\infty}(G))$ is called a fundamental solution of the Cauchy problem for the operator $L(\partial/\partial t, \mathcal{D})$ if

$$L(\partial/\partial t, \mathcal{D})E(t,x) = 0,$$

$$E(0,x) = 0, \ldots, \quad E^{(m-2)}(0,x) = 0, \quad E^{(m-1)}(0,x) = \delta(x).$$

As corollary of Theorem 8.3.2 we have

Proposition 8.3.1. *For any PD-operator*

$$L(\partial/\partial t, \mathcal{D}) \equiv \frac{\partial^m}{\partial t^m} + \sum_{k=0}^{m-1} \frac{\partial^k}{\partial t^k} A_k(t, \mathcal{D})$$

and for any $G \subset \mathbf{R}^n$ there exists a unique fundamental solution $E(t, x) \in$ $C^m(\mathbf{R}^1, H^{-\infty}(G))$.

Remark. It is well known that the significance of Definition 8.3.1 is that knowing the fundamental solution enables us to give the solution of the general Cauchy problem by means of the operators of convolution and differentiation.

8.4. Examples

1. THE CAUCHY PROBLEM FOR A HOMOGENEOUS EQUATION

Let $x \in \mathbf{R}^1, \mathcal{D} = \frac{1}{i}\frac{d}{dx}$. We investigate the problem

$$\sum_{k=0}^m a_k \frac{\partial^k}{\partial t^k} \mathcal{D}^{m-k} u(t, x) = 0 \quad (a_m \neq 0) \tag{8.4.1}$$

$$\frac{\partial^k u}{\partial t^k}(0, x) = \phi_k(x), \quad k = 0, 1, \ldots, m-1. \tag{8.4.2}$$

Putting $\mathcal{D} \leftrightarrow \xi$, $\xi \in \mathbf{R}^1$, we find a system of functions $u_j(t, \xi)$, $j = 0, 1, \ldots$ $\ldots, m-1$, as a system of solutions of the problem

$$\sum_{k=0}^m a_k \xi^{m-k} u_j^{(k)}(t, \xi) = 0$$

$$u_j^{(k)}(0, \xi) = \delta_{kj} \quad (k, j = 0, 1, \ldots, m-1). \tag{8.4.3}$$

We suppose for simplicity that the characteristic polynomial of (8.4.3)

$$P_m(\lambda, \xi) = \sum_{k=0}^m a_k \xi^{m-k} \lambda^k$$

has only simple roots $\lambda_1, \ldots, \lambda_m$. Taking into account the homogeneity of $P_m(\lambda, \xi)$, we have

$$\lambda_k = \omega_k \xi, \quad k = 1, \ldots, m,$$

where the ω_k are the roots of the polynomial $P_m(\lambda, 1)$.

After simple calculations we obtain

$$u_j(t,\xi) = \sum_{k=1}^{m} \frac{V_{j+1,k}}{V\xi^j} \exp(t\omega_k\xi),$$

where $V = V(\omega_1, \ldots, \omega_m)$ is the Vandermonde determinant and the $V_{j+1,k}$ are its cofactors.

Since for any $j = 0, 1, \ldots, m-1$

$$\sum_{k=1}^{m} \omega_k^s V_{j+1,k} = 0 \quad (s = 0, 1, \ldots, m-1),$$

it follows that the solutions $u_j(t,\xi)$ are entire functions of $\xi \in \mathbf{R}^1$ and, consequently, they determine the differential operators of infinite order

$$u_j(t,D) = \sum_{k=1}^{m} \frac{V_{j+1,k}}{VD^j} \exp(t\omega_k D).$$

Thus the solution of the Cauchy problem (8.4.1), (8.4.2) can be represented in the form

$$u(t,z) = \sum_{j,k=0}^{m-1} \frac{V_{j+1,k+1}}{VD^j} \exp(t\omega_{k+1}D)\phi_j(x).$$

By definition every function $\phi_j(x) \in H^\infty(G)$ is entire, so that

$$\exp(t\omega_{k+1}D)\phi_j(x) = \phi_j(x - it\omega_{k+1}).$$

Consequently, the above solution is given by the formula

$$u(t,x) = \frac{1}{V}\sum_{k=0}^{m-1} V_{1,k}\phi_0(x - it\omega_{k+1}) + \frac{1}{V}\sum_{j=1}^{m-1}\frac{1}{(j-1)!}\sum_{k=0}^{m-1} V_{j+1,k+1} \times$$

$$\times \int_0^{x-it\omega_{k+1}} (x - it\omega_{k+1} - \tau)^{j-1}\phi_j(\tau)d\tau. \qquad (8.4.4)$$

Remark. Let us consider the special case of purely imaginary roots $\omega_1, \ldots, \omega_m$. This means that the original equation (8.4.1) is a strictly hyperbolic equation. In this case all the shifts and integrations in (8.4.4) are real and, since the imbedding $H^\infty(\mathbf{R}^1) \subset L_2(\mathbf{R}^1)$ is dense, formula (8.4.4) can be extended to arbitrary initial data $\phi_j(x) \in L_2(\mathbf{R}^1)$. Consequently, we obtain the well known result: for any strictly hyperbolic equation the Cauchy problem is well-posed in the sense of $L_2(\mathbf{R}^1)$.

On the other hand, it is obvious that in the case of complex or purely real roots $\omega_1, \ldots, \omega_m$ the following shifts cannot be extended to functions $\phi_j(x)$ of any finite smoothness. Therefore, the Cauchy problem for the class of homogeneous equations (8.4.1) is well-posed in the sense of Hadamard-Petrovskii, if the equation is precisely of hyperbolic type.

2. SPECIAL CASE. LAPLACE EQUATION

We now consider a special case of the problem (8.4.1), (8.4.2). More precisely, the solution of the Cauchy problem

$$\frac{\partial^2 u}{\partial t^2} - a^2 \frac{\partial^2 u}{\partial x^2} = 0, \quad a \in \mathbf{C}^1, \ |\alpha| = 1, \qquad (8.4.5)$$

$$u(0, x) = \phi(x), \quad \frac{\partial u}{\partial t}(0, x) = \psi(x), \qquad (8.4.6)$$

is given by the formula

$$u(t, x) = \left[\cosh at \frac{d}{dt}\right] \phi(x) + \left[\frac{\sinh at \frac{d}{dx}}{a \frac{d}{dx}}\right] \psi(x). \qquad (8.4.7)$$

If $\phi(x) \in H^\infty(\mathbf{R}^1)$ and $\psi(x) \in H^\infty(\mathbf{R}^1)$, then this formula is the classical d'Alembert formula

$$u(t, x) = \frac{\phi(x + at) + \phi(x - at)}{2} + \frac{1}{2a} \int_{x-at}^{x+at} \psi(\xi) d\xi$$

for any $a \in \mathbf{C}^1$, that is, for any type of equation (8.4.5). For $a = i$ we have the solution of the Cauchy problem for the Laplace equation.

Looking in this case at the d'Alembert formula

$$u(t, x) = \frac{\phi(x + it) + \phi(x - it)}{2} + \frac{1}{2i} \int_{x-it}^{x+it} \psi(\xi) d\xi$$

it is not difficult to give a function-theoretic explanation of the ill-posedness (in the sense of Hadamard-Petrovskii) of the Cauchy problem for the Laplace equation. In fact, this question, clearly, is equivalent to the existence of analytic continuations of functions $\phi(x)$ and $\psi(x)$ to the complex plane. However, not every function belonging to the spaces of finite smoothness has such a continuation.

By contrast, as we have seen, if $a = 1$ (wave equation) such a continuation to $\mathbf{C}^1$ is not required.

Remark. The complex d'Alembert formula can be used for extending the solution $u(x) \in H^\infty(\mathbf{R}^1)$ to a solution belonging to the corresponding Sobolev space of infinite order. Indeed, let us write the formula (8.4.7) in the form of two series

$$u(t, x) = \sum_{n=0}^\infty \frac{(at)^{2n}}{(2n)!} \phi^{(2n)}(x) + \frac{1}{a} \sum_{n=0}^\infty \frac{(at)^{2n}}{(2n+1)!} \psi^{(2n)}(x). \qquad (8.4.8)$$

We set

$$W^\infty\left\{\frac{T^n}{(2n)!}, q\right\} = \left\{\phi(x) : \sum_{n=0}^\infty \frac{T^{2n}}{(2n)!} \|\phi^{(2n)}(x)\|_q < \infty\right\},$$

where $T > 0$ and $\|\cdot\|_q$ is the norm in $L_q(\mathbf{R}^1)$.

It is clear that for any $\phi(x), \psi(x) \in W^\infty\{\frac{T}{(2n)!}, q\}$ the formula (8.4.8) gives the L_q-solution of the problem (8.4.5), (8.4.6) for $|t| \leq T$.

Using Parseval's equality, we can define the more natural space of the initial data as the space of functions $\phi(x), \psi(x) \in L_2(\mathbf{R}^1)$ for which

$$\cosh(at\xi)\tilde{\phi}(\xi) \in L_2(\mathbf{R}^1), \quad \frac{\cosh(at\xi)}{\xi}\tilde{\psi}(\xi) \in L_2(\mathbf{R}^1), \quad |t| \leq T.$$

For such initial data the Cauchy problem is well-posed in the L_2-sense for $|t| \leq T$ also.

Finally we note that the fundamental solution of the Cauchy problem for operator

$$L\left(\frac{\partial}{\partial t}, \frac{\partial}{\partial x}\right) \equiv \frac{\partial^2}{\partial t^2} - a^2 \frac{\partial^2}{\partial x^2}$$

(that is, the solution of the problem

$$L\left(\frac{\partial}{\partial t}, \frac{\partial}{\partial x}\right) E(t, x) = 0, \quad E(0, x) = 0, \quad E'(0, x) = \delta(x))$$

is given by the formula (see (8.4.7))

$$E(t, x) = \frac{\sin(at\mathcal{D})}{a\mathcal{D}} \delta(x).$$

By virtue of Example 5 in §8.2, this functional may be represented as

$$E(t, x) = \frac{1}{2a}\left[\theta(x + at) - \theta(x - at)\right],$$

where $\theta(x \pm at)$ are (in general, complex) shifts of the Heaviside function. In particular, the fundamental solution of the Cauchy problem for the Laplace operator is

$$E(t, x) = \frac{1}{2i}\Big[\theta(x + it) - \theta(x - it)\Big].$$

3. THE CAUCHY PROBLEM FOR THE HEAT EQUATION

As before, let $a \in \mathbf{C}^1$, $|a| = 1$. Evidently, the solution of the Cauchy problem

$$\frac{\partial u}{\partial t} - a\frac{\partial^2 u}{\partial x^2} = 0, \quad u(0, x) = \phi(x) \tag{8.4.9}$$

is given by the formula

$$u(t, x) = \exp\{at\frac{d^2}{dx^2}\}\phi(x).$$

This formula has the non-formal sense when $\phi(x)$ belongs to $H^\infty(\mathbf{R}^1)$ or the wider space

$$W_q^\infty = \Big\{\phi(x) : \sum_{n=0}^\infty \frac{|t|^n}{n!}\|\phi^{(2n)}(x)\|_q < \infty\Big\}$$

or

$$H_{\exp}^\infty = \{\phi(x) : \exp(at\xi^2)\tilde\phi(\xi) \in L_2(\mathbf{R}^1)\}.$$

According to the example in §7.1, the solution $u(t, x)$ of the problem (8.4.9) can be written in the form

$$u(t, x) = \frac{1}{2\sqrt{\pi at}} \int_{-\sqrt{a\,\mathrm{sgn}\,t}\infty}^{\sqrt{a\,\mathrm{sgn}\,t}\infty} e^{-\frac{s^2}{4at}} \phi(x - s)ds.$$

Now let $\phi(x) = \delta(x)$ be the delta-function. Then the following solution (fundamental solution of the Cauchy problem) given by

$$E(t, x) = \exp\{at\frac{d^2}{dx^2}\}\delta(x)$$

is an analytic functional determined by the contour $(-\sqrt{a\,\mathrm{sgn}\,t}\infty, \sqrt{a\,\mathrm{sgn}\,t}\infty)$ and the function $\exp(-s^2/4at)$ (see Example 4 in §8.2).

4. ONE EQUATION WITH A SHIFT

We consider the Cauchy problem for the equation

$$\frac{\partial u}{\partial t}(t, x) + u(t, x + a) = 0 \quad (t \in \mathbf{R}^1, x \in \mathbf{R}^n) \tag{8.4.10}$$

$$u(0, x) = \phi(x), \quad x \in \mathbf{R}^n. \tag{8.4.11}$$

Here $a \in \mathbf{R}^n$ is a shift vector, therefore equation (8.4.10) is a differential equation with a shift.

Obviously, if $\mathcal{D} \equiv (\partial/\partial x_1, \ldots, \partial/\partial x_n)$, then

$$u(t, x + a) = \sum_{|\alpha|=0}^{\infty} \frac{a^\alpha}{\alpha!} \mathcal{D}^\alpha u(t, x) = \exp\{a\mathcal{D}\}u(x),$$

hence equation (8.4.10) is

$$\frac{\partial u}{\partial t}(t, x) + \exp\{a\mathcal{D}\}u(t, x) = 0.$$

From this we have that the solution of (8.4.10), (8.4.11) is

$$u(t, x) = \exp\{-t \exp\{a\mathcal{D}\}\phi(x)\},$$

where $\phi(x) \in H^{\pm\infty}(\mathbf{R}^n)$ is arbitrary. Consequently, using the Taylor expansion, we obtain

$$u(t, x) = \sum_{n=0}^{\infty} \frac{(-t)^n}{n!} \exp\{na\mathcal{D}\}\phi(x) = \sum_{n=0}^{\infty} \frac{(-t)^n}{n!}\phi(x + na).$$

In particular, the fundamental solution of the Cauchy problem for the operator (8.4.10) is given by

$$E(t, x) = \sum_{n=0}^{\infty} \frac{(-t)^n}{n!}\delta(x + na).$$

5. QUASI-POLYNOMIAL SOLUTIONS

We obtain the solution of the problem

$$\frac{\partial u}{\partial t} - A(\mathcal{D})u = 0, \quad t \in \mathbf{R}^1, \ x \in \mathbf{R}^n, \tag{8.4.12}$$

$$u(0, x) = P(x), \tag{8.4.13}$$

where $A(\mathcal{D})$ is any homogeneous differential operator, that is,

$$A(\mathcal{D}) \equiv \sum_{|\alpha|=m} a_\alpha \mathcal{D}^\alpha \quad (a_\alpha \in \mathbf{C}^1)$$

($m \geq 1$ is an integer) and $P(x)$ is a polynomial. In accordance with the general theory, the solution is given by

$$u(t, x) = \exp\{tA(\mathcal{D})\}P(x) = \sum_{n=0}^{\infty} \frac{t^n A^n(\mathcal{D})}{n!} P(x).$$

It is clear that this series is, in fact, a finite sum and therefore

$$u(t, x) = Q(t, x)$$

is a polynomial in the variables t and x. Similarly, if the initial function $\phi(x)$ is a quasi-polynomial, that is, $\phi(x) = \exp(i\lambda x)P(x)$, then simple calculations show that the solution of the Cauchy problem (8.4.12), (8.4.13) is also a quasi-polynomial

$$u(t, x) = e^{t\beta + i\lambda x} Q(x),$$

where $\beta = A(\lambda)$ and $Q(x)$ is a polynomial, the degree of which is the same as $P(x)$.

6. ONE BOUNDARY VALUE PROBLEM IN A STRIP

We consider the strip $\{-\frac{1}{2} \leq t \leq \frac{1}{2}, -\infty < x < \infty\}$ in the (t, x)-plane. We study the solvability of the problem

$$\frac{\partial^2 u}{\partial t^2} + a^2 \frac{\partial^2 u}{\partial x^2} = 0, \quad a \in \mathbb{C}^1, \ |a| = 1, \tag{8.4.14}$$

$$u(\tfrac{1}{2}, x) = \phi_+(x), \quad \frac{\partial u}{\partial t}\left(-\tfrac{1}{2}, x\right) = \phi_-(x). \tag{8.4.15}$$

Putting $\mathcal{D} \equiv \frac{1}{i} \frac{\partial}{\partial x} \leftrightarrow \xi$, we have the following two boundary value problems for the ordinary differential equation (in t) with parameter $\xi \in \mathbb{R}^1$:

$$u_{\pm}''(t, \xi) - a^2 \xi^2 (u_{\pm}(t, \xi) = 0,$$

$$u_+(\tfrac{1}{2}, \xi) = 1, \quad u_+'(-\tfrac{1}{2}, \xi) = 0$$

and

$$u_-(\tfrac{1}{2}, \xi) = 0, \quad u_-'(-\tfrac{1}{2}, \xi) = 1.$$

Solving these problems we obtain

$$u_+(t, \xi) = \frac{\cosh a(t + \tfrac{1}{2})\xi}{\cosh a\xi}, \quad u_-(t, \xi) = \frac{\sinh a(t - \tfrac{1}{2})\xi}{a\xi \cosh a\xi},$$

therefore, the original problem (8.4.14), (8.4.15) has a solution of type

$$u(t, x) = \frac{\cosh a(t + \frac{1}{2})D}{\cosh aD}\phi_+(x) + \frac{\sinh a(t - \frac{1}{2})D}{aD \cosh aD}\phi_-(x),$$

or, what is the same,

$$u(t, x) = \left[\cosh a(t + \tfrac{1}{2})D\right]v_+(x) + \left[\frac{\sinh a(t - \tfrac{1}{2})D}{aD}\right]v_-(x),$$

where the functions $v_\pm(x)$ are the solutions of the differential equations of infinite order

$$[\cosh aD]v_\pm(x) = \phi_\pm(x), \quad x \in \mathbf{R}^1.$$

Remark. Let us study the behaviour of the symbol $\cosh a\xi \equiv \cos ai\xi$ on $a = \sigma + i\tau$, $\sigma \in \mathbf{R}^1$, $\tau \in \mathbf{R}^1$. For $\sigma = 0$ we have $\cos ai\xi = \cos \tau\xi$ and consequently, $\cos ai\xi = 0$ if and only if $\tau\xi = \pi/2 + k\pi$, $k = 0, \pm 1, \ldots$. In this case the symbol of the operator $I/\cos \tau D$ is analytic in the domain

$$G = \mathbf{R}^1 \backslash \{\tau^{-1}(\pi/2 + k\pi)\}.$$

Thus, the problem (8.4.15) for the wave equation ($a = i$) has a unique solution for any $\phi_\pm(x) \in H^\infty(G)$, that is, for any $\phi_\pm(x) \in L_2(\mathbf{R}^1)$ such that $\operatorname{supp} \tilde{\phi}_\pm(\xi) \subset G$. We note that from the classical point of view this problem is ill posed.

If $\sigma \neq 0$, then $|\cosh z| \sim \cosh \sigma\xi$ for any z on the ray $z = a\xi$. Therefore if $\sigma \neq 0$, then the problem (8.4.14) (8.4.15) is well posed not only in $H^\infty(G)$ but in the whole L_2-scale of spaces, having the weight $\cosh \sigma\xi$. In particular, if $\sigma \neq 0$, this problem is well posed in the Sobolev spaces $H^m(\mathbf{R}^n)$, that is, in the classical sense.

7. THE BOUNDARY VALUE PROBLEM IN THE CYLINDER

Let $Q = \{-\frac{1}{2} < t < \frac{1}{2}, \ x \in \Omega \subset \mathbf{R}^n\}$ be a cylinder with lateral surface S. In this cylinder we consider the following boundary value problem

$$-\frac{\partial^2 u}{\partial t^2} + (-\Delta)^m u = 0, \quad (t, x) \in Q, \tag{8.4.16}$$

$$u\left(\pm\tfrac{1}{2}, x\right) = \phi_\pm(x), \quad x \in \Omega, \tag{8.4.17}$$

$$D^\omega u\big|_S = 0, \quad |\omega| \leq m - 1. \tag{8.4.18}$$

As before, setting $(-\Delta)^m \leftrightarrow |\xi|^{2m}$ and solving the corresponding ordinary differential problems, we find that the formal solution of (8.4.16)–(8.4.18) can be written in the form

$$u(t,x) = \frac{\sinh(t+\frac{1}{2})(-\Delta)^{m/2}}{(-\Delta)^{m/2}} u_+(x) + \frac{\sinh(t-\frac{1}{2})(-\Delta)^{m/2}}{(-\Delta)^{m/2}} u_-(x), \quad (8.4.19)$$

where $u_\pm(x)$ are the solutions of the Dirichlet problem of infinite order

$$\frac{\sinh(-\Delta)^{m/2}}{(-\Delta)^{m/2}} u_\pm(x) = \phi_\pm(x), \quad x \in \Omega,$$

$$\mathcal{D}^\omega \Delta^{mn} u_\pm(x)\Big|_{\partial\Omega} = 0, \quad n = 0,\, 1,\ldots,\, |\omega| < n$$

($\partial\Omega$ is the boundary of Ω). The last conditions arise naturally from the boundary conditions (8.4.18) and the possibility of applying an arbitrary power of $-\Delta$, as an operator with homogeneous Dirichlet data.

For the proof of the fact that formula (8.4.19) gives a non-formal solution, one must repeat the same calculations as before. We must note, however, that the corresponding version of the theory of test and generalized functions is of a "discrete" nature because the Dirichlet problem

$$(-\Delta)^m v(x) = \lambda v(x), \quad \mathcal{D}^\omega v(x)\big|_{\partial\Omega} = 0,$$

has a discrete spectrum. This means, in particular, that the following test-function space H^∞ is the inductive limit of the sequence of linear spans of the Dirichlet eigenfunctions.

8. THE DIRICHLET PROBLEM IN A DISC. POISSON INTEGRAL

In this subsection we study the Dirichlet problem in the unit disc in the (x,y)-plane for the Laplace equation

$$\frac{\partial^2 u}{\partial x^2} + \frac{\partial^2 u}{\partial y^2} = 0, \quad x^2 + y^2 < 1, \tag{8.4.20}$$

$$u(x,y) = f(x,y), \quad x^2 + y^2 = 1, \tag{8.4.21}$$

assuming that $f(x,y)$ is a continuous function on the unit circle.

In polar coordinates $x = \rho\cos\phi,\ y = \rho\sin\phi$ we have

$$\frac{\partial^2 u}{\partial \rho^2} + \frac{1}{\rho}\frac{\partial u}{\partial \rho} + \frac{1}{\rho^2}\frac{\partial^2 u}{\partial \phi^2} = 0, \quad 0 < \rho < 1, \tag{8.4.22}$$

$$u\big|_{\rho=1} = f(\phi), \quad 0 \leq \phi < 2\pi, \tag{8.4.23}$$

where $u \equiv u(\rho, \phi)$ is the unknown function and $f(\phi) \equiv f(\rho \cos \phi, \rho \sin \phi)$ is the boundary value as a continuous periodic function of the argument ϕ.

In order to solve the problem (8.4.22), (8.4.23) we set, as usual, $\frac{1}{i} \frac{\partial}{\partial \phi} \leftrightarrow \xi$. We then obtain the ordinary differential equation

$$U''(\rho, \xi) + \frac{1}{\rho} U'(\rho, \xi) - \frac{\xi^2}{\rho^2} U(\rho, \xi) = 0, \quad 0 < \rho < 1,$$

under the following conditions:

$$|U(0, \xi)| < \infty, \quad U(1, \xi) = 1 \text{ for any } \xi \in \mathbf{R}^1.$$

It is easy to see that $U(\rho, \xi) = \rho^{|\xi|}$ and, consequently, the solution

$$u(\rho, \phi) = U(\rho, \mathcal{D}) f(\phi) = \rho^{|\mathcal{D}|} f(\phi),$$

where $\rho^{|\mathcal{D}|}$ is the PD-operator with the symbol $\rho^{|\xi|}$, acting according to the formula

$$\rho^{|\mathcal{D}|} = e^{|\mathcal{D}| \ln \rho} = \sum_{n=0}^{\infty} \frac{\ln^n \rho}{n!} |\mathcal{D}|^n.$$

Taking into account the action of the operator $|\mathcal{D}|$ (see subsection 3 of §8.1) we obtain

$$\rho^{|\mathcal{D}|} \exp in\phi = \rho^{|n|} \exp in\phi. \tag{8.4.24}$$

Remark. It should be noted that the formula (8.4.24) is valid only for $n \neq 0$, since $\sup \tilde{1} = \{0\}$. Nevertheless, it is natural to put $|\mathcal{D}| \cdot 1 = 0$, which implies the equality $\rho^{|\mathcal{D}|} \cdot 1 = 1$.

Finally, if

$$f(\phi) = \sum_{n=-\infty}^{+\infty} c_n \exp in\phi$$

is the Fourier expansion of the boundary function $f(\phi)$, then the solution $u(\rho, \phi)$ is given by the formula

$$u(\rho, \phi) = \rho^{|\mathcal{D}|} f(\phi) = \sum_{n=-\infty}^{+\infty} c_n \rho^{|n|} \exp in\phi.$$

In conclusion we note that standard arguments give the integral representation of this solution

$$u(\rho, \phi) = \frac{1}{2\pi} \int_0^{2\pi} \frac{f(\theta) d\theta}{1 - \rho^2 - 2\rho \cos(\phi - \theta)}$$

(see, for example, Petrovskii [1]).

9. THE DIRICHLET PROBLEM IN THE HALF-PLANE. CAUCHY INTEGRAL

We consider the problem: to find the solution $u(t,x)$, $t > 0$, $x \in \mathbf{R}^1$, of the Laplace equation

$$\frac{\partial^2 u}{\partial t^2} + \frac{\partial^2 u}{\partial x^2} = 0 \qquad (8.4.25)$$

with the boundary conditions

$$u(0,x) = \phi(x), \quad x \in \mathbf{R}^1, \qquad (8.4.26)$$

$$|u(t,x)| \le M < \infty \ (t \to +\infty), \qquad (8.4.27)$$

where $\phi(x)$ is some function and $M > 0$ is a constant.

As usual, we set $\mathcal{D} \equiv \frac{1}{i}\frac{\partial}{\partial x} \leftrightarrow \xi$ and find the solution $U(t,\xi)$ of the ordinary differential equation

$$U''(t,\xi) - \xi^2 U(t,\xi) = 0, \quad U(0,\xi) = 1, \quad |U(t,\xi)| \le M.$$

It is not difficult to calculate that

$$U(t,\xi) = \exp(-t|\xi|);$$

hence

$$u(t,x) = \exp(-t|\mathcal{D}|)\phi(x).$$

Clearly the symbol $\exp(-t|\xi|)$ is analytic in $G = \mathbf{R}^1 \backslash \{0\}$, therefore for any $\phi(x) \in H^\infty(G)$,

$$\exp(-t|\mathcal{D}|)\phi(x) = \exp(-t\mathcal{D})\phi_+(x) + \exp(t\mathcal{D})\phi_-(x),$$

where

$$\phi_\pm(x) = \frac{1}{2\pi} \int_{\mathbf{R}_\pm^1} \tilde{\phi}(\xi)e^{ix\xi}\,d\xi$$

(the notation is clear). In other words, for any $\phi(x) \in H^\infty(G)$ the solution of the problem (8.4.25)–(8.4.27) can be written as

$$u(t,x) = \phi_+(x + it) + \phi_-(x - it).$$

Substituting here the expression (8.4.23), we obtain the classical Cauchy integral (see, for example, Shilov [1], p.26).

10. ANALYTIC CONTINUATION OF A PAIR OF FUNCTIONS DEFINED ON $\mathbf{R}^1$

Let $\phi_1(x)$ and $\phi_2(x)$ be functions defined on the whole real axis. Our aim is to find an entire function $\phi(z) = u(x,y) + iv(x,y)$ such that

$$u(x,0) = \phi_1, \quad v(x,0) = \phi_2(x).$$

For this we introduce the Sobolev space of infinite order (see §4.3)

$$W^\infty\left\{\frac{|y|^n}{n!},2\right\}(\mathbf{R}^1) = \left\{\phi(x): \sum_{n=0}^\infty \frac{|y|^n}{n!}\|\phi^{(n)}(x)\|_{L_2(\mathbf{R}^1)} < \infty\right\},$$

where $y \in \mathbf{R}^1$, and note (for the calculations below) that the operators

$$\cos\left(y,\frac{\partial}{\partial x}\right) \text{ and } \sin\left(y\frac{\partial}{\partial x}\right)$$

are defined on $W^\infty\{\frac{|y|^n}{n!},2\}(\mathbf{R}^1)$ and the maps

$$\cos\left(y\frac{\partial}{\partial x}\right) : W^\infty\left\{\frac{|y|^n}{n!},2\right\}(\mathbf{R}^1) \to L_2(\mathbf{R}^1)$$

and

$$\sin\left(y\frac{\partial}{\partial x}\right) : W^\infty\left\{\frac{|y|^n}{n!},2\right\}(\mathbf{R}^1) \to L_2(\mathbf{R}^1)$$

are continuous.

Proposition 8.4.1. *If for any $y \in \mathbf{R}^1$ the functions $\phi_1(x)$ and $\phi_2(x)$ belong to $W^\infty\{\frac{|y|^n}{n!},2\}(\mathbf{R}^1)$, then there exists a unique entire function $\phi(z) = u(x,y) + iv(x,y)$ such that*

$$u(x,0) = \phi_1(x), \quad v(x,0) = \phi_2(x)$$

and

$$u(x,y) = \cos\left(y\frac{\partial}{\partial x}\right)\phi_1(x) - \sin\left(y\frac{\partial}{\partial x}\right)\phi_2(x),$$

$$v(x,y) = \sin\left(y\frac{\partial}{\partial x}\right)\phi_1(x) + \cos\left(y\frac{\partial}{\partial x}\right)\phi_2(x).$$

Proof. In fact, setting $\frac{1}{i}\frac{\partial}{\partial x} \leftrightarrow \xi$, we obtain from the Cauchy-Riemann conditions

$$\frac{\partial u}{\partial x} = \frac{\partial v}{\partial y}, \quad \frac{\partial u}{\partial y} = -\frac{\partial v}{\partial x}$$

the system of the ordinary differential equations

$$\frac{\partial u}{\partial y}(y,\xi) = -i\xi v(y,\xi)$$

$$\frac{\partial v}{\partial y}(y,\xi) = i\xi u(y,\xi),$$

where $\xi \in \mathbf{R}^1$ is a parameter. Solving this system, we find the fundamental system of solutions

$$u_1 = \cos iy\xi, \quad v_1 = -\sin iy\xi,$$

$$u_2 = \sin iy\xi, \quad v_2 = \cos iy\xi.$$

Hence, making the inverse substitution $\xi \leftrightarrow \frac{1}{i}\frac{\partial}{\partial x}$, we immediately find that

$$\begin{pmatrix} u \\ v \end{pmatrix} = \begin{pmatrix} \cos y\frac{\partial}{\partial x} & -\sin y\frac{\partial}{\partial x} \\ \sin y\frac{\partial}{\partial x} & \cos y\frac{\partial}{\partial x} \end{pmatrix} \begin{pmatrix} \phi_1(x) \\ \phi_2(x) \end{pmatrix},$$

as required. The proposition is proved.

Remark. Proposition 8.4.1 was proved by Casanova [1].

8.5. Quantum relativistic particle with zero spin

As was noted before, the PD-equation

$$i\hbar\frac{\partial u}{\partial t} = mc^2\sqrt{I - \frac{\hbar^2}{m^2c^2}\Delta}\,u \tag{8.5.1}$$

is used by Björken and Drell [1] as a natural analogue of the Schrödinger equation for a relativistic free particle with zero spin. In this section we use the spaces $H^{\pm\infty}(G)$ of test and generalized functions for an investigation of this equation and consider (8.5.1) as a PD-equation in the sense of Definition 8.1.2. From this standpoint one can use group-theoretic methods to calculate the various asymptotics and prove the solvability of the Cauchy problem.

I must point out that the results of this section are due to K.L. Samarov and in part to A.B. Aksenov (see Samarov [1],[2],[3] and Samarov and Aksenov [1]). These results are published with the authors' permission.

1. Derivation of the Schrödinger equation

Let us give the derivation of equation (8.5.1) based on the standard quantum mechanical scheme. Namely, we shall say that space-time continuum is, by definition, $(n+1)$-dimensional affine space $A^{n+1}(n > 1)$, the points of which are the world points or events. The group $\mathbf{R}^{n+1}$ acts in A^{n+1} as a transitive group. Moreover, two events $A_1 \in A^{n+1}$ and $A_2 \in A^{n+1}$ define one vector $\overrightarrow{A_1 A_2} \in \mathbf{R}^{n+1}$ with pseudo-Euclidean structure as the space $\mathbf{R}_1^{n+1}$, the coordinates in $\mathbf{R}_1^{n+1}$ being chosen such that the pseudo-Euclidean non-generate symmetric bilinear form has the form

$$\langle x, y \rangle = x^0 y^0 - \sum_{i=1}^{n} x^i y^i,$$

where

$$x = (x^0, x^1, \ldots, x^n) \in \mathbf{R}^{n+1},$$
$$y = (y^0, y^1, \ldots, y^n) \in \mathbf{R}^{n+1}.$$

Then the corresponding quadratic form is the space-time interval between the events.

For any smooth curve $x = x(\tau)$, $a \leq \tau \leq b$, in the space A^{n+1} satisfying the condition

$$\left\langle \frac{dx}{d\tau}, \frac{dx}{d\tau} \right\rangle \geq 0,$$

one can define the length of the curve as

$$l = \int_a^b \sqrt{\langle \frac{dx}{d\tau}, \frac{dx}{d\tau} \rangle} d\tau$$

and then define the world line of the free relativistic material point with mass $m > 0$ as the extremal of the action functional

$$S = -\int_a^b mc \sqrt{\left\langle \frac{dx}{d\tau}, \frac{dx}{d\tau} \right\rangle} d\tau$$

(c is the velocity of light in a vacuum).

If the charge of the particle is equal to e, then the operation of "including the electromagnetic field" is defined by passing from the functional S to the functional

$$S^* = -\int_a^b mc \left\langle \frac{dx}{d\tau}, \frac{dx}{d\tau} \right\rangle + \frac{e}{c} \sum_{i=1}^{n} M_i dx^i,$$

where $M_i = M_i(x)$ are the components of some $(n+1)$-dimensional vector (see, for example, Dubrovin, Novikov and Mishchenko [1]).

Choosing the parameter τ as $\tau = x^0 c^{-1} = t$, we see that the functional S^* may be written in the form

$$S^* = \int_a^b \left[-mc^2 \sqrt{1 - c^{-2} \sum_{i=1}^n (\dot{x}^i)^2} + \frac{e}{c} \sum_{i=1}^n M_i \dot{x}^i + e M_0 \right] dt$$

with Lagrangian

$$L = -mc^2 \sqrt{1 - c^{-2} \sum_{i=1}^n (\dot{x}^i)^2} + \frac{e}{c} \sum_{i=1}^n M_i \dot{x}^i + e M_0,$$

where $M_0 = M_0(x)$ is the electrostatic potential and the vector $(M_1, \ldots, M_n) \equiv (M_1(x), \ldots, M_n(x))$ is the magnetical potential.

We introduce the momenta via the Legendre substitution

$$p_i = \frac{\partial L}{\partial \dot{x}^i}, \quad i = 0, 1, \ldots, n.$$

Then
$$p_0 = 0,$$
$$p_i = m\dot{x}^i \left(1 - c^{-2} \sum_{i=1}^n (\dot{x}^i)^2 \right)^{-1/2} + \frac{e}{c} M_i, \quad i = 1, \ldots, n.$$

Then we have

$$\sum_{i=1}^n \dot{x}^i \frac{\partial L}{\partial \dot{x}^i} - L = \frac{mc^2}{\left(1 - c^{-2} \sum_{i=1}^n (\dot{x}^i)^2 \right)^{1/2}} - e M_0,$$

$$1 + m^{-2} c^{-2} \sum_{i=1}^n (p_i - \frac{e}{c} M_i)^2 = \frac{1}{\left[1 - c^{-1} \sum_{i=1}^n (\dot{x}^i)^2 \right]^{1/2}}, \qquad (8.5.2)$$

and from (8.5.1) we obtain the Hamiltonian

$$H_1 = mc^2 \sqrt{1 + m^{-2} c^{-2} \sum_{i=1}^n (p_i - \frac{e}{c} M_i)^2} - e M_0 \qquad (8.5.3)$$

and the Hamiltonian

$$H_2 = -mc^2 \sqrt{1 + m^{-2} c^{-2} \sum_{i=1}^n (p_i - \frac{e}{c} M_i)^2} - e M_0. \qquad (8.5.4)$$

The Hamiltonian (8.5.4) reduces to (8.5.3) after the substitution

$$H_2 \to -H_2, \quad M_0 \to -M_0 \tag{8.5.5}$$

or the substitution

$$H_2 \to -H_2, \quad e \to -e, \quad M_i \to -M_i, \quad i = 1,\dots,n, \tag{8.5.6}$$

but the standard non-relativistic Hamiltonian

$$H_3 = \frac{1}{2m} \sum_{i=1}^{n} (p_i - \frac{e}{c} M_i)^2 - e M_0 \tag{8.5.7}$$

is arrived at only for the Hamiltonian (8.5.3) by passing to the limit

$$\lim_{c \to \infty} (H_1 - mc^2).$$

Nevertheless, there are no mathematical reasons for preferring (8.5.3) to (8.5.4) or vice versa.

Converting (8.5.3) and (8.5.4) into the quantum equations by means of the formal rules

$$p_k \to -\hbar \frac{\partial}{\partial x^k}, \quad H \to i\hbar \frac{\partial}{\partial t} \tag{8.5.8}$$

we obtain the equation

$$i\hbar \frac{\partial u}{\partial t} = \left[mc^2 \sqrt{I + m^{-2} c^{-2} \sum_{k=1}^{n} \left(i\hbar \frac{\partial}{\partial x^k} - \frac{e}{c} M_k \right)^2} - e M_0 \right] u \tag{8.5.9}$$

and the equation

$$i\hbar \frac{\partial u}{\partial t} = \left[-mc^2 \sqrt{I + m^{-2} c^{-2} \sum_{k=1}^{n} \left(-i\hbar \frac{\partial}{\partial x^k} - \frac{e}{c} M_k \right)^2} - e M_0 \right] u. \tag{8.5.10}$$

In accordance with (8.5.5) and (8.5.6) equation (8.5.9) is transposed to equation (8.5.10) by means of the substitution

$$t \to -t, \quad M_0 \to -M_0$$

or the substitution

$$t \to -t, \quad e \to -e, \quad M_k \to -M_k, \quad k = 1,\dots,n.$$

Equation (8.5.9) is the Schrödinger PD-equation for a quantum relativistic particle with zero spin and moving in an electromagnetic field.

For a relativistic free particle this equation has the form

$$i\hbar\frac{\partial u}{\partial t} = mc^2\sqrt{I = \frac{\hbar}{m^2c^2}\Delta u}. \tag{8.5.11}$$

2. FUNDAMENTAL SOLUTION OF THE CAUCHY PROBLEM

We consider the Cauchy problem for the PD-equation (8.5.11), that is,

$$i\frac{\partial u}{\partial t} = \frac{m}{\omega}\sqrt{I - \omega^2\Delta}u, \quad x \in \mathbf{R}^3, \ t > 0 \tag{8.5.12}$$

$$u(0, x) = \phi(x) \tag{8.5.13}$$

$(\omega = \hbar/mc)$, where the operator $\sqrt{I - \omega^2\Delta}$ is regarded as the PD-operator with symbol $\sqrt{1 + \omega^2\xi^2} \in \mathcal{O}(G)$, $G = \mathbf{R}^3$.

As follows from the general theory (see §8.3), for any function $u(x) \in H^\infty(G)$ the operator $\sqrt{I - \omega^2\Delta}$ acts locally (with respect to the dual variables $\xi \in \mathbf{R}^3$) as a differential operator of infinite order in accordance with a local element of analyticity of $\sqrt{1 + \omega^2\xi^2}$ or, what is the same,

$$\sqrt{I - \omega^2\Delta}u(x) = \frac{1}{(2\pi)^3}\int_{\mathbf{R}^3}(1 + \omega^2\xi^2)^{1/2}\exp ix\xi\tilde{u}(\xi)d\xi,$$

where

$$\tilde{u}(\xi) = \int_{\mathbf{R}^3}u(x)\exp(-ix\xi)dx.$$

Then, as follows from the general theory the Cauchy problem (8.5.12), (8.5.13) is well-posed and for any $\phi(x) \in H^\infty(\mathbf{R}^3)$ (or $\phi(x) \in H^{-\infty}(\mathbf{R}^3)$) the solution is given by formula

$$u(t, x) = \exp\{-itc\omega^{-1}\sqrt{I - \omega^2\Delta}\}\phi(x), \tag{8.5.14}$$

where $u(t, x) \in C^1(\mathbf{R}^1; H^\infty(\mathbf{R}^3))$ (or $u(t, x) \in C^1(\mathbf{R}^1; H^{-\infty}(\mathbf{R}^3))$). In particular, the fundamental solution of the Cauchy problem is

$$E(t, x) = \exp\{-itc\omega^{-1}\sqrt{I - \omega^2\Delta}\}\delta(x).$$

Our aim is to describe $E(t, x)$ in terms of Bessel functions.

Theorem 8.5.1. *The fundamental solution $E(t, x)$ can be written in the form*

$$E(t, x) = \begin{cases} \dfrac{1}{4\pi r c} \dfrac{\partial}{\partial r} \dfrac{\partial}{\partial t} [iY_0(\omega^{-1}\sqrt{c^2t^2 - r^2}) - \pi J_0(\omega^{-1}\sqrt{c^2t^2 - r^2})] & (ct \geq r) \\[4mm] -\dfrac{i}{2\pi r c} \dfrac{\partial}{\partial r} \dfrac{\partial}{\partial t} K_0(\omega^{-1}\sqrt{r^2 - c^2t^2}) & (ct \leq r); \end{cases}$$

where J_0 and Y_0 are the Bessel functions of order zero (of the first and second types) and K_0 is the modified Bessel function, $r = |x|$.

Proof. For any $N > 0$ we put

$$\widetilde{\chi}_N(\xi) = \begin{cases} 1 & \text{if } |\xi| \leq N; \\ 0 & \text{if } |\xi| > N. \end{cases}$$

Clearly the functions

$$\chi_N(x) = \frac{1}{(2\pi)^3} \int_{|\xi| \leq N} e^{ix\xi} d\xi$$

belong to $H^\infty(\mathbf{R}^3)$ and consequently, the Cauchy problem for the equation (8.5.12) with initial condition

$$u(0, x) = \chi_N(x)$$

has the unique solution

$$E_N(t, x) = \frac{1}{(2\pi)^3} \int_{|\xi| \leq N} \exp\{ix\xi - \frac{itc}{\omega}\sqrt{1 + \omega^2\xi^2}\}d\xi \tag{8.5.15}$$

which belongs to $C^1(\mathbf{R}^1; H^\infty(\mathbf{R}^3))$.

It is easy to see that $E_N(t, x) \to E(t, x)$ in the sense of $C^1(\mathbf{R}^1; H^\infty(\mathbf{R}^3))$.

Let us study the function $E_N(t, x)$ in detail. For this we consider in the space $\mathbf{R}^3$ of $\xi = (\xi_1, \xi_2, \xi_3)$ the orthogonal transformation T_1 such that for a fixed vector x

$$x = (x_1, x_2, x_3) \to T_1x = (0, 0, |x|).$$

Denoting this substitution by $\eta = T_1\xi$, we have

$$E_N(t, x) = \frac{1}{(2\pi)^3} \int_{|\eta| \leq N} \exp\{i|x|\eta_3 - \frac{itc}{\omega}\sqrt{1 + \omega^2\eta^2}\}d\eta. \tag{8.5.16}$$

To transform the triple integral into a single integral we make the substitution

$$T_2 : \eta \to \zeta = (\zeta_1, \zeta_2, \zeta_3)$$

according to the formulae

$$\eta_1 = (\zeta_1^2 - \zeta_3^2)^{1/2} \cos\zeta_2, \quad \eta_2 = (\zeta_1^2 - \zeta_2^2)^{1/2} \sin\zeta_2, \quad \eta_3 = \zeta_3.$$

This map takes the ball $|\eta| \leq N$ into the set

$$\{0 \leq \zeta_1 \leq N, \ 0 \leq \zeta_2 \leq 2\pi, \ -\zeta_1 \leq \zeta_3 \leq \zeta_1\}$$

and its Jacobian is equal to ζ_1. Now from (8.5.16) we have

$$E_N(t, x) = \frac{1}{2\pi^2 |x|} \int_0^N \zeta_1 \exp(-itc\omega^{-1}\sqrt{1 + \zeta_1^2\omega^2}) \sin(|x|\zeta_1)d\zeta_1.$$

Setting $\zeta_1 = \omega^{-1}\sinh z$, we have

$$E_N(t, x) = -\frac{i}{4\pi^2 rc} \frac{\partial}{\partial r} \frac{\partial}{\partial t} \int_{-\sinh^{-1}N\omega}^{\sinh^{-1}N\omega} \exp\left\{-\frac{itc}{\omega}\cosh z + \frac{ir}{\omega}\sinh z\right\}dz$$

and passing to the limit as $N \to \infty$, we obtain

$$E(t, x) = -\frac{i}{4\pi^2 rc} \frac{\partial}{\partial r} \frac{\partial}{\partial t} \int_{-\infty}^{\infty} \exp\left\{-\frac{itc}{\omega}\cosh z + \frac{ir}{\omega}\sinh z\right\}dz.$$

Then in order to represent the function

$$\Phi(t, r) = \int_{-\infty}^{\infty} \exp\left\{-\frac{itc}{\omega}\cosh z + \frac{ir}{\omega}\sinh z\right\}dz$$

in terms of Bessel functions, we consider the following domains:

$$Q_1 = \{(t, x) : ct > r\} \ (\text{interior of the light cone})$$

and

$$Q_2 = \{(t, c) : ct < r\} \ (\text{exterior of the light cone})$$

and obtain the desired representation in each domain separately:

1) in Q_1 we have

$$\Phi(t, r) = \int_{-\infty}^{\infty} \exp\left\{-\frac{i}{\omega}\sqrt{c^2t^2 - r^2}\left(\frac{ct}{\sqrt{c^2t^2 - r^2}}\cosh z - \right.\right.$$

$$\left.\left. -\frac{r}{\sqrt{c^2t^2 - r^2}}\sinh z\right)\right\}dz \qquad (8.5.17)$$

and if α is such that

$$\cosh\alpha = \frac{ct}{\sqrt{c^2t^2 - r^2}}, \quad \sinh\alpha = \frac{r}{\sqrt{c^2t^2 - r^2}},$$

then (8.5.17) implies that

$$\Phi(t,r) = \int_{-\infty}^{\infty} \exp\left\{-\frac{i}{\omega}\sqrt{c^2t^2 - r^2}\cosh(z - \alpha)\right\}dz =$$
$$= -\pi\left(Y_0(\omega^{-1}\sqrt{c^2t^2 - r^2}) + iJ_0(\omega^{-1}\sqrt{c^2t^2 - r^2})\right);$$

2) in Q_2 we have

$$\Phi(t,r) = \int_{-\infty}^{\infty} \exp\left\{-\frac{i}{\omega}\sqrt{r^2 - c^2t^2}\left(\frac{ct\cosh z}{\sqrt{r^2 - c^2t^2}} - \frac{r\sinh z}{\sqrt{r^2 - c^2t^2}}\right)\right\}dz \quad (8.5.18)$$

and if real β is such that

$$\cosh\beta = \frac{r}{\sqrt{r^2 - c^2t^2}}, \quad \sinh\beta = \frac{ct}{\sqrt{r^2 - c^2t^2}},$$

then (8.5.18) implies that

$$\Phi(t,r) = \int_{-\infty}^{\infty} \exp\left\{\frac{i}{\omega}\sqrt{r^2 - c^2t^2}\sinh(z - \beta)\right\}dz = 2K_0(\omega^{-1}\sqrt{r^2 - c^2t^2})$$

(see formulae 9.6.21 and 9.6.23 in Abramovitz and Stegun [1]).

Summing these results we obtain the required formulae. The theorem is proved.

Corollary 8.5.1. *For any $\phi(x) \in H^\infty(\mathbf{R}^3)$ the solution of the Cauchy problem (8.5.12), (8.5.13) has the form*

$$u(t,x) = E(t,x) * \phi(x).$$

3. LORENTZ INVARIANCE

In this subsection we show that the solutions of the Schrödinger equation are invariant with respect to the Lorentz transformation.

For simplicity we change the scale of the independent variables and write the Schrödinger equation in the form

$$\left(i\frac{\partial u}{\partial t} - \sqrt{I - \Delta}\right)u(t,x) = 0$$

$$x = (x_1, \ldots, x_n) \in \mathbf{R}^n, \ t \in \mathbf{R}^1. \tag{8.5.19}$$

For some m, $1 \leq m \leq n$, we consider the one-parameter subgroup of Lorentz transformations ($a \in \mathbf{R}^1$ is the parameter)

$$L_m(a) : (u, x, t) \rightarrow (u', x', t') \qquad (8.5.20)$$

where

$$
\begin{aligned}
u' &= u \\
x'_l &= x_l, \quad 1 \leq l \leq n, \ l \neq m, \\
x'_m &= x_m \cosh a + t \sinh a \\
t' &= x_m \sinh a + t \cosh a.
\end{aligned}
\qquad (8.5.21)
$$

Theorem 8.5.2. *Let $u(t, x)$ be a solution of the equation (8.5.19). Then for any $a \in \mathbf{R}^1$ the function $u' \equiv u(t', x')$ is also a solution of the Schrödinger equation (8.5.19).*

Proof. The formulae (8.5.21) imply that as $a \rightarrow 0$

$$
\begin{aligned}
\frac{\partial}{\partial t} &= \frac{\partial}{\partial t'} + a \frac{\partial}{\partial x'_m} + o(a) \\
\frac{\partial}{\partial x_m} &= \frac{\partial}{\partial x'_m} + a \frac{\partial}{\partial t'} + o(a) \\
\frac{\partial}{\partial x_l} &= \frac{\partial}{\partial x'_l}, \quad l \neq m.
\end{aligned}
\qquad (8.5.22)
$$

Then (the notation is clear) as $a \rightarrow 0$

$$\Delta = \Delta' + 2a \frac{\partial^2}{\partial x'_m \partial t} + o(a)$$

$$\Delta^l = (\Delta')^l + 2la \frac{\partial^2}{\partial x'_m \partial t}(\Delta')^{l-1} + o(a), \quad l \in N. \qquad (8.5.23)$$

We show that as $a \rightarrow 0$

$$\left(i\frac{\partial}{\partial t} - \sqrt{I - \Delta} \right) u(t', x') = u(t', x') \cdot o(a). \qquad (8.5.24)$$

Indeed, using (8.5.22), (8.5.23), we obtain from (8.5.19):

$$\left(i\frac{\partial}{\partial t}-\sqrt{I-\Delta}\right)u(t',x')=\left(i\frac{\partial}{\partial t}-\sum_{k=0}^{\infty}(-1)^k C_{1/2}^k \Delta^k\right)u(t',x')=$$

$$=\left(i\frac{\partial}{\partial t'}+ia\frac{\partial}{\partial x'_m}-\sum_{k=0}^{\infty}(-1)^k C_{1/2}^k\left[(\Delta')^k+2ka\frac{\partial^2}{\partial x'_m \partial t'}(\Delta')^{k-1}+o(a)\right]\right)u(t',x')=$$

$$=\left[i\frac{\partial}{\partial t'}-(I-\Delta')^{1/2}+a\frac{\partial}{\partial x'_m}\left(i-2\frac{\partial}{\partial t'}\frac{\partial}{\partial \Delta'}(I-\Delta')^{1/2}+o(a)\right)\right]u(t',x')=$$

$$=\left[i\frac{\partial}{\partial t'}-(I-\Delta')^{1/2}+a\frac{\partial}{\partial x'_m}\left(i+\frac{\partial}{\partial t'}(I-\Delta')^{-1/2}+o(a)\right]u(t',x')=$$

$$=\left[i\frac{\partial}{\partial t'}-(I-\Delta')^{1/2}-ia\frac{\partial}{\partial x'_m}(i-\Delta')^{-1/2}\left(i\frac{\partial}{\partial t'}-(I-\Delta')^{1/2}\right)+o(a)\right]u(t',x')=$$

$$=u(t',x')\cdot o(a)$$

as required.

As is known (see Ovsyannikov [1], Ibragimov [1]) relation (8.5.24) is the criterion for the invariance of equation (8.5.19) with respect to the one-parameter group $L_m(a)$. The theorem is proved.

Remark. For $n=1$ the theorem may be proved without using group theoretic methods. Indeed, if $n=1$, then any solution of the equation (8.5.19) may be represented in the form

$$u(t,x)=\frac{1}{2\pi}\int_{-\infty}^{\infty}F(k)\exp\{i(kx-\sqrt{1+k^2}t)\}dk,$$

where $F(k)$ is the Fourier image of some function.

Let us consider the integral

$$J\equiv\frac{1}{2\pi}\int_{-\infty}^{\infty}F(k)\exp\{i(kx'-\sqrt{1+k^2}t')\}dk,$$

the parameters x' and t' of which are defined for $n=m=1$ from the formulae (8.5.21). Making the substitution

$$k=\sinh(\sin^{-1}\xi+a)$$

and noting that for all a

$$kx'-\sqrt{1+k^2}t'=\xi x-\sqrt{1+\xi^2}t,$$

we find that

$$J = \frac{1}{2\pi} \int_{-\infty}^{\infty} F_1(\xi) \exp\{i(\xi x - \sqrt{1+\xi^2}t)\} d\xi,$$

where

$$F_1(\xi) = \left(\cosh a + \frac{\xi \sinh a}{(1+\xi^2)^{1/2}}\right) F\left(\xi \cosh a + \sqrt{1+\xi^2} \sinh a\right).$$

Moreover, the inverse Fourier transform of $F_1(\xi)$ is a member of $H^\infty(\mathbf{R}^3)$ (or $H^{-\infty}(\mathbf{R}^3)$), if the inverse Fourier transform of $F(\xi)$ does.

4. DESCRIPTION OF LORENTZ-INVARIANT SOLUTIONS

Our aim in this section is to prove that each Lorentz-invariant solution of the Schrödinger PD-equation of the relativistic free particle is an ordinary generalized function (or distribution) and the set of all such solutions is a one-dimensional space. We also give here the representation of the fundamental solution of the Cauchy problem in terms of the Lorentz-invariant solutions.

Let $h(x)$ be a functional on $H^\infty(\mathbf{R}^n)$, that is $h(x) \in H^{-\infty}(\mathbf{R}^n)$. Then the solution of the Cauchy problem

$$\left(i\frac{\partial}{\partial t} - \sqrt{I - \Delta}\right)u = 0 \tag{8.5.25}$$

$$u(0, x) = h(x) \tag{8.5.26}$$

is (for each fixed t) also an element of $H^{-\infty}(\mathbf{R}^n)$ acting according to the formula

$$\langle u(t, x), \phi(x) \rangle = \langle h(x), \exp\{-it\sqrt{I - \Delta}\}\phi(x) \rangle =$$
$$= (2\pi)^{-n} \langle \tilde{h}(\xi), \exp\{-it\sqrt{1+\xi^2}\}\tilde{\phi}(-\xi) \rangle,$$

moreover, $\tilde{h}(\xi)$ is a regular functional defined by some locally square integrable function on $\mathbf{R}^n$.

We now give a description of the Lorentz-invariant solutions. A solution $u(t, x)$ of equation (8.5.25) is Lorentz-invariant if and only if for any m $(1 \le m \le n)$ we have

$$\left(t\frac{\partial}{\partial x_m} + x_m\frac{\partial}{\partial t}\right)u(t, x) = 0. \tag{8.5.27}$$

These equations imply that the solution $u(t, x)$ is a function of the invariant $x^2 - t^2$ of the Lorentz group. Applying the Fourier transformation to (8.5.25) and (8.5.27), we easily see that

$$\left(i\frac{\partial}{\partial t} - \sqrt{1 + \xi^2}\right)\tilde{u}(t, \xi) = 0, \tag{8.5.28}$$

$$\left(i\xi_m t + i\frac{\partial^2}{\partial t \partial \xi_m}\right)\tilde{u}(t, \xi) = 0 \quad (1 \le m \le n) \tag{8.5.29}$$

Substituting the value of $i\tilde{u}'(t, \xi)$ from (8.5.28) into (8.5.29), we have for $m = 1, \ldots, n$

$$\frac{\partial}{\partial \xi_m}\left[it\sqrt{1 + \xi^2} + \ln(\tilde{u}\sqrt{1 + \xi^2})\right] = 0.$$

Consequently,

$$it\sqrt{1 + \xi^2} + \ln(\tilde{u}\sqrt{1 + \xi^2}) \equiv T(t),$$

where $T(t)$ is a function, depending only on t, and hence

$$\tilde{u}(t, \xi) = T(t)\,(1 + \xi^2)^{-1/2}\exp(-it\sqrt{1 + \xi^2}).$$

In view of equation (8.5.28) we find that $T(t) \equiv c$, where $c = \text{const}$, therefore

$$\tilde{u}(t, \xi) = c(1 + \xi^2)^{-1/2}\exp(-it\sqrt{1 + \xi^2}).$$

Setting

$$V_n(t, x) = \frac{1}{(2\pi)^n}\int_{\mathbb{R}^n}(1 + \xi^2)^{-1/2}\exp\{i\xi x - it(1 + \xi^2)^{1/2}\}d\xi$$

we obtain the following theorem.

Theorem 8.5.3. *Every Lorentz-invariant solution of equation (8.5.25) has the form*

$$u(t, x) = cV(t, x), \quad c \in \mathbb{C}^1,$$

where for any $\phi(x) \in H^\infty(\mathbb{R}^n)$

$$\langle u(t, x), \phi(x)\rangle = \frac{c}{(2\pi)^n}\int_{\mathbb{R}^n}(1 + \xi^2)^{-1/2}e^{-it(1 + \xi^2)^{1/2}}\tilde{\phi}(-\xi)d\xi.$$

Remark. From this representation it is easy to see the connection between the fundamental solution $E(t, x)$ of the Cauchy problem for equation (8.5.25) and the Lorentz-invariant solutions. Namely,

$$E(t, x) = i\frac{\partial}{\partial t}V(t, x).$$

5. RECURRENCE FORMULAE FOR THE LORENTZ-INVARIANT SOLUTIONS

In this subsection we show the connection between the Lorentz-invariant solutions of different dimensions.

Theorem 8.5.4. *For any $n \geq 1$ we have the recurrence formula*

$$V_{n+2}(t, x) = -\frac{1}{2\pi r} \frac{\partial}{\partial r} V_n(t, x) \tag{8.5.30}$$

where $r = |x|$ and

$$V_1(t, x) = \begin{cases} -\frac{1}{2} Y_0(\sqrt{t^2 - r^2}) - \frac{i}{2} \operatorname{sgn} t \cdot J_0(\sqrt{t^2 - r^2}), & r < |t|, \\ \frac{1}{\pi} K_0(\sqrt{r^2 - t^2}), & r > |t|, \end{cases} \tag{8.5.31}$$

$$V_2(t, x) = \begin{cases} -\frac{i}{2\pi} \operatorname{sgn} t \cdot \dfrac{\exp\{-i \operatorname{sgn} t \cdot \sqrt{t^2 - r^2}\}}{\sqrt{t^2 - r^2}}, & r < |t|, \\ \dfrac{1}{2\pi} \dfrac{\exp\{-\sqrt{r^2 - t^2}\}}{\sqrt{r^2 - t^2}}, & r > |t|. \end{cases} \tag{8.5.32}$$

Proof. Making a rotation in the ξ-space $\mathbf{R}^n_\xi$ such that the image of the vector x (as a vector of $\mathbf{R}^n_\xi$) lies on the ξ_1-axis, we obtain

$$V_n(t, x) \equiv \frac{1}{(2\pi)^n} \int_{\mathbf{R}^n} \frac{\exp\{i\xi x - it\sqrt{1 + \xi^2}\}}{\sqrt{1 + \xi^2}} d\xi =$$

$$= \frac{1}{(2\pi)^n} \int_{\mathbf{R}^n} \frac{\exp\{i\xi_1 r - it\sqrt{1 + \xi^2}\}}{\sqrt{1 + \xi^2}} d\xi$$

Hence, in terms of the spherical coordinates

$$\xi_1 = \rho \cos \theta_1, \quad \xi_i = \rho \sin \theta_1 \ldots \sin \theta_{i-1} \cos \theta_i (2 \leq i \leq n - 1),$$

$$\xi_n = \rho \sin \theta_1 \ldots \sin \theta_{n-2} \sin \theta_{n-1}, \quad \rho = |\xi|,$$

$$0 \leq \theta_i \leq \pi \ (1 \leq i \leq n - 2), \quad 0 \leq \theta_{n-1} < 2\pi,$$

we obtain

$$V_n(t, x) = \frac{1}{(2\pi)^n} \int_0^\infty \rho^{n-1} (1 + \rho^2)^{1/2} \exp\{-it\sqrt{1 + \rho^2}\} \times$$

$$\times \left(\int_0^\pi \exp\{i\rho r \cos \theta_1\} \sin^{n-2} \theta_1 d\theta_1 \right) d\rho \cdot C_{n-1}, \tag{8.5.33}$$

where

$$C_{n-1} = 2\pi \int_0^\pi \sin^{n-3}\theta_2 d\theta_2 \ldots \int_0^\pi \sin\theta_{n-2} d\theta_{n-2}$$

(recall that the Jacobian of the transformation to spherical coordinates equals $\rho^{n-1}\sin^{n-2}\theta_1\sin^{n-3}\theta_2\ldots\sin\theta_{n-2}$).

Now in order to satisfy formula (8.5.30) it suffices to note that (integration by parts)

$$-\frac{1}{r}\frac{\partial}{\partial r}\int_0^\pi \exp\{i\rho r\cos\theta\}\sin^{n-2}\theta d\theta =$$

$$= \frac{\rho^2}{n-2}\int_0^\pi \exp\{i\rho r\cos\theta\}\sin^n\theta d\theta,$$

and (elementary calculations)

$$\int_0^\pi \sin^{n-1}\theta d\theta \int_0^\pi \sin^{n-2}\theta d\theta = \frac{2\pi}{n-1}$$

and to compare these formulae with formula (8.5.33).

We now turn to formulae (8.5.31) and (8.5.32). In fact, for $n = 1$, after making the substitution $\rho = \sinh z$ we have

$$V_1(t,x) = \frac{1}{2\pi}\int_{-\infty}^\infty \frac{\exp\{i(\xi x - it\sqrt{1+\xi^2})\}}{\sqrt{1+\xi^2}}d\xi =$$

$$= \frac{1}{2\pi}\int_{-\infty}^\infty \exp\{i(x\sinh z - t\cosh z)\}dz$$

which, in accordance with the calculations in §8.2, implies (8.5.30).

Further, if $n = 2$, then from (8.5.33) we have

$$V_2(t,x) = \frac{1}{(2\pi)^2}\int_0^\infty d\rho\frac{\rho}{\sqrt{1+\rho^2}}\exp\{it\sqrt{1+\rho^2}\}\int_0^\pi \exp\{i\rho r\cos\theta\}d\theta =$$

$$= \frac{1}{4\pi}\int_0^\infty \frac{\rho}{\sqrt{1+\rho^2}}\exp\{it\sqrt{1+\rho^2}\}J_0(\rho r)d\rho.$$

Using the well known Sommerfeld integral (see, for example, Mathematical Encyclopaedia, Vol.2, Moscow, 1979) we immediately obtain (8.5.32). The theorem is proved.

6. NON-RELATIVISTIC LIMIT AND FACTORIZATION OF KLEIN-GORDON-FOCK OPERATOR

We make the substitution

$$u(t,x) = \exp\{-it\hbar^{-1}mc^2 v(t,x)\}.$$

Then the basic equation

$$L_1(u) \equiv \left(i\hbar \frac{\partial}{\partial t} - mc^2 \sqrt{I - \frac{\hbar^2}{m^2 c^2} \Delta} \right) u(t, x) = 0 \qquad (8.5.34)$$

is transformed into the equation

$$\left(i\hbar \frac{\partial}{\partial t} - mc^2 \sqrt{I - \frac{\hbar^2}{m^2 c^2} \Delta} + mc^2 \right) v(t, x) = 0, \qquad (8.5.35)$$

for which we consider the Cauchy problem with initial condition

$$v(0, x) = \phi(x), \quad \phi(x) \in H^\infty(\mathbf{R}^3). \qquad (8.5.36)$$

We denote the solution of (8.5.35), (8.5.36) (for fixed $\hbar$ and m) by $v(t, x; c)$. Then we have the following result.

Theorem 8.5.5. *If $c \to +\infty$, then $v(t, x; c)$ converges in the topology of $C(\mathbf{R}^1; H^\infty(\mathbf{R}^3))$ to the solution of the Cauchy problem for the non-relativistic Schrödinger equation*

$$\left(i\hbar \frac{\partial}{\partial t} + \frac{\hbar^2}{2m} \Delta \right) w(t, x) = 0 \qquad (8.5.37)$$

$$w(0, x) = \phi(x). \qquad (8.5.38)$$

Proof. Since $\phi(x) \in H^\infty(\mathbf{R}^3)$, the solution $v(t, x; c)$ can be represented in the form

$$v(t, x; c) = \frac{1}{(2\pi)^3} \int_{\mathbf{R}^3} \exp\left\{ ix\xi - \frac{imc^2 t}{\hbar} \left(1 - \sqrt{1 + \frac{\hbar^2}{m^2 c^2} \xi^2} \right) \right\} \tilde{\phi}(\xi) d\xi.$$

Further, since $\tilde{\phi}(\xi) \in L_2(\mathbf{R}^3)$ has a compact support, we can pass to the limit under the integration sign:

$$\lim_{c \to \infty} v(t, , x; c) = \frac{1}{(2\pi)^3} \int_{\mathbf{R}^3} \exp\left\{ ix\xi + \frac{i\hbar t}{2m} \xi^2 \right\} \tilde{\phi}(\xi) d\xi =$$

$$= \exp\left\{ \frac{i\hbar t}{2m} \Delta \right\} \phi(x).$$

It remains to note that this function is the unique solution of the Cauchy problem (8.5.37), (8.5.38). The theorem is proved.

Remark. A similar proof can also be given for the initial data $\phi(x) \in H^{-\infty}(\mathbf{R}^3)$.

Let us now consider the Klein-Gordon-Fock equation (see, for example, Landau and Lifshits [1])

$$L_6(u) \equiv \left[\hbar^2 \frac{\partial^2}{\partial t^2} + m^2 c^4 \left(I - \frac{\hbar^2}{m^2 c^2}\Delta\right)\right] u(t, x) = 0 \qquad (8.5.39)$$

and the equation

$$L_7(u) \equiv \left(i\hbar \frac{\partial}{\partial t} + mc^2 \sqrt{I - \frac{\hbar^2}{m^2 c^2}\Delta}\right) u(t, x) = 0, \qquad (8.5.40)$$

which is the complex conjugate of the Schrödinger PD-equation.

Theorem 8.5.6. *Every solution of the Cauchy problem for equation (8.5.39) with initial conditions*

$$u(0, x) = \phi_1(x) \in H^{\pm\infty}(\mathbf{R}^3), \quad \frac{\partial u}{\partial t}(0, x) = \phi_2(x) \in H^{\pm\infty}(\mathbf{R}^3) \qquad (8.5.41)$$

is a linear combination of certain solutions of the Cauchy problems for equations (8.5.34) and (8.5.40).

Proof. Clearly, $L_6(u) = -L_1(L_7(u))$. Hence the solution of the Cauchy problem (8.5.39), (8.5.41) may be written in the form

$$u(t, x) = \tfrac{1}{2} e^{\Lambda t}(\phi_1 + \Lambda^{-1}\phi_2) + \tfrac{1}{2} e^{-\Lambda t}(\phi_1 - \Lambda^{-1}\phi_2),$$

where

$$\Lambda = -imc^2 \hbar^{-1}\sqrt{I - \frac{\hbar^2}{m^2 c^2}\Delta}.$$

On the other hand, the function

$$u_1(t, x) = e^{\Lambda t}(\phi_1 + \Lambda^{-1}\phi_2)$$

is the solution of the equation (8.5.34) with initial datum $\phi_1 + \Lambda^{-1}\phi_2$, while the function

$$u_2(t, x) = e^{-\Lambda t}(\phi_1 - \Lambda^{-1}\phi_2)$$

is the solution of equation (8.5.40) with initial datum $\phi_1 - \Lambda^{-1}\phi_2$, as required. The theorem is proved.

References

M. Abramowitz, and I.A. Stegun (eds.):

[1] *Handbook of mathematical functions with formulae, graphs and mathematical tables*, National Bureau of Standards, Washington, DC, 1964.

J.D. Bjorken, and S.D. Drell:

[1] *Relativistic quantum mechanics*, McGraw-Hill, New York-Toronto-London, 1965.

F.E. Browder:

[1] "Non-linear elliptic boundary value problem", *Bull. Amer. Math. Soc.*, **69**:6 (1963), 862–874.

R. Casanova:

[1] "Existencia y unicidad de la combinación analitica de funciones definidas en el eje real", *Revista Ciencas Mat.* V:3 (1984), 23–26.

Yu.A. Dubinskii:

[1] "The algebra of PD-operators with analytic symbols and its applications to mathematical physics", *Uspekhi Mat. Nauk* **37**:5 (1982), 97–137 (in Russian).

[2] "The Cauchy problem for partial differential equations with complex variables", *Dokl. Akad. Nauk SSSR* **264**:5 (1982), 1045–1048 (in Russian).

[3] "PD-operators with complex variables and their applications", *Dokl. Akad. Nauk SSSR* **268**:5 (1983), 1046–1050 (in Russian).

[4] "Fourier transformation of analytic functions. The complex Fourier method", *Dokl. Akad. Nauk SSSR* **275**:3 (1984), 533–536 (in Russian).

[5] *Sobolev spaces of infinite order and differential equations*, Reidel, Dordrecht-Boston-Lancaster-Tokyo, 1986.

[6] *Sobolev spaces of infinite order and differential equations*, Teubner, Leipzig, 1986.

[7] "Analytic PD-operators and applications", Acta Math. Vietnam. **11**:1 (1986), 3–61 (in Russian).

[8] "The Cauchy problem for PD-equations with variable analytic symbols", *Dokl. Akad. Nauk SSSR* **289**:1 (1986), 24–27 (in Russian).

[9] "On the well-posedness of the Cauchy problem in classes of entire functions of finite order", *Dokl. Akad. Nauk SSSR* **301**:6 (1988), 1305–1308 (in Russian).

[10] "On the necessity of the Kovalevskaya conditions and their generalizations", *Dokl. Akad. Nauk SSSR* **304**:2 (1990), 26–28 (in Russian).

[11] "The Cauchy problem and PD-operators in a complex domain", *Uspekhi Mat. Nauk* **45**:2 (1990), 115–142 (in Russian).

B.A. Dubrovin, S.P. Novikov and A.T. Mishchenko:

[1] *Modern geometry*, Nauka, Moscow, 1979.

L. Ehrenpreis:

[1] *Fourier analysis in several complex variables*, Wiley Interscience, New York, 1970.

I.M. Gel'fand, and G.E. Shilov:

[1] *Generalized functions*, Fizmatgiz, Moscow, 1959.

E. Holmgren:

[1] "Über systeme von linearen partiellen Differentialgleichungen", *Ofversigt Kgl. Vetenskaps Acad. Förhandl.* **58** (1901), 91–103.

L. Hörmander:

[1] *An introduction to complex analysis in several variables*, van Nostrand, Princeton, 1966.

A.Yu. Hrennikov:

[1] "Differential equations in locally convex spaces and PD-equations of evolution", *Differentsial'nye Uravneniya* **22**:9 (1986), 1596–1602.

[2] "Some equations with infinite dimensional PD-operators", *Dokl. Akad. Nauk SSSR* **267**:6 (1982), 1313–1319.

[3] "Functional superanalysis", *Uspekhi Mat. Nauk* **43**:2 (1988), 87–114.

N.G. Ibragimov:

[1] *Transformation groups in mathematical physics*, Nauka, Moscow, 1978.

V.I. Kachalov, and S.A. Lomov:

[1] "Analytic qualities of differential equations with singular points", *Dokl. Akad. Nauk SSSR* **304**:1 (1989), 15–19.

[2] "The smoothness of the solutions of differential equations with respect to a singular parameter", *Dokl. Akad. Nauk SSSR* **299**:4 (1988), 1214–1217.

E. Kamke:

[1] *Differentialgleichungen. Lösungsmethoden und Lösungen*, Leipzig, 1959.

L.D. Landau, and E.M. Lifshits:

[1] *Field theory*, Fizmatgiz, Moscow, 1962.

A.F. Leont'ev:

[1] *Exponential series*, Nauka, Moscow, 1976.

J. Leray, and J. Lions:

[1] "Quelques résultats de Vishik sur les problèmes elliptiques non linéaires par les méthodes de Minty-Browder", *Bull. Soc. Math. France* **93**:1 (1965), 97–107.

A. Martineau:

[1] "Sur les fonctionelles analytiques et la transformation de Fourier-Borel", *J. Analyse Math.* **9** (1963), 1–163.

G. Minty:

[1] "Monotone (non-linear) operators in Hilbert spaces", *Duke Math. J.* **29**:3 (1962), 341–346.

S. Misochata:

[1] "On the Kovalevskaya systems", *Uspekhi Mat. Nauk* **29**:2 (1974), 216–224 (in Russian).

V.V. Napalkov:

[1] *Convolution equations in multidimensional space*, Nauka, Moscow, 1982.

O.V. Odinokov:

[1] "On the well-posedness of the Cauchy problem for the operators with smooth symbols", *Izv. Akad. Armyan. SSR* **23**:1 (1988), 65–75.

[2]   "The Cauchy problem for operators with smooth symbols", *Uspekhi Mat. Nauk* **42**:4 (1987), 163 (in Russian).

O.A. Oleinik, and V.P. Palamodov:

[1]   "Systems of partial differential equations", In: Petrovskii [1] (in Russian).

L.V. Ovsyannikov:

[1]   *Group analysis of differential equations*, Nauka, Moscow, 1978.

I.G. Petrovskii:

[1]   *Selected papers (Systems of partial differential equations)*, Nauka, Moscow, 1980 (in Russian).

G. Pólya:

[1]   "Untersuchungen über Lüken und singularitaten von Potenzreichen", *Math. Z.* **19** (1929), 119–125.

Ya.V. Radyno:

[1]   "Vectors of exponential type and functional calculus", *Dokl. Akad. Nauk SSSR* **27**:10 (1983), 875–878.

[2]   "Vectors of exponential type in operator calculus and differential equations", *Differentsial'nye Uravneniya* **21**:9 (1985), 1559–1565.

A.P. Robertson, and W.J. Robertson:

[1]   *Topological vector spaces*, Cambridge Univ. Press, 1964.

L.I. Ronkin:

[1]   *Introduction to the theory of entire functions of many complex variables*, Nauka, Moscow, 1971.

K.L. Samarov:

[1]   "On the solution of the Cauchy problem for the Schrödinger equation of a free relativistic particle", *Dokl. Akad. Nauk SSSR* **271**:2 (1983), 334–337.

[2]   "On the Schrödinger PD-equation", *Dokl. Akad. Nauk SSSR* **279**:1 (1984), 83–87 (in Russian).

[3]   "On the solution of PD-equations based on the Bessel operator", *Dokl. Akad. Nauk SSSR* **293**:1 (1987).

K.L. Samarov, and A.V. Aksenov:

[1] "Lorentz invariance of solutions of the Schrödinger PD-equation ", *Differentsial'nye Uravneniya* **26**:2 (1990), 268–271.

H. Schaefer:

[1] *Topological vector spaces*, Macmillan, New York, 1966.

G.E. Shilov:

[1] *Mathematical analysis* (2nd special part), Nauka, Moscow, 1965.

E.M. Stein, and G. Weiss:

[1] *Introduction to Fourier analysis in Euclidean spaces*, Princeton Univ. Press, 1971.

S. Steinberg:

[1] "The Cauchy problem for differential equations of infinite order", *J. Differential Equations* **9**:1 (1971), 501–519.

B.A. Taylor:

[1] "Some locally convex spaces of entire functions", *Proc. Symp. Pure Math., XI, Amer. Math. Soc., Providence, RI*, 1968, 431–467.

A.N. Tikhonov:

[1] "Some uniqueness theorems for the heat equation", *Mat. Sb.* **42**:2 (1935), 199–216 (in Russian).

Tran Duk Van:

[1] "On PD-operators with real analytic symbols and their applications", *J. Fac. Sci. Univ. Tokyo, Sect. IA Math.* **36** (1989), 803–825.

[2] "On PD-operators with analytic symbols and their applications", In: *UTYO-MATH*, Univ. Tokyo, No. 88-19, 1988.

[3] "On functional equations and functional differential equations", (to appear).

Tran Duk Van, and Dinh Nho Hào:

[1] "PD-operators with real analytic symbols and approximation methods for PD-equations", (to appear).

Tran Duk Van, Dinh Nho Hào, and R. Gorenflo:

[1] "Approximating the solution to the Cauchy problem and the boundary value problem for the Laplace equation", (to appear).

Tran Duk Van, Dinh Nho Hào, Trinh Ngoc Minh, and R. Gorenflo:

[1] "On the Cauchy problem for systems of partial differential equations with distinguished variables", Preprint *FB Mathematik* A88–05, FU Berlin, 1988.

Tran Duk Van, Nguyen Duy Thai Son, and Dinh Zung:

[1] "Approximate solution of the Cauchy problem for the wave equation by the method of differential operators of infinite order", *Acta Math. Vietnam.* **13** (1988), 2.

F. Treves:

[1] "The Ovsyannikov theorem and hyperdifferential equations", *Notas Mat., Inst. Mat. Pura Appl., Rio de Janeiro* **46** (1968), 1–237.

Trinh Ngoc Minh, and Tran Duk Van:

[1] "Cauchy problem for systems of partial differential equations with distinguished variables", *Dokl. Akad. Nauk SSSR* **32**:2 (1985), 562–565.

Trinh Ngoc Minh:

[1] "Linear differential operators of infinite order and their applications", *Acta Math. Vietnam.* **12**:1 (1987), 101–124 (in Russian).

S.R. Umarov:

[1] "Some spaces of infinite order and their applications to operator equations, *Dokl. Akad. Nauk SSSR* **275**:2 (1984), 313–317 (in Russian).

[2] "Boundary value problems for differential operator equations and their applications to PD-equations", *Izv. Akad. UzSSR* **4** (1986), 38–42.

A. Zygmund:

[1] *Trigonometric series*, Vols. I, II, Cambridge Univ. Press, 1959.

Author Index

Index of Basic Formulae

Subject Index